Managing Water in River and Reservoir Systems

Other Titles of Interest

Water Resources Systems Analysis through Case Studies: Data and Models for Decision Making, edited by David W. Watkins, Jr. (ASCE/EWRI 2013). This book contains 10 case studies suitable for classroom use to demonstrate engineers' use of widely available modeling software in evaluating complex environmental and water resources systems. (ISBN 978-0-7844-1287-9)

Adventures in Managing Water: Real-World Engineering Experiences, edited by Daniel Peter Loucks and Laurel Saito (ASCE/EWRI 2019). This book provides valuable insight for students and professionals into real-world experiences of water engineers and other water management professionals. (ISBN 978-0-7844-1533-7)

Water Engineering with the Spreadsheet: A Workbook for Water Resources Calculations Using Excel, by Ashok Pandit (ASCE Press 2016). Pandit provides more than 90 problems, from easy to difficult, in the areas of fluid mechanics, hydraulics, hydrology, and stormwater management. Using the accompanying spreadsheets, students will gain a solid foundation on the topics along with hands-on experience with Excel. (ISBN 978-0-7844-1404-0)

River Training and Sediment Management with Submerged Vanes, by A. Jacob Odgaard (ASCE Press 2009). Odgaard provides a complete guide to help engineers determine when submerged vanes would be an effective design solution for problems associated with sedimentation and river training and how the vanes should be planned and installed. (ISBN 978-0-7844-0981-7)

Energy Production and Reservoir Water Quality: A Guide to the Regulatory, Technical, and Theoretical Basis for Required Studies, edited by James Martin, John Higgins, John Edinger, and John Gordon (ASCE/Energy Division 2007). This report examines the inextricable link between energy production and reservoir water quality, which poses an ongoing challenge for civil engineers. (ISBN 978-0-7844-0896-4)

Managing Water in River and Reservoir Systems

Water Resources, Institutional Practices, and Constructed Infrastructure in Texas

Ralph A. Wurbs, Ph.D., P.E., Hon.D.WRE

ASCE PRESS

Library of Congress Cataloging-in-Publication Data

LCCN 2023055214
Cataloging-in-Publication Data on file with the Library of Congress

---

Published by American Society of Civil Engineers
1801 Alexander Bell Drive
Reston, Virginia 20191-4382
www.asce.org/bookstore | ascelibrary.org

***Errata:* Errata, if any, can be found at https://dx.doi.org/10.1061/9780784485385**

Front Cover Photograph: Whitney Dam and Reservoir on the Brazos River in Central Texas, Courtesy of USACE Fort Worth District

ISBN 978-0-7844-8538-5 (PDF)
ISBN 978-0-7844-8539-2 (ePub)

Manufactured in the United States of America.

28 27 26 25 24 1 2 3 4 5

Dedication

The love and support of my wife of 52 years, Keri, our family, and our Christian faith are the foundation of my life. Water is also essential but in a different way. None of our adult children or their spouses have chosen a professional career in water management, but they enjoy drinking water regularly and recreating at the lake occasionally. Perhaps this book will inspire one or more of our grandchildren to pursue civil engineering and water management careers. Keri and I would greatly enjoy seeing some or all of our grandchildren studying at Texas A&M University in the future, following in the footsteps of their grandparents, parents, aunts, and uncles.

Ralph A. Wurbs
Texas A&M University
January 2024

Contents

List of Figures

List of Tables

Preface

Water management in Texas is driven by dramatic spatial and temporal hydrologic variability, continual rapid population growth, declining groundwater supplies, and intensifying demands on river and reservoir systems. Dams, reservoirs, and other constructed facilities, along with effective planning, water allocation, and resource management capabilities are essential for providing reliable municipal, industrial, and agricultural water supplies and reducing flood damages. River and reservoir system water management also plays an important role in recreation, ecosystem protection, electrical energy production, and land development. Numerous dam and reservoir projects, conveyance facilities, and flood control levee and channel improvements, most constructed during the 1940s to 1980s era of large-scale water project construction nationwide, are operated throughout Texas to regulate extremely variable river flows for beneficial purposes. Other projects are in various stages of planning and development. Effective water management requires integration of improvements in both operation of constructed facilities and institutional capabilities for planning, allocation, and management of both water and other related resources.

This book explores (1) institutional capabilities for water management, (2) river system hydrology, (3) operation of dams and reservoirs and other facilities statewide to deal with extreme hydrologic variability, (4) planning and development of additional new projects, (5) allocation of limited streamflow and reservoir storage capacity, (6) ecosystem preservation, and (7) other aspects of integrated water management. Collaborative endeavors of international, federal, interstate, state, river basin, and local agencies, private sector enterprises, and an entire water management community are a central theme of the book.

The Texas experience provides an excellent case study of river and reservoir system water management relevant nationwide and worldwide. Diverse climate, hydrology, socioeconomic conditions, and water management practices are found in the various regions of Texas. Most regions of the nation and world are similar in relevant ways to some region of Texas. Water management policies, institutional programs, constructed facilities, and operating practices are crucial to the prosperity of the state and of necessity must be comprehensive and effective. Consequently, Texas has pursued progressive endeavors in statewide and regional planning, infrastructure development, water allocation, environmental flow protection, water availability modeling, and other aspects of river and reservoir system management. Over a century of experience that includes past, recent, and still evolving water management accomplishments in Texas can inform similar efforts throughout the United States and the world.

This book is written for professional water resources planners and managers, civil engineers, environmental scientists, elected and appointed officials at all levels of government, private sector industrial and agricultural water users, university students, researchers, and the public. Everyone benefits from effective water management. Interests in rivers and lakes are broad.

Acronyms and Abbreviations

AAWRE	American Academy of Water Resources Engineers
AWRA	American Water Resources Association
BRA	Brazos River Authority
CRM	Conditional reliability modeling
CRMWD	Canadian River Municipal Water District
CRMWD	Colorado River Municipal Water District
CWMS	Corps Water Management System
EFS	Environmental Flow Standards
EPA	Environmental Protection Agency
FEMA	Federal Emergency Management Agency
FERC	Federal Energy Regulatory Commission
FIRO	Forecast-informed reservoir operation
FWD	Fort Worth District of USACE
GAM	Groundwater availability model
GBRA	Guadalupe-Blanco River Authority
HCFCD	Harris County Flood Control District
HEC	Hydrologic Engineering Center of USACE
HEC-DSS	Data storage system (developed by the USACE HEC)
IBWC	International Boundary and Water Commission
IH	Interstate highway
IWRM	Integrated water resources management
LCRA	Lower Colorado River Authority
msl	Mean sea level (ft above mean sea level)
MWD	Municipal Water District
NDSP	National Dam Safety Program
NFIP	National Flood Insurance Program
NID	National Inventory of Dams
NOAA	National Oceanic and Atmospheric Administration
NPDES	National Pollutant Discharge Elimination System
NRCS	Natural Resource Conservation Service
NTMWD	North Texas Municipal Water District
NWIS	National Water Information System

NWS	National Weather Service
PMF	Probable Maximum Flood
PMP	Probable Maximum Precipitation
PMS	Probable Maximum Storm
RA	River Authority
SB	Senate Bill
SRA	Sabine River Authority
SWIFT	State Water Implementation Fund for Texas
SWPA	Southwestern Power Administration
TAMU	Texas A&M University
TCEQ	Texas Commission on Environmental Quality
TIFP	Texas Instream Flow Program
TPWD	Texas Parks and Wildlife Department
TRA	Trinity River Authority
TRWD	Tarrant Regional Water District
TSSWCB	Texas State Soil and Water Conservation Board
TWDB	Texas Water Development Board
TWRI	Texas Water Resources Institute
USACE	United States Army Corps of Engineers
USBR	United States Bureau of Reclamation
USEPA	United States Environmental Protection Agency
USFWS	United States Fish and Wildlife Service
USGS	United States Geological Survey
WAM	Water availability model
WAPA	Western Area Power Administration
WCID	Water Control and Improvement District
WRAP	Water Rights Analysis Package

Unit Conversions

1.0 acre-foot (ac-ft) = 43,560 cubic feet (ft^3) = 325,850 gallons (gal.) = 1,234 cubic meters (m^3)

1.0 cubic foot per second (ft^3/s) = 1.9835 acre-feet/day (ac-ft/day) = 0.02832 cubic meters per second (m^3/s)

1.0 square mile (sq. mi) = 640 acres (ac) = 2.590 square kilometers (km^2)

1.0 acre (ac) = 43,560 square feet (ft^2) = 4,047 square meters (m^2)

1.0 cubic foot (ft^3) = 0.02832 cubic meters (m^3)

1.0 mile (mi) = 5,280 feet (ft) = 1.6093 kilometers (km)

1.0 inch (in.) = 2.540 centimeters (cm) = 25.40 millimeters (mm)

1.0 milligram per liter (mg/L) = 1.0 gram per cubic meter (g/m^3)

Chapter 1

Water Resources Management in Texas

Water management consists of planning, development, regulation, conservation, protection, allocation, and use of water in streams, rivers, lakes, reservoirs, aquifers, and other components of the natural and human-modified hydrologic cycle. Water use sectors include municipal, industrial, agricultural irrigation, rural domestic, livestock, hydroelectric power generation, steam-electric power generation, mining, navigation, recreation, and environmental protection. Flood risk mitigation and stormwater management are fundamental aspects of water management. Water quality, erosion and sedimentation, fish and wildlife, and ecosystems are important. Water management activities include policy formulation; resource assessments; water allocation; environmental regulation; formulation and implementation of management strategies; and planning, design, construction, maintenance, rehabilitation, and operation of constructed facilities. For brevity, the term *water management* is adopted in this book to refer to any of these endeavors or any combinations thereof.

Although this introductory first chapter begins from a broader water management perspective, the focus of the book is on managing water in major rivers and reservoirs. The terms *streams* and *rivers* are used interchangeably, although river implies a larger stream. The terms *lakes* and *reservoirs* are also used interchangeably, although reservoir implies impoundment by a constructed dam. Lakes include natural as well as constructed storage projects. All major lakes in Texas are impounded by constructed dams.

The objectives of the book are to contribute to understanding (1) water management in Texas, and (2) fundamental concepts and practices applicable nationwide and worldwide. Understanding water management in Texas requires an understanding of water management concepts in general as applied throughout the nation and world as well as specific endeavors within Texas. Conversely, studying river and reservoir water management experiences in Texas is an effective way to expand knowledge regarding water management in general.

Water Management throughout Texas and the United States

This book explores river/reservoir system water management in Texas. The Texas experience provides an excellent case study relevant to managing water in river and reservoir systems throughout the United States and the world. Climate, hydrology, economic development, water use, and water management practices vary dramatically across the state of Texas from the arid or semiarid desert of the western one-third of Texas to humid eastern forests, from sparsely populated rural regions to the metropolitan areas of El Paso, San Antonio, Austin, Houston, Fort Worth, and Dallas shown in Figure 1-1. The climate of Texas is representative of

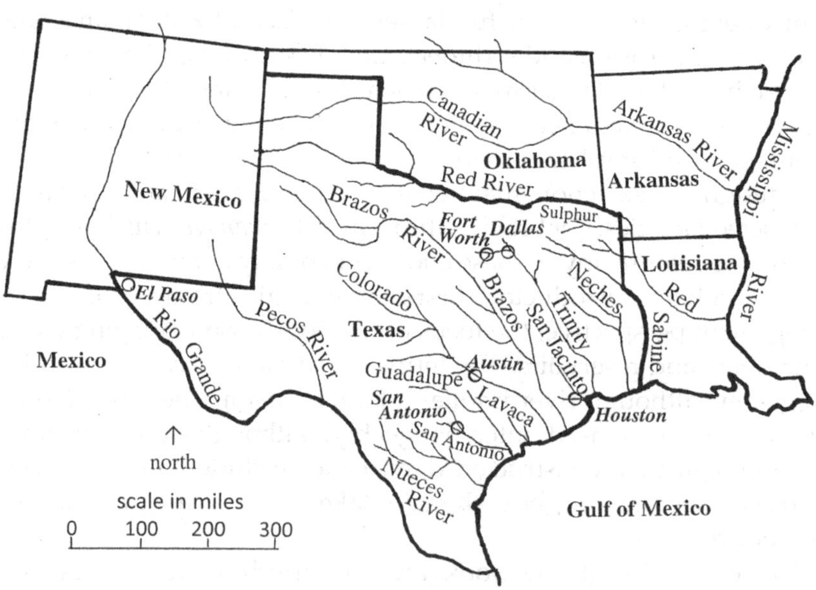

Figure 1-1. Major rivers and largest cities of Texas.

both drier western states and wetter eastern states of the United States. Likewise, diverse water management issues and practices in Texas are representative of other diverse regions of the nation and the world.

Texas has the second largest gross national product of any state in the United States and as a sovereign country would be the ninth or tenth largest economy of the nations of the world. Texas has a larger land area than two-thirds of the countries in the world. The population of Texas is larger than 187 of 235 nations of the world. Texas has the second largest population and second largest land area of the 50 states in the United States. Texas has had one of the largest, often the largest, annual population increases of any state in the United States over the last many years.

River/reservoir system water management in the 15 major river basins and eight coastal basins of Texas is representative of the entire United States in many respects. Texas is a notable leader in pursuing innovative new water management endeavors that can be beneficially replicated in other states of the United States and other countries.

The Texas experience reflects actual past and present water management nationwide particularly with respect to federal, local governmental, and private sector endeavors. Widely shared endeavors include federal planning and implementation of large-scale water projects during the 1940s to 1980s nationwide construction era of water development; establishment of federal environmental protection policies and programs during the 1960s to 1980s; and traditional local responsibilities for water supply, sanitation, and stormwater management. Many of the large water projects constructed across the United States in the past by federal agencies are in Texas. The state also has the full spectrum of many smaller communities, large growing cities and metroplexes, and diverse commercial enterprises, along with associated water needs, issues, and management practices.

The role of state agencies in water management has expanded significantly during the last several decades. Texas has been a leader in establishing structured institutional processes for regional and statewide water planning, water allocation, environmental flow preservation, drought management, water availability modeling, and other state-level activities. The problems, needs, and opportunities addressed by progressive state agency endeavors in Texas are relevant nationwide and worldwide. Strategies and practices established over the last three decades by state agencies in Texas provide an experience base for implementing similar programs elsewhere.

Texas encompasses portions of the international and interstate Rio Grande Basin and five other interstate river basins. Nine other major river basins and eight coastal basins are encompassed entirely or almost entirely within the state. Several of the river authorities manage water resources of entire river basins. Reservoir operation and water allocation practices requiring collaboration of neighboring states and countries are aspects of Texas water management also relevant elsewhere.

Water Management Community

People and organizations manage water resources within an institutional framework of traditions, laws, policies, programs, practices, and administrative processes. The water management community consists of many people and organizations with different concerns and responsibilities working together on collaborative endeavors. Water management institutions relevant to planning, development, and management of river/reservoir systems in Texas are discussed in Chapter 2. Several key federal, state, regional, and local government agencies; private companies; and other entities that comprise this water management community are briefly introduced here in Chapter 1.

US Congress and Texas Legislature

The US Congress and Texas legislature enact laws that establish water management policies, authorize projects and programs, and appropriate funding. Many national and state laws enacted over many decades have shaped water management policies and practices in Texas. Examples of federal and state legislation particularly relevant to this book are highlighted as follows.

The Flood Control Act of 1936 and amendments thereto established the nationwide flood control responsibilities of the US Army Corps of Engineers (USACE), which later grew to encompass comprehensive multiple-purpose water development. The Water Supply Act of 1958 authorized inclusion of municipal water supply storage in federal reservoirs subject to all costs allocated to water supply being reimbursed to the federal government by nonfederal project sponsors. Under authority of Section 404 of the Clean Water Act of 1972 and amendments thereto, the Corps of Engineers administers a permit program regulating construction and other activities involving dredging and/or filling in rivers, streams, and wetlands.

The Water Rights Adjudication Act enacted by the Texas legislature in 1967 created the water right permit system administered by the Texas Commission on Environmental Quality (TCEQ) to allocate the waters of the river/reservoir systems of the state to water management entities (Caroom and Maxwell 2013). The Texas legislature traditionally assigns the designation "Senate Bill 1" during each legislative session to legislation of special importance. Motivated by severe drought conditions during 1995 to 1996, the 1997 Senate Bill 1 (SB1) created a regional and statewide water planning process and authorized development of a water availability modeling system to support both the planning process and administration of the water rights permit system. Primary responsibility for the SB1 planning process was assigned to the Texas Water Development Board (TWDB). TCEQ led the multiple-agency development of a water availability modeling system.

Senate Bill 2 (SB2), enacted into law by the Texas legislature in 2001, established the Texas Instream Flow Program (TIFP) jointly administered by TCEQ, TWDB, and the Texas Parks and Wildlife Department (TPWD) to develop capabilities for preserving stream flows to protect ecosystems. Studies to improve understanding of environmental flow needs in the river systems of the state are continuing under TIFP. The Texas legislature in its 2007 Senate Bill 3 (SB3) created an accelerated process for establishing environmental flow standards for selected priority river systems using existing information and the best available science.

The National Flood Insurance Program (NFIP) was created by the National Flood Insurance Act of 1968, expanded by the Flood Protection Act of 1973, and further refined by subsequent legislation including the 2012 Flood Insurance Reform Act. The Federal Emergency Management Agency (FEMA) administers NFIP, as in NFIP at the federal level. Local communities are responsible for floodplain management requirements of NFIP. Private insurance companies sell flood insurance to property owners.

Flood mitigation has historically been a local and federal responsibility, with almost no state involvement. The Texas legislature in its 2019 Senate Bill 8 authorized TWDB creation of a state and regional flood planning process with regions based on river basins. The new state-level flood control planning process is like the 1997 SB1 planning process that focuses on water supply.

International and Interstate Water Managers

Pursuant to a 1906 convention, 1944 treaty, and other agreements, the Mexico and US sections of the International Boundary and Water Commission (IBWC), headquartered in Juarez and El Paso, respectively, manage the waters of the Rio Grande, including operation of International Amistad and Falcon Reservoirs. Five interstate river compact commissions representing Texas and neighboring states administer allocation between the states of the waters of the Rio Grande and Canadian, Red, Sabine, and Pecos Rivers. TCEQ provides staff support for the interstate river commissioners.

Federal Water Agencies

USACE, through its Civil Works Program, has constructed and now owns and operates more reservoir projects nationwide and in Texas than any other entity. USACE has constructed and now owns and operates 27 multiple-purpose lakes in Texas that contain both water supply and flood control storage, two flood control reservoirs that have no water supply storage, and a brine control dam. These Corps of Engineers reservoirs contain about 27.9%, 78.3%, and 44.0%, respectively, of the conservation, flood control, and total storage capacity of the major reservoirs of Texas. USACE is responsible for flood control operations. River authorities, water districts, and cities contract for the water supply storage capacity of the USACE reservoirs and reimburse the federal government all construction, operation, and maintenance costs allocated to water supply.

The US Bureau of Reclamation (USBR) has constructed many large reservoir projects in the 17 western states, which includes Texas. The five reservoir projects constructed by USBR in Texas have been turned over to nonfederal sponsors for operation and maintenance. USACE is responsible for flood control operations at multipurpose reservoirs constructed by USBR.

The Natural Resource Conservation Service (NRCS) of the US Department of Agriculture has constructed 11,850 flood-retarding dams in rural tributary watersheds nationwide, including 2,040 dams in Texas. NRCS flood-retarding dams have ungated outlet structures that do not require operation by people. After construction on private land with federal funding, these projects are maintained by nonfederal sponsors at nonfederal expense.

State Agencies

TWDB and TCEQ play important roles in river and reservoir system management, but unlike other entities mentioned in the preceding and later paragraphs do not construct, own, and/or operate reservoir projects or other constructed facilities.

TWDB with about 400 employees is responsible for statewide planning, which includes working with 16 regional planning groups pursuant to the 1997 SB1 to periodically update regional water plans which are consolidated into a statewide plan. TWDB also administers more than 10 different grant and revolving fund loan programs to assist local and regional entities in financing water projects. The State Water Implementation Fund for Texas administered by TWDB provides low-interest loans with favorable terms for any political subdivision of Texas or nonprofit water supply corporation for implementation of projects included in the most recently adopted state water plan. Projects may include developing reservoirs or well fields, building new pipelines, desalination treatment plants, purchasing water rights, or other endeavors recommended in the state and regional water plans.

Reports documenting the 2002, 2007, 2012, 2017, and 2022 regional and state water plans are available at the TWDB website. Completion by TWDB and 16 regional planning groups of the next regional and state water plans is scheduled for 2027. TWDB and regional flood planning groups are scheduled to complete the state's first statewide flood mitigation plan in 2024.

TCEQ has about 2,800 employees in its Austin central office and 16 regional offices, making it the largest state environmental regulatory agency in the United States. TCEQ administers the National Pollutant Discharge Elimination in Texas under the oversight of the federal Environmental Protection Agency as well as other state water quality and environmental protection programs. TCEQ has a dam safety program for inspecting dams and enforcing safety regulations. TCEQ administers two versions of a water rights permit system: one for the Texas side of the Rio Grande downstream of Fort Quitman and the other version for the remainder of Texas.

Partnering and consensus water planning are key concepts in Texas. State water programs are characterized by collaboration between TCEQ, TWDB, and TPWD and between these agencies and the water management community. TPWD responsibilities intersect TCEQ and TWDB activities particularly regarding recreation management and protecting and enhancing fish, wildlife, and ecosystems. TWDB is the lead agency for the SB1 regional and statewide planning endeavors with TCEQ and TPWD playing supporting roles. TCEQ administers the water right system, but consistency with regional plans is a requirement for approval of water right permit applications. TCEQ maintains a water availability modeling system developed pursuant to the 1997 SB1. TWDB and SB1 regional planning groups routinely apply the modeling system. TCEQ, TWDB, and TPWD are partners in implementing the 2001 SB2 TIFP and 2007 SB3 procedures for establishing environmental flow standards in collaboration with stakeholders and the scientific community.

Reservoir recreation is extremely popular throughout Texas. The major federal and nonfederal reservoirs are used extensively for recreation along with other primary project purposes. A number of smaller reservoirs have been constructed specifically for recreation. Reservoir owners, TPWD, and other entities manage recreation areas around the reservoirs.

Regional and Local Governmental and Private Entities

Water management infrastructure is constructed and maintained by numerous cities, municipal water districts, irrigation districts, river authorities, and utility companies. Groundwater conservation districts provide leadership in managing groundwater. River authorities are responsible for development and management of the water resources of all or portions of major river basins. River authorities construct and operate their own reservoir projects to provide municipal, industrial, and/or agricultural water supply and contract for storage capacity in federal reservoirs.

Private companies also construct, own, and operate reservoir projects and other water management facilities. Many of the private reservoirs are owned by electric power utilities and used for thermal-electric power plant cooling water.

Local communities are responsible for stormwater and floodplain management. FEMA has nationwide responsibilities for NFIP with its local floodplain management requirements and for disaster relief operations after floods and other natural disasters. The recently initiated TWDB statewide flood control planning program is accomplished in collaboration with local entities.

Numerous civil engineering firms provide technical expertise to support the endeavors of the water management agencies discussed throughout this book. The many consulting firms active in the Texas water management community range in size from a single registered professional engineer to many thousands of employees. Professional consulting services are employed in planning studies, preparing water right permit applications, delineating floodplains, designing facilities, and various other activities. Likewise, the water agencies hire construction contractors to build and rehabilitate constructed facilities.

Water users, recreationists, environmentalists, residential and commercial developers, and property owners are integrally involved in a full spectrum of water-related activities. Everyone uses water. Public involvement is an important fundamental aspect of water management. Local as well as state and federal elected and appointed political officials play roles in water management. University research and education are important in addressing water issues.

Water Use in Texas

Regional and statewide water planning by TWDB staff and 16 regional planning groups and their consultants is introduced earlier in this chapter and discussed further in Chapter 2. Sixteen updated regional water plans are consolidated into a statewide plan in a 5-year planning cycle. Most of the following information is from the most recent statewide water plan (TWDB 2022b), which is found at a TWDB website listed in Table 4-1 of Chapter 4.

The state population increased from about three million people in 1900 to 9.6 million in 1960 to 20.9 million in 2000 and 29.7 million in 2020. State Water Plan (TWDB 2022b) projections of the future population of Texas in millions of people are as follows: 33.9 million in 2030, 38.1 in 2040, 42.3 in 2050, 46.8 in 2060, and 51.5 million in 2070. The population tripled between 1960 and 2020 and is projected to increase 73% between 2020 and 2070.

Water supplies are lowest and water demands highest during droughts. TWDB staff in collaboration with the 16 regional planning groups estimate annual water demands in specified years (2020, 2030, 2040, 2050, 2060, 2070) over a 50-year planning period and capabilities for supplying the demands based on the premise of a hypothetical repeat of the most hydrologically severe drought on record. Additional unmet water needs are estimated as the difference between water demands and available water supplies. Annual water demands, available supplies, and additional water needs are expressed in units of ac-ft/year (TWDB 2022b). An acre-foot is a volume equivalent to covering an area of one acre to a depth of 1 ft. An acre-foot is 43,560 ft^3 or 325,850 gal.

Estimated annual water needs during severe drought conditions for the various types of water use are summarized in Table 1-1. The TWDB also compiled estimates of water supply capabilities provided from currently available water supply sources. Estimated demands exceed supplies. The unmet water demands are summarized in Table 1-2. These needs for additional water are shortages in supplying demands assuming extreme drought conditions and no new water supply

Table 1-1. Projected Annual Water Demand in Acre-Feet by Water Use Category.

Category	2020	2030	2040	2050	2060	2070
Irrigation	9,448,000	9,383,000	8,703,000	8,154,000	7,737,000	7,594,000
Municipal	5,223,000	5,826,000	6,440,000	7,089,000	7,783,000	8,507,000
Steam-electric	931,000	935,000	935,000	935,000	935,000	935,000
Manufacturing	1,339,000	1,531,000	1,531,000	1,531,000	1,531,000	1,531,000
Mining	407,000	409,000	365,000	323,000	287,000	281,000
Livestock	332,000	343,000	353,000	363,000	374,000	382,000
Total	17,680,000	18,427,000	18,327,000	18,395,000	18,647,000	19,230,000

Source: TWDB (2022b).

Table 1-2. Projected Unmet Annual Water Need in Acre-Feet by Water Use Category.

Category	2020	2030	2040	2050	2060	2070
Irrigation	2,396,000	3,319,000	3,280,000	3,188,000	3,094,000	3,046,000
Municipal	215,000	802,000	1,371,000	1,912,000	2,502,000	3,144,000
Steam-electric	187,000	192,000	196,000	199,000	201,000	203,000
Manufacturing	159,000	264,000	275,000	286,000	295,000	301,000
Mining	119,000	123,000	111,000	102,000	96,000	101,000
Livestock	40,000	44,000	48,000	54,000	60,000	63,000
Total	3,116,000	4,744,000	5,281,000	5,741,000	6,248,000	6,858,000

Source: TWDB (2022b).

projects. The statewide quantities in Tables 1-1 and 1-2 are aggregations of quantities compiled for each of the 16 SB1 planning regions of the state (TWDB 2022b).

Except for cooling water for steam-electric power plants, the quantities in Tables 1-1 and 1-2 represent annual volumes of water to be withdrawn from groundwater aquifers or river/reservoir systems, reused return flows from municipal and industrial wastewater treatment plants, or otherwise supplied. The quantities represent water to be supplied as contrasted with consumptive use excluding return flow. In the case of steam-electric cooling water, most of the water withdrawn is returned to lakes and streams. Therefore, the steam-electric quantities in Tables 1-1 and 1-2 and TWDB planning reports are limited to consumptive evaporative losses excluding return flows. Quantities representing total diversions from water supply sources for steam-electric power plants would be far larger.

The quantities in Table 1-2 are the portions of the demands of Table 1-1 not supplied by currently existing facilities with a hypothetical repeat of the most severe drought of record. Without additional supplies being developed through the strategies and projects recommended in the regional and statewide planning reports, estimated shortages of 3,116,000 ac-ft/year would have occurred in 2020 under extreme drought conditions, and shortages of 6,858,000 ac-ft/year will occur in 2070 with extreme drought conditions. Population growth leads to municipal water users statewide potentially facing water shortages during severe drought that are 14.4 times larger in 2070 than in 2020 unless recommended strategies and projects are implemented.

The future unmet water needs in Table 1-2 reflect decreases in water supply capabilities of existing reservoirs and groundwater well pumping facilities as well as the increases in water demands shown in Table 1-1. Decreases in water supply capabilities of surface water reservoirs result from loss of storage capacity over time due to sedimentation. Decreasing well pumping capabilities are due to the lowering of aquifer storage levels resulting from decades of pumping exceeding natural recharge.

The 16 regional water plans and statewide plan recommend more than 5,800 water management strategies and projects for supplying increasing water needs over the next several decades. Construction of 23 new major reservoir projects is included in these proposed plans. Strategies for modifying the operations of existing reservoirs to address intensifying demands on limited resources are explored in the planning studies and in this book.

The water demands outlined in the preceding paragraphs and Table 1-1 are to be supplied by diversions from sources of supply. Hydroelectric power generation, recreation, and environmental flow needs involve instream use rather than withdrawals from supply sources. Hydroelectric power facilities are operated at 26 reservoirs in Texas. Except for Lake Texoma, hydropower generation is usually limited to releases for downstream water supply diversions. Recreation is popular at most of the major reservoirs. Appropriating streamflow pursuant to the 2007 SB3 to protect environmental instream flow requirements represents an important type of water need.

Water Resources of Texas

Water resources include precipitation, soil moisture, groundwater aquifers, and seawater as well as water in streams and lakes. Water needs of natural vegetation, lawns and landscaping plants, and agricultural crops are supplied directly by rainfall, which may in many cases be supplemented by irrigation. Both groundwater aquifers and surface water streams and lakes are important components of Texas water resources. Both ground and surface water are discussed in this chapter. Subsequent chapters focus on surface water.

The use of seawater from the Gulf of Mexico for municipal and agricultural water supply in Texas to date has been negligible due to the high economic cost of treatment processes to remove salts. Some industrial use of seawater occurs without treatment. A substantial number of municipal water supply systems include desalination of brackish surface and groundwater. Desalination of seawater and additional desalination of brackish surface and groundwater are strategies for developing additional water supplies in the future as increasing demands exceed available supplies.

Hydroclimate of Texas

Precipitation, reservoir surface evaporation, streamflow, and reservoir storage throughout Texas are explored in detail in later chapters. Precipitation and evaporation are climatic drivers of river flows, including inflows to reservoirs, and significantly affect agricultural and urban lawn irrigation and other water use needs. Most of the rain falling on a watershed is returned to the atmosphere through evapotranspiration. Snowfall and sleet are minimal throughout Texas. Precipitation consists mainly of rainfall.

Statewide databases of precipitation and reservoir evaporation rates are introduced in Chapter 4 and analyzed in Chapter 5. Losses of stored water by evaporation is a significant factor affecting reservoir storage contents and regulated river flows. Mean annual precipitation increases uniformly from west to east across Texas from less than 9 in. at El Paso to over 56 in. in the lower Sabine River Basin, with a statewide average of 28.0 in./year. Mean annual reservoir surface evaporation rates range from 70 in./year in west Texas to 40 in./year in northeast Texas, with a statewide average of 59.4 in./year. The statewide average monthly and annual precipitation depth plots of Figures 1-2 and 1-3 illustrate temporal variability.

Droughts and floods are drivers of the development and management of river and reservoir systems. The hydrologically most severe drought since before 1900 for much of Texas began gradually in 1950 and ended in April 1957 with one of the greatest floods on record. Major droughts in the 1910s and 1930s also affected large areas of the state. More recent droughts were much more economically costly due to population and economic growth. The 2010 to 2014 drought is comparable in hydrologic severity to the 1950 to 1957 drought in major areas of

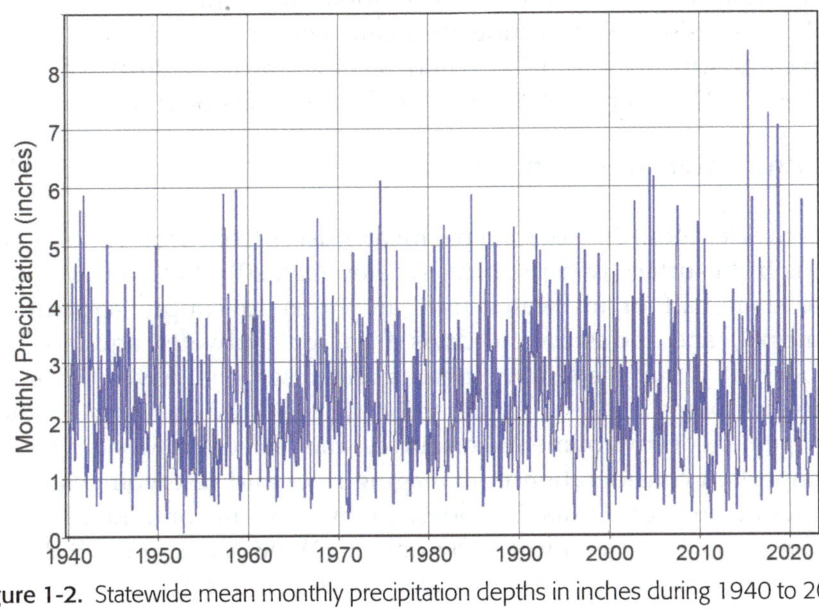

Figure 1-2. Statewide mean monthly precipitation depths in inches during 1940 to 2022.

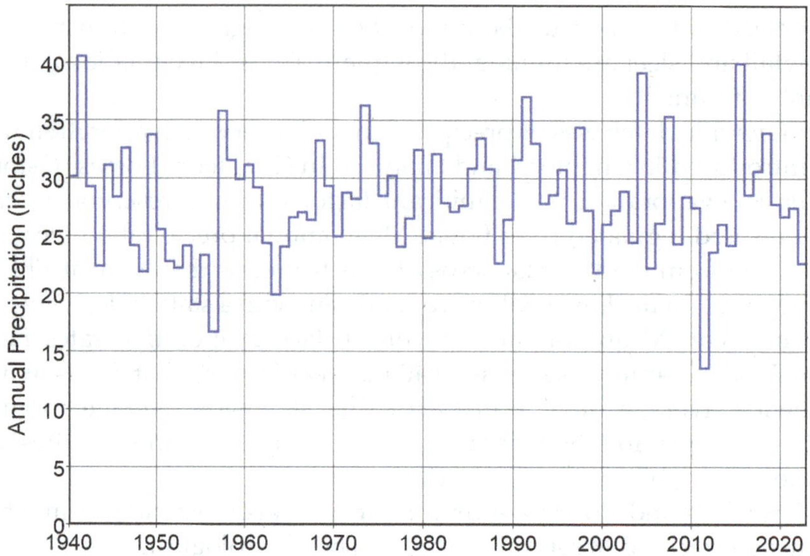

Figure 1-3. Statewide mean annual precipitation depths in inches during 1940 to 2022.

the state (Winters 2013). For more than half of Texas, 2011 had the lowest annual precipitation since the beginning of official precipitation records in 1895 (Nielsen-Gammon 2012). On the other extreme, 2015 was one of the wettest years on record, with multiple major floods. Several very costly floods since 2015 include Hurricane Harvey in 2017 (ASCE 2018), Tropical Storm Imelda in 2019, and several storms in 2020. The years 2022 and 2023 are among the hottest and driest years on record for Texas.

Groundwater and Surface Water

Nationwide, about 26% of total freshwater use is from groundwater, and 74% is from surface water sources (Dieter et al. 2018). Currently, a little more than half of all water used in Texas to supply agricultural, municipal, and industrial needs is from groundwater aquifers, and almost half is supplied from streams and reservoirs. River flows and reservoir storage are also important for hydroelectric energy generation, management of environmental resources, and recreation. Flooding is also an important consideration in managing river and reservoir systems.

About 80% of Texas is underlain by nine major and 22 minor aquifers, which are delineated by online TWDB maps (TWDB 2022a). Agricultural irrigation currently accounts for about 80% of groundwater use. Municipalities account for about 15% of groundwater use, which supplies about 35% of municipal water demands. Declining groundwater supplies combined with increased water demands accompanying population growth in Texas have intensified demands on limited surface water resources. The Ogallala, Gulf Coast, and Edwards Aquifers have experienced particularly notable declines in groundwater supplies.

Although agricultural operations throughout Texas rely upon irrigation from groundwater to varying extents, the majority of the irrigation supplied from groundwater occurs in the upper Canadian, Red, and Brazos River Basins of northern Texas supplied by the Ogallala Aquifer. Development of surface water is limited in this flat semiarid region. Historically, irrigated agriculture in the High Plains Region of northern Texas supplied by the Ogallala Aquifer accounted for a very large portion of total statewide water use. However, water level declines in the Ogallala Aquifer combined with scarce surface water supply alternatives have resulted in significant decreases in irrigation and associated crop production that are projected to continue.

Prior to the 1970s, Houston and its vicinity relied almost totally on the Gulf Coast Aquifer with only minimal use of surface water. Several decades of overpumping of groundwater resulted in major land subsistence problems. Consequently, Houston and neighboring cities and industries have been shifting from groundwater toward the goal of near-total reliance on surface water.

The city of San Antonio and adjacent smaller cities rely almost totally on the Edwards Aquifer for municipal and industrial water supply. San Antonio is the largest city in the nation relying totally on groundwater. Regulation of the Edwards has been driven by preservation of spring flows and ecological resources.

Municipal and industrial water use by one-fourth of the Texas population that resides in more than 70 cities of the Dallas and Fort Worth metropolitan area is supplied from surface water stored in reservoirs on the Trinity River and its tributaries, with little use of groundwater. Austin and nearby smaller cities as well as farmers in the lower Colorado Basin are supplied primarily from reservoirs on the Colorado River.

Ground and surface water are quite different from the perspectives of both hydrologic processes and water management (Wurbs 2015, Young et al. 2018). Although potential benefits of conjunctive ground and surface water management have long been recognized, combining and coordinating the use of water from surface and groundwater sources is difficult due to differences in water rights associated with surface water versus groundwater.

Water in the rivers and lakes of Texas is owned by the state, and its use is regulated through a statewide water rights system administered by TCEQ. Like most states in the western United States, surface water is regulated through a prior appropriation permit system that protects senior water users from junior (more recent) appropriators diminishing their supply reliability. Seniorities (priorities) are based on the dates that water is first used or water right permits are granted.

Groundwater rights in Texas have historically been based on the common law rule allowing landowners to pump as much water as they wish from under their land (Caroom and Maxwell 2013, Porter 2014). Increased state regulation of groundwater has evolved primarily through the establishment of local groundwater conservation districts. The legislature passed laws in 1949 and 1985 authorizing creation of groundwater conservation districts with county-level voter approval. Twelve districts existed prior to 1985. As of 2023, 98 groundwater districts have been created covering about 70% of the land area of the 254 counties of the state. Sixty districts each cover a single county and the other 38 encompass more than one county. The primary purposes of the districts are to encourage water conservation and protect water quality. Legislatively mandated duties of groundwater conservation districts include permitting water wells, developing comprehensive management plans, and adopting necessary rules to implement management plans.

Most groundwater districts focus on prevention of waste, water conservation education, recharge projects, and data collection. Some regulate pumping. The districts tread a narrow path between private ownership of groundwater and state responsibility to protect the water resource. Texans are reluctant to allow anyone to tell them how much water they can pump from under their own land. Governmental regulation of pumping has been driven by necessity as depleting aquifers resulted in major problems.

The Harris–Galveston Coastal Subsidence District and Edwards Underground Water Authority have developed the strongest permitting systems in the state to regulate groundwater use. The Subsidence District was created in 1957 in response to severe subsidence in the low-lying, urbanized coastal region containing the cities of Houston and Galveston due to over-drafting groundwater. Creation of the Edwards Aquifer Authority in 1993 was motivated by a federal court ruling protecting endangered species under the Endangered Species Act. The Edwards is a limestone aquifer shared by San Antonio, several smaller cities, and extensive irrigated farming interests. Springs fed by the Edwards Aquifer maintain the flow of several rivers and support ecosystems that include endangered species.

River Basins and Streamflow

Texas has an area of 268,600 sq. mi that includes 15 major river basins and eight coastal basins located between the major rivers as delineated in Figure 1-4. The Arkansas and Red Rivers are tributaries of the Mississippi River, which discharges into the Gulf of Mexico, as shown in Figure 1-1. The Canadian River is a tributary of the Arkansas River. Although the Sulphur River and Cypress Creek are tributaries of the Red River, their upper basins are treated as separate Texas river basins. With the exceptions of the Canadian and Red Rivers and their tributaries, the major river basins and coastal basins of Figure 1-4 discharge directly into the Gulf of Mexico. The Guadalupe River and San Antonio River join to form the combined GSA Basin a short distance above the Guadalupe River's outlet to the Gulf of Mexico.

River/reservoir system operations are governed by extended droughts and intense floods. Water management consists of managing hydrologic variability and uncertainties regarding the future. Rivers throughout Texas exhibit great variability including severe multiple-year droughts and major floods along with year-to-year, seasonal, and continuous fluctuations. Streamflow characteristics statewide are explored in Chapter 5 using databases introduced in Chapter 4.

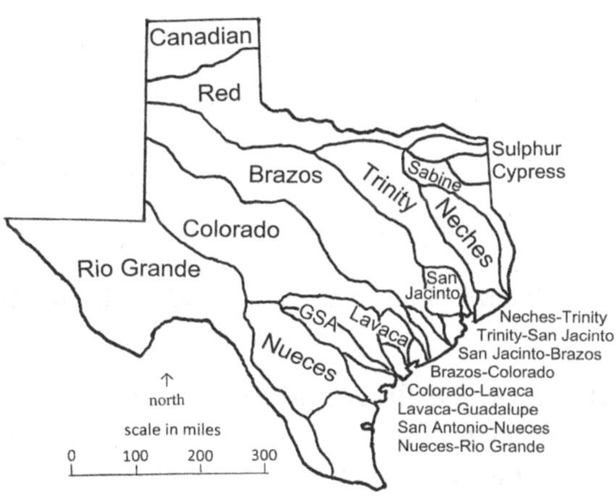

Figure 1-4. Fifteen major river basins and eight coastal basins of Texas.

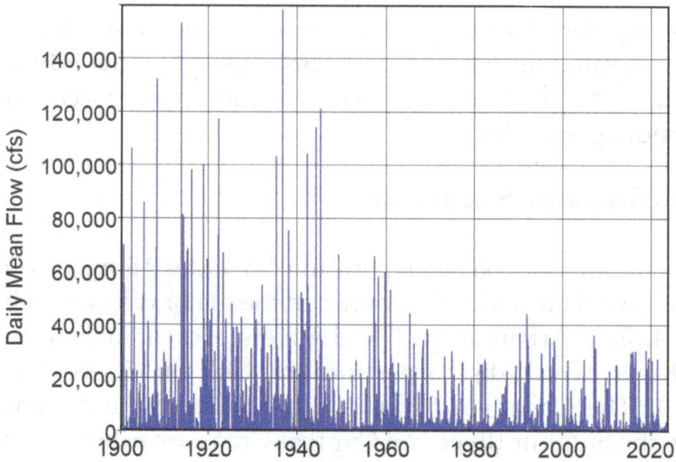

Figure 1-5. Daily mean flows of the Brazos River at Waco during January 1900 through December 2023.

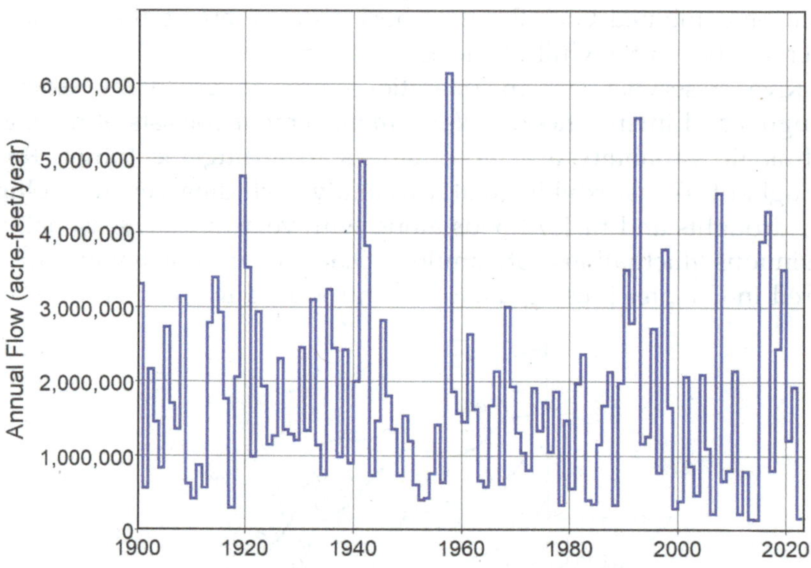

Figure 1-6. Annual flows of the Brazos River at Waco during 1900 to 2023.

Observed flows of the Brazos River at the US Geological Survey gauge at the city of Waco in Central Texas are plotted in Figures 1-5 and 1-6. Figure 7-2 in Chapter 7 is a map of the Brazos River Basin that includes this gauge site. Mean flows in cubic feet per second (ft^3/s)

during each day from January 1900 through December 2023 are plotted in Figure 1-5. Annual flow volumes in acre-feet during each of the 124 years of 1900 to 2023 are plotted in Figure 1-6.

The plots of Figures 1-5 and 1-6 illustrate the great variability characteristic of river flows throughout Texas. Large volumes of reservoir storage are required to reduce flooding, develop reliable water supplies, and otherwise deal with the extreme streamflow variability.

Analyses of flows at multiple gauge sites presented in Chapter 5 indicate that long-term trends of decreases in certain flow metrics are evident at some locations, flow increases have occurred in other cases, and some gauge sites exhibit no evident long-term trends of flow changes. Flood flows are affected differently than low flows by reservoirs. Figures 1-5 and 1-6 illustrate the different effects of flood control reservoirs on daily versus annual downstream river flows. Effects of multiple-purpose reservoirs with flood control pools constructed upstream of this site during the 1950s to 1960s are apparent in the daily mean flows of Figure 1-5, but effects on annual flows of Figure 1-6 are negligible. Flow retarding effects of reservoir flood control storage reduces daily high flows without affecting the long-term average flow.

Dams and Reservoirs

The National Inventory of Dams (NID) maintained by USACE in support of dam safety programs is accessed through a website cited in Table 4-1 of Chapter 4. In January 2024, NID included 91,843 dams in the United States meeting criteria explained in Chapter 2. This includes 7,384 dams in Texas. The TCEQ-administered water rights system includes about 3,400 reservoirs with conservation storage capacities of 200 ac-ft or greater (Wurbs 2022). Water rights are not required for dams providing only flood control. Flood control dams are included in NID along with all other types of dams but are not included in the 3,400 reservoirs authorized by water right certifications.

Of the 3,400 reservoirs authorized by water rights, 98% of the conservation storage capacity is contained in about 210 reservoirs that have authorized conservation storage capacities of at least 5,000 ac-ft. The remaining 2% of the storage volume is in other (over 3,000) reservoirs with capacities ranging from 200 to 5,000 ac-ft. Storage facilities of less than 200 ac-ft in size are not included in the water rights inventory. Eighty-one reservoirs with capacities of at least 50,000 ac-ft account for 92% of the total conservation storage capacity of the approximately 3,400 licensed reservoirs.

Most of the storage capacity in the numerous storage facilities throughout Texas is contained in a small number of the largest major reservoirs. The major reservoirs discussed in this book are exclusive of thousands of small farm ponds, urban ponds created for aesthetics, water supply storage facilities, urban stormwater detention basins, and rural flood-retarding structures.

Conservation and Flood Control Storage Capacity

Reservoir storage capacity is categorized as (1) conservation storage for beneficial purposes, (2) gate-controlled flood control, and (3) surcharge. Water is stored in a conservation pool for municipal, industrial, and/or agricultural water supply, hydroelectric energy generation, thermal-electric cooling water, recreation, and/or other beneficial purposes. An inactive portion of conservation storage with no releases may facilitate recreation and provide head for hydropower. Flood control storage capacity is maintained empty except during and immediately following flood events. The water surface level rises into surcharge storage whenever the reservoir is full to its conservation and flood control storage capacity and inflows exceed its outflow capacity. Encroachment into surcharge storage results in uncontrolled spills over the spillway of the dam.

For a reservoir with a gated outlet structure, storage in the flood control pool may be regulated by gate operations. For a reservoir with only ungated outlets, flood inflows are attenuated by surcharge storage resulting from limited spillway capacity, uncontrolled by human gate operating decisions. Flood flows are reduced by outlet structure discharge capacities, without human-operated gates, at more than 2,000 ungated flood-retarding structures in Texas constructed by NRCS in smaller rural watersheds. Likewise, stormwater flow attenuation in thousands of urban detention facilities results from limited discharge capacities of ungated outlet structures. Most NRCS rural flood-retarding structures and local community urban stormwater detention basins are much smaller than 5,000 ac-ft.

Major Reservoirs

A major reservoir is typically viewed by the water agencies in Texas as having a storage capacity of 5,000 ac-ft or larger at its targeted operating level or top of conservation pool. Since storage capacities decrease over time with sedimentation and may not necessarily be precisely measured, the application of this criterion for categorizing reservoirs is uncertain for reservoirs with storage capacities a little more or less than 5,000 ac-ft. As an aid in visualizing this volume, a reservoir with a surface area of 500 ac and an average depth of 10 ft contains a volume of 5,000 ac-ft.

The definition of major reservoir adopted for this book considers gate-controlled flood control storage capacity as well as conservation capacity. A major reservoir is defined as having at least 5,000 ac-ft of conservation storage capacity or at least 5,000 ac-ft of designated gate-controlled flood control storage capacity or both. Releases from gate-controlled flood control storage are made by gate operations implementing human operating decisions. Conservation storage is most often associated with municipal, industrial, and agricultural water supply, thermal-electric and hydroelectric energy production, fisheries, and recreation. However, several brine control reservoirs in Texas are also included in this book in the category of conservation storage.

An inventory of existing major dams and reservoirs located wholly or partially in Texas provided in Appendixes A and B is comprised of descriptive information for 195 reservoirs. This includes 192 reservoirs with conservation storage capacities of 5,000 ac-ft or greater, of which 33 also have flood control pools. Three other reservoirs have gated flood control storage capacities of at least 5,000 ac-ft but no conservation storage capacity.

The 195 major reservoirs described in the appendixes range in size as shown in Table 1-3. Ranges are specified in the table alternatively by the total storage capacity below the top of flood control pool, total capacity below the top of conservation pool, and total capacity of the flood control pool. The flood control capacity is the volume between the top of conservation and the top of flood control. The conservation storage capacity is the volume below the top of conservation pool.

The summation of the total controlled storage capacity exclusive of surcharge storage of each of the 195 major reservoirs is 58,872,709 ac-ft, comprised of 40,129,620 ac-ft of conservation storage in 192 reservoirs and 18,743,089 ac-ft of flood control storage in 36 of the 195 reservoirs. The 69 reservoirs with capacities of at least 100,000 ac-ft account for 95.15% of the total capacity of the 195 reservoirs.

TWDB maintains an array of information regarding reservoirs conveniently accessed online as described in Chapter 4. Most reservoirs included in the TWDB compilation meet the 5,000 ac-ft of conservation storage capacity criterion for being categorized as a major reservoir. The

Table 1-3. Ranges of Storage Capacities of the 195 Major Reservoirs Described in the Appendixes.

Range (acre-feet)	Number of Reservoirs and Storage Capacity (acre-feet)					
	Total Storage		Conservation Storage		Flood Control	
≥1,000,000	13	35,025,436	8	19,781,089	5	9,734,312
500,000–999,999	15	10,664,533	11	7,554,475	6	3,677,434
100,000–499,999	41	10,328,080	39	9,426,633	19	4,994,085
50,000–99,999	9	638,739	15	1,050,770	4	307,980
10,000–49,999	78	1,927,081	79	2,019,854	2	29,278
5,000–9,999	39	288,840	40	296,799	0	0
Total	195	58,872,709	192	40,129,620	36	18,743,089

TWDB inventory of existing major reservoirs includes 188 water supply reservoirs and 20 other reservoirs serving recreation, hydropower, or other purposes without providing water supply. As previously noted, the 2022 state water plan includes proposals for the construction of 23 additional reservoir projects in the future (TWDB 2022b).

All major reservoirs in Texas, including Caddo Lake, are impounded by constructed dams. Caddo Lake on Cypress Bayou on the Texas and Louisiana border is the only natural lake in Texas with a storage capacity greater than 5,000 ac-ft. However, a dam was constructed by a private company in 1914 to raise the water level and reconstructed by USACE in 1968 to 1971 to preserve the lake. Caddo Lake has a storage capacity of about 129,000 ac-ft.

The photographs in Figures 1-7 and 1-8 illustrate ecosystem and terrain differences between major reservoirs in different regions of Texas. Figure 1-7 shows an area of Caddo Lake consisting of an inundated forest of bald cypress trees draped with Spanish moss. The shoreline of Possum Kingdom Reservoir pictured in Figure 1-8 consists of steep rock cliffs with scattered juniper, mesquite, and oak trees on top. Possum Kingdom Reservoir in northcentral Texas is included in Tables 1-4 and 1-5 and Figure 1-9 as the 28th largest reservoir.

Largest Reservoirs

The 28 reservoirs located partially or completely in Texas with storage capacities of 500,000 ac-ft or larger are listed in Tables 1-4 and 1-5 in order of total storage capacity. The locations of the dams are shown in Figure 1-9.

Figure 1-7. Caddo Lake on Cypress Bayou in northeast Texas.
Source: Courtesy of Texas Water Resources Institute.

Figure 1-8. Possum Kingdom Reservoir on the Brazos River in northcentral Texas.
Source: Courtesy of Texas Water Development Board.

Table 1-4. Reservoirs with Total Storage Capacities of 500,000 Acre-Feet or Greater.

Map No.	Reservoir (Dam)	River of Dam	Watershed Area (square miles)	Owner
1	Texoma (Denison Dam)	Red River	33,783	USACE Tulsa District
2	International Amistad	Rio Grande	145,040	IBWC
3	Toledo Bend	Sabine River	7,178	Sabine River Authority
4	Sam Rayburn	Angelina River	3,449	USACE Fort Worth
5	International Falcon	Rio Grande	164,482	IBWC
6	Wright Patman	Sulphur River	3,443	USACE Fort Worth
7	Whitney	Brazos River	26,606	USACE Fort Worth
8	Travis (Mansfield Dam)	Colorado River	38,130	Lower Colorado RA
9	Livingston	Trinity River	16,616	Trinity River Authority
10	Meredith (Sanford Dam)	Canadian River	16,048	Canadian River MWD
11	Richland-Chambers	Richland Creek	1,957	Tarrant Regional WD
12	Belton	Leon River	3,570	USACE Fort Worth
13	Ray Roberts	Elm Fork Trinity	692	USACE Fort Worth
14	Lewisville	Elm Fork Trinity	1,660	USACE Fort Worth
15	Buchanan	Colorado River	11,900	Lower Colorado RA
16	Tawakoni (Iron Bridge)	Sabine River	756	Sabine River Authority
17	Lake O' the Pines	Cypress Creek	850	USACE Fort Worth
18	Lavon	East Fork Trinity	770	USACE Fort Worth
19	Canyon	Guadalupe River	1,425	USACE Fort Worth
20	Waco	Bosque River	1,670	USACE Fort Worth
21	Choke Canyon	Frio River	4,667	City of Corpus Christi
22	Cedar Creek	Cedar Creek	1,007	Tarrant Regional WD
23	Lake Fork	Lake Fork Creek	493	Sabine River Authority
24	Twin Buttes	South Concho	3,724	City of San Angelo
25	Stillhouse Hollow	Lampasas River	1,318	USACE Fort Worth
26	Kemp	Wichita River	2,086	City of Wichita Falls
27	O. H. Ivie	Colorado River	12,647	Colorado River MWD
28	Possum Kingdom (Morris Sheppard Dam)	Brazos River	13,310	Brazos River Authority

Table 1-5. Water Surface Areas and Storage Capacities of the 28 Largest Reservoirs.

Map No.	Reservoir	Began Storage	Area (acres)	Storage Capacity (ac-ft)		
				Conservation	Flood Control	Total
1	Texoma	1943	74,690	2,516,000	2,877,000	5,393,000
2	International Amistad	1968	66,460	3,276,000	1,734,000	5,010,000
3	Toledo Bend	1966	181,600	4,477,000	–	4,477,000
4	Sam Rayburn	1965	112,600	2,876,000	1,122,000	3,998,000
5	International Falcon	1953	85,200	2,647,000	501,000	3,148,000
6	Wright Patman	1956	18,250	97,900	2,556,000	2,654,900
7	Whitney	1951	23,220	554,000	1,445,000	2,000,000
8	Travis	1940	19,300	1,135,000	787,000	1,922,000
9	Livingston	1969	83,280	1,742,000	–	1,742,000
10	Meredith	1965	16,410	818,000	590,000	1,408,000
11	Richland–Chambers	1987	43,400	1,113,000	–	1,113,000
12	Belton	1954	12,140	435,000	663,000	1,098,000
13	Ray Roberts	1987	28,600	788,000	276,000	1,065,000
14	Lewisville	1954	27,200	599,000	383,000	982,000
15	Buchanan	1937	22,140	887,000	–	887,000
16	Tawakoni	1960	37,300	872,000	–	872,000
17	Lake O' the Pines	1957	16,900	241,000	601,000	842,000
18	Lavon	1953	20,600	409,000	339,000	748,000
19	Canyon	1964	8,310	379,000	362,000	741,000
20	Waco	1965	8,190	190,000	537,000	726,000
21	Choke Canyon	1982	26,440	663,000	–	663,000
22	Cedar Creek	1965	32,870	645,000	–	645,000
23	Lake Fork	1979	26,890	637,000	–	637,000
24	Twin Buttes	1962	9,080	186,000	446,000	632,000
25	Stillhouse Hollow	1968	6,480	228,000	403,000	630,000
26	Kemp	1922	15,360	245,000	323,000	568,000
27	O. H. Ivie	1990	19,150	554,000	–	554,000
28	Possum Kingdom	1941	7,100	538,000	–	538,000
	Totals		1,049,160	29,747,900	15,945,000	45,692,900

Integer identifiers of dam sites on the map reference the first column of Tables 1-4 and 1-5. Watershed drainage areas in square miles above the dams are tabulated in the fourth column of Table 1-4. The reservoir name is followed by the name of the dam in parenthesis in the second column of Table 1-4 if the names are different.

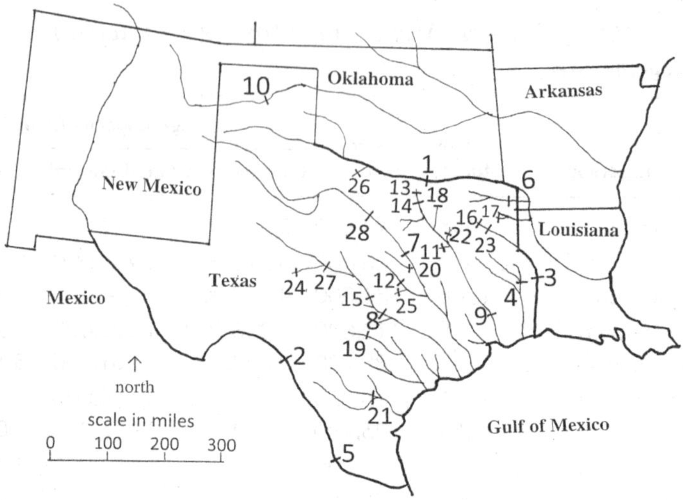

Figure 1-9. Locations of the 28 largest reservoirs in Texas listed in Tables 1-4 and 1-5.

The third column of Table 1-5 is the year that outlet structure gates were initially closed to impound water. Water surface areas for full conservation pools are listed in the fourth column of Table 1-5. Conservation, operator-controlled flood control, and total storage capacities are tabulated in the last three columns. These 28 largest reservoirs have conservation, flood control, and total storage capacities totaling 29,747,900, 15,945,000, and 45,692,900 ac-ft, respectively, which represents 74.13%, 85.07%, and 77.61% of the totals for the 195 major reservoirs located partially or completely in Texas. The total water surface area at the top of conservation pool for the 28 reservoirs is 1,049,160 ac (1,640 sq. mi). Major portions of the storage capacity of International Amistad and Falcon Reservoirs on the Rio Grande, Lake Texoma on the Red River, and Toledo Bend Reservoir on the Sabine River are controlled by management entities and water users in Mexico, Oklahoma, and Louisiana as discussed in Chapter 2.

Storage capacities and water surface areas shown for the 28 reservoirs in Table 1-5 and the information in Appendixes A and B for the 195 major reservoirs are from multiple sources, the most prevalent being TWDB and USACE information accessible at websites referenced in Table 4-1. Information for USACE reservoirs is from USACE sources. An array of information presented throughout the book is available online from TWDB websites. Other agencies owning reservoirs also provide

Figure 1-10. Toledo Bend Dam and Reservoir on the Sabine River.
Source: Courtesy of Texas Water Development Board.

information about their reservoirs on their websites. Storage capacities are subject to variations over time due to sedimentation and other factors discussed in Chapters 3 and 6.

Toledo Bend Reservoir shown in Figure 1-10 has the largest conservation storage capacity of any reservoir located wholly or partially in Texas. Lake Texoma has the largest combined total conservation and flood control storage capacity of any reservoir located wholly or partially in Texas. Lake Whitney shown in the photo on the front cover is the seventh largest reservoir.

For comparison, the two largest reservoirs in the United States are Lakes Mead and Powell impounded by Hoover and Glen Canyon Dams on the Colorado River in Nevada, Arizona, and Utah constructed and operated by the USBR. The third, fourth, and fifth largest are Sakakawea, Oahe, and Fort Peck Reservoirs impounded by Garrison, Oahe, and Fort Peck Dams on the Missouri River in North Dakota, South Dakota, and Montana constructed and operated by the USACE. These five reservoirs have combined flood control and conservation storage capacities of 28.2, 25.2, 23.8, 23.5, and 18.7 million ac-ft, respectively.

Lake Texoma in Tables 1-4 and 1-5, with a total capacity of 5.393 million ac-ft, is the largest reservoir in either Texas or Oklahoma and the 12th largest in the United States.

Mansfield Dam on the Colorado River impounding Lake Travis is the tallest dam in Texas. Figure 7-10 is a photograph of Mansfield Dam. The concrete gravity dam has a maximum height of 266 ft, top of dam elevation of 750 ft above mean sea level, and length including earth and rockfill section at either end of 7,100 ft. The maximum dam height above streambed is included in the data tabulated in Appendix A for each of the 195 major reservoirs.

Oldest and Newest Reservoirs

McDaniels (1964), Dowell and Breeding (1967), and Wurbs (1985, 1987) discuss the early history of developing dam and reservoir projects in Texas. As previously noted, Caddo Lake is the only natural lake in Texas with a storage capacity of 5,000 ac-ft or greater, but a dam was constructed in 1914 and rehabilitated during 1968 to 1971 to preserve the lake. Although a few small dams were constructed in Texas before 1900, except for Caddo Lake, Eagle Lake is the oldest of the major reservoirs still in existence. Eagle Lake, with impoundment beginning in 1900, is a 9,600 ac-ft irrigation reservoir in the Colorado River Basin. The 35 major reservoirs in operation in 1935 were small projects constructed for irrigation, municipal and industrial water supply, and/or hydroelectric power. Most of the present reservoir storage capacity in Texas was created between 1935 and 1985, with 1960 to 1970 being the period of greatest addition of storage capacity resulting from completion of new projects.

Billington et al. (2005) describe the history of large federal dam construction throughout the United States. Mansfield Dam and Lake Travis on the Colorado River was the first of the large multiple-purpose projects constructed in Texas by the federal government (McDaniels 1964). The project was constructed by USBR from 1937 to 1942 and is now owned and operated by the Lower Colorado River Authority (LCRA). Denison Dam and Lake Texoma on the Red River was the first USACE project in the state. Construction was initiated and completed in 1939 and 1943. The USACE Tulsa District owns and operates the project. Denison Dam was the largest rolled earth-fill dam in the United States at the time of construction. Lake Texoma is still the largest reservoir in Texas in terms of total flood control and conservation storage capacity (Table 1-5).

The newest major reservoir project in Texas for which impoundment of water has begun as of February 2024 is the Bois d' Arc Dam and Reservoir on Bois d' Arc Creek in the Red River Basin in northeast Texas along with a water treatment plant and pipeline conveyance facilities for transporting water to the upper Trinity River Basin. Construction began in 2018 with some remaining work underway during 2024. Impoundment of water began in April 2021, and the reservoir had filled to 72% of capacity by February 2024. The project is owned and operated by the North Texas Municipal Water District, which serves 13 member cities. Bois d' Arc Reservoir has a water supply storage capacity of 368,000 ac-ft and water surface area when full of 16,640 ac.

Since construction of the Arbuckle and Ralph Hall projects is not yet completed to the stage of actually storing water, these two reservoirs are not included in the 195 existing reservoirs listed in the appendixes. Construction of Arbuckle Reservoir owned by LCRA was substantially completed in 2019, but additional remedial work was still required as of January 2024 to mitigate seepage problems before water can be stored. LCRA expects Arbuckle Reservoir to become operational during 2024. The off-channel reservoir located in Wharton County will have a storage capacity of 40,000 ac-ft and cover an area of 1,100 ac.

Construction of the Lake Ralph Hall municipal water supply project owned by the Upper Trinity Regional Water District began in June 2021 with water delivery expected by 2026. This lake on the North Sulphur River in northeast Texas will have a conservation storage capacity of 180,000 ac-ft and a surface area of 7,600 ac.

References

ASCE. 2018. *Addressing flood risk: A path forward for Texas after Hurricane Harvey*. Austin, TX: ASCE Texas Section Task Committee on Post-Hurricane Harvey Recommendations.

Billington, D. P., D. C. Jackson, and M. V. Melosi. 2005. *The history of large federal dams: Planning, design, and construction*. Denver: US Bureau of Reclamation.

Caroom, D. G., and S. M. Maxwell. 2013. *Texas water law overview*. Austin, TX: Texas Water Law Institute.

Dieter, C. A., M. A. Maupin, R. R. Caldwell, M. A. Harris, et al. 2018. *Estimated use of water in the United States in 2015*. Circular 1441. Reston, VA: US Geological Survey.

Dowell, C. L., and Breeding, S. D. 1967. *Dams and reservoirs in Texas, historical and descriptive information*. Rep. 48. Austin, TX: Texas Water Development Board.

McDaniels, L. L. 1964. *Conservation storage reservoirs in Texas: Some aspects and chronology of surface-water development*. Bulletin 6404. Austin, TX: Texas Water Commission.

Nielsen-Gammon, J. W. 2012. "The 2011 Texas drought." *Texas Water J.* 3 (1): 59–95.

Porter, C. R. 2014. *Sharing the common pool: Water rights in the everyday lives of Texans.* College Station, TX: Texas A&M University Press.

TWDB. 2022a. "Major aquifers." Accessed November 27, 2023. https://www.twdb.texas.gov/groundwater/aquifer/major.asp.

TWDB (Texas Water Development Board). 2022b. *Water for Texas: 2022 state water plan*. Austin, TX: TWDB.

Winters, K. E. 2013. *A historical perspective on precipitation, drought severity, and streamflow in Texas during 1951–56 and 2011*. Scientific Investigations Rep. 2013–5113. Reston, VA: US Geological Survey.

Wurbs, R. A. 1985. *Reservoir operation in Texas*. TR-135. College Station, TX: TWRI.

Wurbs, R. A. 1987. "Reservoir management in Texas." *J. Water Resour. Plan. Manag.* 113 (1): 130–150.

Wurbs, R. A. 2015. "Sustainable statewide water resources management in Texas." *J. Water Res. Plan. Mgmt.* 141 (12): A4014002-1-10.

Wurbs, R. A. 2022. "Institutional framework for modeling water availability and allocation." *J. Water*. 2022 (12): 2767.

Young, S. C., R. F. Mace, and C. Rubinstein. 2018. "Surface water-groundwater interaction issues in Texas." *Texas Water J.* 9 (1): 129–149.

Chapter 2

Institutional Framework for River and Reservoir System Water Management

The welfare and prosperity of the citizens of Texas, like elsewhere throughout the United States, depend on the capabilities of a myriad of people and organizations responsible for collaboratively managing water and related land and environmental resources. Planning, development, and management of the water resources of rivers and reservoirs are accomplished within a setting of evolving institutional responsibilities and capabilities. The institutional framework for river and reservoir system water management is embedded within a more general water management institutional setting, which is comprised of an array of interconnected organizations, laws, policies, customs, infrastructure operational practices, and resource management strategies.

Constructed facilities and natural resources are managed to serve human needs while protecting the environment. Hydrologic variability necessitates large volumes of reservoir storage to provide reliable water supplies during periods of drought or low flows. Conversely, excessive rainfall and streamflow result in socioeconomic damages and loss of life due to flooding. Reservoir storage reduces flood damages. Sanitation and pollution control are essential. The vitality of natural ecological systems is dependent on human stewardship of water resources.

Water has played a fundamental role in the development of civilizations since ancient times. Population growth, urban and industrial development, and expansion of irrigated agriculture have resulted in dramatically increased demands on water resources and water managers during the past century worldwide and in Texas.

Institutional Responsibilities and Capabilities

The institutional framework for both water management in general and river/reservoir system water management is shaped by the following types of responsibilities:

- Responsibilities for specific types of water-related needs, problems, and opportunities (municipal, industrial, and agricultural water supply; wastewater collection and treatment; hydroelectric energy; flood control; floodplain management; urban stormwater management; fish and wildlife conservation; recreation; aesthetics; water quality management; navigation);
- Functional responsibilities (policy formulation; financing; planning at statewide, regional, river basin, and project levels; project design and construction; maintenance and operation of constructed facilities; water allocation; protection of the environment and public health and safety; research; education);
- Responsibilities best suited to different levels of government or the private sector (international, federal, state, regional, or local government; private industry; nonprofit entities); and
- Interconnections between the water sector and other sectors (energy, environment, transportation, land development, and regional and urban development).

The main entities responsible for accomplishing the responsibilities outlined above are listed in Table 2-1. Individual agencies or organizations typically have multiple types of responsibilities that span the various categories of water management responsibilities in diverse ways. A wealth of information can be accessed through the websites maintained by the federal and state agencies that are included in Table 2-1. Other organizations also have websites that provide an array of information.

The history of dams and other water development projects dates back thousands of years and closely follows the rise and decline of civilizations, especially in cultures highly dependent on agricultural

Table 2-1. Texas Water Management Community.

Federal Agencies		
International Boundary and Water Commission (IBWC) http://www.ibwc.gov		
Corps of Engineers (USACE) http://www.usace.army.mil		
Bureau of Reclamation (USBR) http://www.usbr.gov		
Natural Resource Conservation Service (NRCS) http://www.nrcs.usda.gov		
Geological Survey (USGS) http://www.usgs.gov		
National Weather Service (NWS) http://www.nws.noaa.gov		
Federal Emergency Management Agency (FEMA) http://www.fema.gov		
Fish and Wildlife Service (USFWS) http://www.fws.gov		
Environmental Protection Agency (EPA) http://www.epa.gov		
State Agencies		
Texas Water Development Board (TWDB) http://www.twdb.state.tx.us		
Texas Commission on Environmental Quality (TCEQ) http://www.tceq.state.tx.us		
Texas Parks and Wildlife Department (TPWD) http://www.tpwd.state.tx.us		
Texas Water Resources Institute (TWRI) http://twri.tamu.edu		
Local and Regional	*Private Sector*	*Other*
Cities and towns	Individual citizens	US Congress
Municipal water districts	Industrial water users	Texas Legislature
Irrigation districts	Private water suppliers	Advocacy groups
Groundwater districts	Consulting firms	Volunteer service groups
Other types of districts	Construction contractors	Professional associations
River authorities	Equipment suppliers	Universities

irrigation (Jansen 1988, Schnitter 1994, Jackson 1997). Ancient water projects and associated water management institutions were developed in the cradles of civilization of Babylonia, Egypt, India, Persia, and Asia. Many ancient water control structures and water management customs and practices are still functioning today.

Viessman and Welty (1985) discuss development of water technology and institutions in the United States focusing on the period from 1900 through 1980. A diverse array of issues is addressed in collections of papers edited by Reuss (1993), Haynes and Barnes (2009), and Boes (2011). Dworsky et al. (2020) assess current and future institutional issues in federal water resources planning and management in the United States. Executive Order 13956 articulates federal policy initiatives in modernizing water management nationwide (Executive Office of the President 2020).

The institutional framework for river/reservoir system water management in Texas has been evolving for more than 200 years and is still being refined and expanded in response to population growth, economic development, changes in priorities at all levels, and technological advances. Texas became a republic in 1836 and a state in the United States in 1845. The evolution of water rights, irrigation practices, and other aspects of water management in Texas predates the Republic of Texas era. Most of the constructed facilities and institutional capabilities functioning today were created during the twentieth century. Although the construction of new large-scale water projects has diminished since the 1980s, new facilities continue to be constructed. Institutional responsibilities and capabilities are growing during the 21st century.

The history of dam construction in Texas is briefly introduced in Chapter 1. A few dams, conveyance channels, and other types of water management facilities were constructed in Texas before 1900. Eagle Lake completed in 1900 is the oldest dam and reservoir project in Texas that is still functioning. Most of the major dams in Texas were constructed between 1935 and 1985, with 1945 to 1975 being the most active period of dam construction.

The 1940s to 1970s nationwide era of constructing large-scale water projects transitioned into a greater focus on operation, maintenance, and rehabilitation of a massive inventory of aging constructed facilities concurrently with dramatic growth in regulations to protect the environment (Wurbs 2017b). Management strategies are now relying more on using limited water resources more efficiently, better managing floodplain land use, optimizing operations of existing reservoir systems, and integrating diverse water management practices. Water right systems and other water allocation mechanisms continue to grow in importance with intensifying demands on limited resources. A shift from federal and local dominance to increased state-level responsibilities has occurred. Advances in computer-based decision-support technologies are improving capabilities for managing hydrologic variability and future uncertainty. For example, water supply reliability assessments and flood forecasting techniques are being applied to improve reservoir system operations.

Institutional changes are driven by politics, economics, and technology. Floods and droughts also motivate institutional change. For example, devastating flooding in the Mississippi River Basin in 1935 resulted in the US Congress establishing, through the Flood Control Act of 1936, the USACE responsibility for controlling floods on major rivers (Arnold 1988). The USACE subsequently constructed many multipurpose reservoirs with flood control, flood-control-only reservoirs, levees, and channel improvements throughout the nation, including many in Texas. The creation of TWDB in 1957 and other significant water management activities in Texas were driven by the 1950 to 1957 drought. Drought conditions again from 1996 to 1997 encouraged enactment of the 1997 Senate Bill 1 by the Texas legislature that created the regional water planning process discussed in Chapter 1 and water availability modeling (WAM) system described in Chapter 4. Recent flooding including Hurricane Harvey flooding in 2017 motivated creation of a regional flood control planning program within the TWDB.

Reservoir Project Developers, Owners, and Managers

Development of a reservoir project requires many years of preconstruction planning feasibility studies, followed by detailed design studies along with political project authorization and financing and obtaining of permits from regulatory agencies, followed by multiple years of construction. The constructed project requires continual maintenance and operation and occasional replacement or rehabilitation of project components. For many of the large reservoirs in Texas and throughout the United States, the time from conceptualization of the reservoir project in an initial planning study to completion of construction spanned several decades. Many people and organizations contribute in many ways to the development process. Reservoir owners are not necessarily always the same entity that originally planned and developed the project. However, for many reservoirs, the present owner and operator and the original primary developer are the same agency.

Inventory of Existing Major Reservoirs

Basic information for each of the 195 existing major reservoirs located wholly or partially in Texas is found in Appendixes A and B. Appendix A is a list of the 195 reservoirs grouped by river basin with the following information for each reservoir: primary owner or manager, project purposes, date of initial impoundment, dam height, and water surface area at the top of conservation pool and former or alternative names. Appendix B provides additional information including location and storage capacities for each of the 195 major reservoirs.

Thirty-three of the major reservoirs have both flood control and conservation pools. Three of the 195 reservoirs (Addicks, Barker, and Olmos) contain only flood control. Most have conservation storage for municipal and/or industrial supply and recreation. Agricultural irrigation is also supplied from many of the conservation pools. Twenty-six hydropower plants generate electricity using reservoir releases or water conveyance flow. Several reservoirs were constructed for the deposal of salt brine. One project stores wastewater from steel production.

The 195 major reservoirs are grouped by type of primary owner or controlling entity in Table 2-2. The number of reservoirs in each owner

Table 2-2. Reservoir Storage Capacity by Type of Owner and Operator as Percentage of Total.

Primary Reservoir Owner/Operator	Number Projects	Conservation (%)	Flood Control (%)	Total Capacity (%)
IBWC, United States and Mexico, Rio Grande	2	14.76	11.93	13.86
USACE FWD—Natural Caddo Lake	1	0.321	0.00	0.219
USACE FWD—Multipurpose Reservoirs	25	20.32	58.65	32.52
USACE Galveston—Flood Control Dams	2	0.000	2.183	0.695
USACE Tulsa—Multipurpose Reservoirs	3.	7.352	17.45	10.57
USACE Tulsa—Truscott Brine Control	1	0.267	0.00	0.182
US Fish & Wildlife—Buffalo Lake	1	0.045	0.00	0.031
US Forest Service—Coffee Mill Lake	1	0.020	0.00	0.014
TPWD—Shelton Reservoir	1	0.014	0.00	0.009
USBR Construction, Nonfederal Owners	5	7.375	9.723	8.123
River Authorities	22	26.12	0.067	17.83
Water Districts	42	15.14	0.00	10.32
Cities	47	5.444	0.00	3.711
Counties	5	0.137	0.00	0.093
Electric Utility Companies	24	2.233	0.00	1.522
Other Private Companies	13	0.450	0.00	0.307
Totals for 195 Major Reservoirs	195	100.0	100.0	100.0
Total Storage Capacity (ac-ft)		40,129,620	18,743,089	58,872,709

category is shown in Table 2-2, along with the storage capacity for reservoirs in each group expressed as a percentage of the total storage capacity of all 195 reservoirs for (1) conservation storage capacity, (2) flood control storage capacity, and (3) total storage capacity. The storage capacities for each owner group are expressed as a percentage of the volumes in the last line of Table 2-2. The total storage capacity in acre-feet shown in the last line of Table 2-2 is also found in the last line of Table 1-3 of the preceding chapter.

Surcharge storage with inflows spilling through spillways or outlet conduits is not included in the quantities discussed here. Allocation of storage between flood control and conservation purposes varies with season or other factors in some reservoirs. The 195 reservoirs have combined conservation and flood control storage capacities totaling about 58,873,000 ac-ft with about 31.84% designated for flood control and the other 68.16% serving primarily conservation storage purposes. Storage capacities of each reservoir are included in Appendix B.

The 195 major reservoir projects were constructed and/or are owned, maintained, and operated by IBWC, USACE, USBR, Fish and Wildlife Service (USFWS), US Forest Service, eight river authorities, 36 water districts, 42 cities, two county agencies, a state agency (TPWD), and 23 electric utility companies and other private companies. Some entities own and operate systems of multiple reservoirs. Other entities contract for the conservation storage capacity, water supplied, or electrical energy generated at the reservoir projects. Recreation lands adjacent to the reservoirs are managed by the reservoir owners, TPWD, and/or other organizations. Government agencies that maintain and operate reservoirs are called reservoir owners here even though the projects may be legally owned by federal, state, or city governmental jurisdictions representing the citizens.

Flood Control Storage in Federal Reservoirs

All reservoirs incidentally store streamflow in their conservation pools when floods occur during times when the conservation pools have been drawn down. The term *flood control storage* as used here refers to designated gate-controlled flood control pools.

All but one of the major reservoirs containing flood control pools were constructed by the USACE, IBWC, or USBR. Olmos Reservoir, the smallest of the major reservoirs with a primary flood control purpose, was constructed and is owned by the San Antonio River Authority.

The federal government has borne all costs associated with flood control incurred with federally planned, designed, and constructed reservoirs. Conversely, local urban stormwater management and drainage, like municipal and industrial water supply, is a local responsibility.

IBWC maintains and operates International Amistad and Falcon, which are the second and fifth largest reservoirs in Texas (Table 1-5, Figure 1-9). USACE maintains and is responsible for flood control operations of the other federal reservoirs in Texas. USACE Galveston District maintains and operates Addicks and Barker in Houston, which are large flood-control-only reservoirs that remain empty except during and following floods. Reservoir lands are used for recreation between floods. USACE Fort Worth District maintains and operates 25 multiple-purpose reservoirs, all including flood control, that contain 32.52% of the total storage capacity of the 195 reservoirs. The Fort Worth District also maintains the dam at Caddo Lake, which has no flood control pool. As noted in Chapter 1, Caddo Lake is the only natural lake in Texas with at least 5,000 ac-ft of storage, but USACE constructed a dam in 1968 to 1971 for the safety and preservation of the historically naturally occurring ecologically unique lake.

The USACE Tulsa District maintains and operates two multipurpose reservoirs (Texoma and Pat Mayse) and the Truscott brine control dam in the Texas portion of the Red River Basin. Lake Texoma has the largest flood control and total storage capacities in Texas. Lake Kemp is owned and maintained by the city of Wichita Falls and Wichita County Water Improvement District No. 2, but USACE Tulsa District enlarged the reservoir to include a flood control pool. USACE Tulsa District now has oversight responsibility for flood control operations of Lake Kemp.

The USBR has planned and constructed five reservoir projects (Travis, Meredith, Twin Buttes, Choke Canyon, and Texana) in Texas, which have been turned over to local sponsors consisting of two river authorities, a water district, and two cities to own, maintain, and operate. USACE has oversight responsibility for flood control operations at three USBR-constructed reservoirs that contain designated flood control storage capacity, which were developed with federal funding. The five USBR-developed, nonfederal-owned reservoirs contain 7.375%, 9.723%, and 8.123% of the conservation, flood control, and total storage capacities of the 195 major reservoirs.

Water Supply Storage

Most of the nonfederal and federal reservoirs include water supply as the primary or one of multiple primary purposes. Of the 195 major reservoirs, 154 containing 49.54% of the conservation storage and 0.067% of the flood control storage (Olmos Reservoir) were constructed and are owned and operated by nonfederal entities. About half of the water supply storage capacity along with almost all flood control storage capacity is contained in the federally constructed reservoirs, of which five are owned and operated by nonfederal sponsors. The water supply storage in all USACE reservoirs is controlled by nonfederal sponsors pursuant to water supply contracts executed in accordance with the Water Supply Act of 1958.

Nonfederal water supply reservoirs are more numerous, but the federal reservoirs tend to be much larger. All 28 reservoirs with total storage capacities exceeding 500,000 ac-ft listed in Tables 1-4 and 1-5 contain large storage volumes for water supply. The 18 reservoirs in Tables 1-4 and 1-5 with flood control storage include two owned by IBWC, 12 owned by USACE, one (Lake Kemp) enlarged by USACE, three constructed by USBR, and the nonfederal Olmos Reservoir owned by the San Antonio River Authority. The 10 reservoirs with no flood control storage were constructed and are operated totally by nonfederal entities with no federal funding. Toledo Bend contains 40.77% of the conservation storage capacity of these 10 largest completely nonfederal reservoirs.

Municipal and industrial storage is included in all but two (Addicks and Barker) of the reservoirs constructed by USACE and USBR, subject to nonfederal sponsorship. Two small federally funded reservoirs constructed by NRCS are owned and operated by cities for water supply. Although municipal and industrial water supply was already being included in federal reservoirs, the Water Supply Act of 1958 and amendments thereto established a uniform policy (Wurbs 1994, USACE 2016).

Under the provisions of the Water Supply Act of 1958, the federal water agencies may provide additional storage capacity for municipal and industrial water supply in reservoirs to be constructed for federal purposes such as navigation, flood control, and irrigation. Costs allocated to water supply must be repaid to the federal government, with interest, by nonfederal sponsors over a period not to exceed 50 years.

Repayment of costs for future water use can be delayed interest-free until the water is first used up to the limit of 10 years after the completion of construction. No more than 30% of the costs of the project may be allocated to storage for future supply. Inclusion of municipal and industrial water supply storage in a federal reservoir requires a contractual agreement with one or more nonfederal sponsors prior to construction.

All costs for the federal projects, including construction, operation and maintenance, and major replacement, are allocated to project purposes. All costs allocated to flood control are borne by the federal government. All costs allocated to municipal and industrial water supply are repaid by nonfederal sponsors, consisting of river authorities, water districts, or cities.

The conservation storage capacity in several of the federal reservoirs is used for agricultural irrigation as well as municipal and industrial water supply. USBR has implemented other cost-sharing arrangements for irrigation water in large federally subsidized irrigation projects in other western states. However, nonfederal sponsorship and cost allocation of storage capacity for agricultural irrigation supply in federal reservoirs constructed by the USBR and USACE in Texas has been handled the same for irrigation and municipal and industrial water users following the provisions of the Water Supply Act of 1958.

Texas and Oklahoma have well over half of the water supply storage capacity incorporated in USACE reservoirs nationwide. The majority of USACE reservoirs located in other states serve purposes of flood control, recreation, hydropower, and/or navigation without including municipal, industrial, or agricultural water supply. USACE reservoir development in Texas and Oklahoma has emphasized the inclusion of water supply in multiple-purpose reservoirs.

Hydroelectric Energy

In the Columbia River Basin of the northwestern United States and other locations in the world, hydropower accounts for a substantial portion of the total electrical energy consumed, and large reservoirs are constructed and operated primarily for hydroelectric power generation. This is not the case in Texas. With some exceptions, the most notable being Lake Texoma, hydropower is usually secondary to municipal, industrial, and agricultural water supply in operating reservoirs in Texas. Releases for water supply diversions at downstream locations pass through hydropower plants to generate electricity without additional releases specifically for hydropower. The energy head reflected in the difference in water surface elevation in the reservoir versus downstream of the hydropower plant is converted from hydraulic to mechanical to electrical energy.

Hydropower is a component of electrical energy generation systems that rely primarily on thermal-electric plants. Hydroelectric energy generation reduces coal and natural gas consumption by the thermal-electric plants in the connected power grid.

Hydroelectric power plant owners are listed in Table 2-3 with the reservoirs at or near where the hydropower plants are located. Hydropower accounts for about 1% of the total electrical energy generated and consumed in Texas.

Godfrey and Dowell (1968) describe the early history of hydroelectric power development in Texas. Hydropower played a key role in developing several major reservoir projects. Construction of many of the hydropower plants occurred concurrently with dam construction or shortly thereafter. In other cases, hydropower plants were constructed many years after the dams. Construction of the Livingston Lake and Dam project was completed with impoundment of water beginning in 1968. Construction of the R. C. Thomas Hydroelectric Power Plant at

Table 2-3. Hydroelectric Power Plants in Texas.

Hydropower Owner	Reservoirs
International Boundary & Water Commission	Amistad and Falcon
US Army Corps of Engineers	Texoma, Whitney, Rayburn, Steinhagen
Lower Colorado River Authority	Buchanan, Inks, Travis, Austin, Granite Shoals, Marble Falls
Guadalupe-Blanco River Authority	Canyon, Dunlap, McQueeney, Meadow, Placid, Nolte, Wood
Tarrant Regional Water District	Arlington
Entergy Texas, Inc.	Toledo Bend
East Texas Electric Coop, Inc	Livingston
Maverick County Water Control and Improvement District No. 1	Eagle Pass
Red Bluff Water Power Control District	Red Bluff
City of Sequin	Sequin
City of Garland	Lewisville

Lake Livingston began in 2015 and was completed in 2021. Conversely, hydroelectric power plants have been decommissioned. The Brazos River Authority (BRA) owned and operated a hydroelectric power plant at Morris Sheppard Dam beginning in 1941 that was decommissioned in 2007 because repairs needed for the then 67-year-old hydropower equipment were determined to not be economically feasible.

Hydropower could be added to other existing reservoirs in the future. Water depths behind dams provide energy head for hydropower. Water is released to empty flood control pools following flood events. Water supply releases are diverted at downstream locations. Opportunities potentially exist to pass these releases through hydropower turbines constructed in the future to generate electricity without adversely affecting water supply or flood control.

Four USACE and two IBWC reservoirs have hydroelectric power plants. Lake Travis constructed by the USBR also has hydropower, but it was added later by the LCRA. The Western Area Power Administration (WAPA) is responsible for marketing the electrical power generated at the two IBWC reservoirs. The Southwestern Power Administration (SWPA) markets the power from USACE projects. WAPA and SWPA are two of the four agencies of the US Department of Energy that market hydroelectric energy generated at federal reservoir projects in various geographical regions of the nation. The power administrations are required by law to market energy in such a manner as to encourage the most widespread use at the lowest possible rates to customers consistent with sound business principles. SWPA and WAPA operate through contracts and agreements with the electric cooperatives, municipalities, and utility companies that buy and distribute the electrical energy.

Hydropower is a primary use of the conservation pool of Lake Texoma. Hydropower generation at Lakes Whitney and International Amistad and Falcon is mainly from releases passed through hydropower turbines on the way to downstream water supply diverters. Releases through the hydropower plant at Sam Rayburn Reservoir are reregulated downstream at B. A. Steinhagen Reservoir. Another hydropower plant is operated at B. A. Steinhagen Reservoir.

Toledo Bend Reservoir, owned by the Sabine River Authorities of Texas and Louisiana, is the largest water supply reservoir in Texas and the southern United States. Hydroelectric power is generated for Entergy Texas, Inc., with releases from Toledo Bend. The Red Bluff Water Power

Control District operates Red Bluff Reservoir on the Pecos River for hydropower and irrigation.

The LCRA is a major Central Texas supplier of both electrical energy and water. LCRA operates the six Highland Lakes on the Colorado River, hydroelectric power plants at each of the six dams, four thermal-electric power plants, and two off-channel reservoirs that provide water for thermal-electric power plants. LCRA passes reservoir releases for downstream water supply diversions through the hydroelectric power plants at the six Highland Lakes whenever feasible but seldom releases water for only hydropower.

The Guadalupe-Blanco River Authority (GBRA) owns and operates six small hydropower reservoirs on the Guadalupe River and contracts with USACE for water supply storage capacity in Canyon Reservoir, which also has a GBRA-operated hydropower plant. Reservoir releases from Canyon Lake flow through the seven downstream hydroelectric power plants. The hydroelectric plant adjacent to the Canyon Dam discharge channel is operated by GBRA to provide electricity through New Braunfels Utilities to the residents of the City of New Braunfels. GBRA operates six hydroelectric plants constructed during 1928 to 1932 further downstream at small dams on the Guadalupe River to generate electricity for the Guadalupe Valley Electric Cooperative.

Recreation at Reservoir Projects

Federal and nonfederal entities are involved to various extents in providing recreation opportunities at the major reservoir projects. Several of the smaller major reservoirs are used only for recreation. Most reservoir-related recreation occurs at larger multipurpose projects.

The federal water resources program has emphasized multiple-purpose development. Federal reservoir projects all include public access and recreation facilities. Prior to 1965, recreation was included in federal projects as a fully federal expense. The Federal Water Recreation Act of 1965 established a policy of development of the full recreational potential at federal reservoirs as a full project purpose subject to nonfederal cost-sharing. Under the provisions of this law, the federal government is authorized to bear 50% of the separable costs of providing recreational facilities and to make federal lands available for the use of nonfederal entities agreeing to operate and maintain the facilities. This authority includes providing additional lands and facilities at existing projects as well as proposed new projects. If nonfederal interests do not accept responsibility for recreational development, USACE or USBR develops at federal expense minimum facilities required for public health and safety and provides access by public roads to the recreation facilities.

The federal reservoirs include a significant amount of project-owned publicly accessible land around the shoreline. Flood control pools are empty most of the time, increasing the area of land adjacent to the water in the conservation pool. Many of the nonfederal reservoirs have privately owned land adjacent to the shoreline most of the way around the lake.

Cities and water districts often include recreational facilities as an incidental use of their water supply reservoirs. Before 1971, most river authorities were not authorized to supply recreation, but were directed not to prevent free public use of their lands. They served the public by making land available to other agencies for recreational development or leasing their land for commercial enterprises. The River Authority Act passed by the Texas legislature in 1971 granted the river authorities the authority and responsibility to develop water resources for public recreation purposes and to acquire and improve parkland near public waters.

The 3,500 permanent employees of the TPWD are responsible for the management of 95 state parks, 51 wildlife management areas, eight fish hatcheries, and numerous field offices. State park lands, wetlands, and recreation facilities are described in the Texas Outdoor Recreation Plan (TPWD 2018), which is updated every five years. Many of the state parks and wildlife management areas are located at reservoirs. TPWD's mission is to manage and conserve the natural and cultural resources of Texas and to provide hunting, fishing, and outdoor recreation opportunities for the use and enjoyment of present and future generations. TPWD was formed in 1963 through a merger of the previously functioning State Parks Board and Game and Fish Commission.

Corps of Engineers Civil Works Program

The Civil Works Program of USACE is the oldest and largest reservoir construction and management organization in the nation, with 537 reservoirs in operation nationwide (Patterson and Doyle 2018). The military history of USACE dates to the founding of the nation. The

Corps of Engineers provided engineering and construction support for the revolutionary army led by General George Washington and continues to provide technical services for the US Army and US Air Force today. The USACE Civil Works Program began with the construction of navigation improvements in the 1800s. The early engineering capabilities of the USACE motivated the US Congress to assign the military organization responsibility for civilian navigation improvements. The Flood Control Act of 1936 established the agency's nationwide flood control responsibilities that later grew to encompass comprehensive multiple-purpose water planning and development.

USACE has about 37,000 employees, most civilian. The Civil Works Program is organized geographically into eight divisions subdivided by river basins into 40 districts. Military design and construction administration services are consolidated into selected district offices. All district offices have civil works responsibilities, and certain offices also provide services to the Army and Air Force. Research and development organizations include the Engineering Research and Development Center in Vicksburg, Mississippi; the Hydrologic Engineering Center in Davis, California; and the Institute for Water Resources in Alexandria, Virginia. The Office of the Chief of Engineers and other headquarters offices are in Washington, DC.

The Corps of Engineers Civil Works Program includes planning, design, construction, and operation of water resources development projects as well as other water research and regulatory activities. The Corps has constructed and now maintains most of the inland navigation facilities in the United States and over 500 major flood control and multipurpose reservoir projects nationwide, including 29 in Texas. The agency has constructed numerous flood control levee and channel improvement projects that have been turned over to local sponsors for maintenance. USACE is also responsible for flood control operations at reservoirs constructed by USBR or other agencies that have flood control pools included at federal expense.

USACE Districts

The USACE Southwestern Division Civil Works Program is comprised of the Fort Worth, Galveston, Tulsa, Little Rock, and Albuquerque Districts. The entire river basin land area assigned to the Fort Worth and Galveston Districts and a portion of the Red River Basin component of the Tulsa District are encompassed within Texas. The Southwestern Division Office is in Dallas. Certain reservoir management personnel work from offices located at each of the reservoirs. Centralized reservoir management activities are housed in the district offices in Fort Worth, Galveston, and Tulsa. The district offices are responsible for administration of the Section 404 dredge and fill permit program within their jurisdictional boundaries.

The Galveston District of the Corps of Engineers encompasses the coastal area of Texas. Responsibilities of the Galveston District Office include dredging and otherwise maintaining the Texas portion of the Gulf Intracoastal Waterway and maintenance and operation of Addicks and Barker Reservoirs in Houston. The 1,100 mi long Gulf Intracoastal Waterway completed in 1949 extends from Florida to Brownsville, Texas, facilitating barge transportation. Addicks and Barker Reservoirs are the only USACE reservoirs in Texas constructed for only flood control, without inclusion of a conservation storage pool. Land areas behind Addicks and Barker Dams are used for parks, sports fields, and recreation activities between major flood events.

The Fort Worth District (FWD) encompasses most of Texas. The FWD Office constructed and now maintains and operates 25 multiple-purpose reservoir projects and the dam at Caddo Lake. The unique Caddo Lake is discussed later in this chapter. FWD has constructed major flood control levee and channel improvement projects including Trinity River floodway improvements in the Dallas-Fort Worth metropolitan area, the multiple-component San Antonio River Project in downtown San Antonio, and several smaller projects in smaller cities.

The Red River Basin is included in the USACE Tulsa District. About 24,300 sq. mi of the 93,450 sq. mi Red River Basin is in Texas. The Tulsa District Office constructed and is responsible for the operation of the multipurpose Lakes Texoma and Pat Mayse and the Truscott brine control dam in Texas as well as other projects located throughout Oklahoma. The Sulphur River and Cypress Creek are tributaries of the Red River, but their upper watersheds in northeast Texas are encompassed by the Fort Worth District.

Clean Water Act Section 404 Permit Program

Under authority of Section 404 of the Clean Water Act of 1972 and amendments thereto, the USACE, including the three district offices with jurisdiction for areas in Texas, administers a permit program regulating construction and other activities involving dredging and/or filling in rivers, streams, and wetlands. The Section 404 permit process ensures that all construction projects and other filling or dredging activities follow all federal environmental protection laws and requirements. Obtaining a Section 404 permit from USACE is an important aspect of private and public development projects throughout the nation. USACE approval of a Section 404 permit requires adherence of the permit applicant whether private sector or governmental to all federal environmental laws and regulations.

USACE Planning and Development Process

In 1925, the Congress directed the Corps of Engineers and Federal Power Commission to jointly prepare a list and estimate the cost of determining where hydropower development might be practical on navigable streams (Viessman and Welty 1985). The idea was to produce general plans for the improvement of these streams for navigation in combination with the development of hydropower, flood control, and irrigation. A list of streams was submitted to Congress in 1927 and printed as House Document 308. A series of comprehensive planning studies for the rivers identified in House Document 308 conducted during the 1930s to 1950s was known as 308 studies. Most of the USACE reservoirs now in operation in Texas were originally conceptualized in these basin-wide 308 studies. More detailed feasibility studies for specific reservoirs or systems of reservoirs followed the 308 studies prior to detailed design and construction.

Numerous reservoir projects were initially identified in the early general basin-wide planning studies. The subsequent pathway to implementation of specific projects was lengthy. The general process followed in planning and implementation of the USACE reservoir and other water resources development projects includes the following major steps:

1. City, state, or other local entities petition their representatives in Congress for federal assistance in addressing water-related problems and needs.
2. A senator or congressional representative requests the appropriate congressional committee to direct the Corps of Engineers to conduct a feasibility study and develop recommendations. Authorization for

the planning study is either a resolution by an appropriate senate or house committee or a congressional act.

3. Congress appropriates funds annually to conduct the feasibility study.
4. The professional staff of the appropriate USACE district office conducts planning studies to formulate and evaluate alternatives to addressing identified problems and needs. These studies are closely coordinated with state and local agencies, stakeholders, advocacy groups, and the public. The feasibility study is documented in detail by a report forwarded through review channels to the Congress. The report may recommend no further federal action, a plan for construction of one or more projects, or some other plan of action.
5. Based on the recommendations of the feasibility report, the Congress authorizes advanced planning, design, and construction of specific projects.
6. Congress appropriates funds annually for advanced planning, design, and construction.
7. The USACE district office performs advanced planning and design studies, awards and administers construction contracts with selected construction companies, and supervises construction.
8. For navigation improvements, flood control reservoirs, and multipurpose reservoirs, USACE maintains and operates the completed project. Local flood protection levee and channel improvement projects are turned over to local sponsors for operation and maintenance.

Construction of most USACE reservoirs in Texas was authorized under this process by omnibus authorization acts passed during the 1940s to 1960s. For example, construction of Hords Creek Reservoir in the Colorado River Basin was authorized by the Flood Control Acts of 1941 and 1944. Construction of Ray Roberts and Joe Pool Reservoirs was authorized in the Flood Control Act of 1965 as components of a comprehensive plan for the Trinity River Basin. Construction of Aquilla Reservoir originally conceptualized in comprehensive planning studies for the Brazos River Basin was authorized in the Flood Control Act of 1968.

This planning and development process is still valid today. However, construction of the most recently constructed USACE reservoir project in Texas (Jim Chapman Reservoir) was completed in 1991. A 1940s to

1980s focus on developing new reservoir projects shifted to operation, maintenance, and rehabilitation of existing projects and other resource management endeavors along with environmental protection and other regulatory endeavors.

USACE, USBR, and other federal agencies developed planning methodologies documented in agency publications (Grigg 1985). The Water Resources Council (WRC) was created by the Water Planning Act of 1965 to coordinate federal water programs and policies. The WRC was dismantled in 1983. *The Principles and Standards for Planning Water and Related Land Resources* (Water Resources Council 1973) was published in 1973, revised in 1979, and replaced in 1982 by a document titled *Economic and Environmental Principles and Guidelines (P&G) for Water and Related Land Resource Implementation Studies*. The P&G outlines planning procedures based on achieving economic development, environmental quality, regional development, and social well-being. Methods for assessing problems and needs and formulating and evaluating solutions are outlined. The Water Resources Development Act of 2007 directed USACE to revise the P&G based on certain considerations. Publication of the revised P&G was announced in the Federal Register in 2014 (Council on Environmental Quality 2014).

Programs and Responsibilities of Other Federal Water Agencies

Each of the federal agencies listed in Table 2-1 has multiple missions. The four agencies that plan and construct dams and other facilities along with other water management responsibilities are IBWC, USACE, USBR, and NRCS. The United States and Mexico Sections of IBWC are federal agencies of the respective countries. USACE and USBR provided IBWC with planning, engineering, and construction management services during the 1940s to 1960s when the two reservoir projects on the Rio Grande were being developed.

The responsibilities of the US Geological Survey (USGS) and National Weather Service (NWS) deal primarily with collecting and disseminating climatic and hydrologic data, performing flood and weather forecasting, and scientific research. FEMA administers the NFIP and emergency relief programs for responding to the full spectrum of disasters including floods, droughts, and hurricanes. The Environmental Protection Agency (EPA) is primarily a regulatory agency but also conducts research. The USACE, through its Section 404 permit program for construction and dredging activities, is also a major environmental regulatory agency. The USFWS is responsible for conserving fish and wildlife resources throughout the rivers, reservoirs, and lands of Texas and the nation.

US Bureau of Reclamation

The USBR of the Department of the Interior was created by the Reclamation Act of 1902 to plan and implement water projects needed to support population and economic growth in the arid West. USBR is responsible for development and management of water resources in the 17 western states, including Texas, for multiple purposes, including irrigation, hydroelectric power, municipal and industrial water supply, pollution abatement, propagation of fish and wildlife, recreation, erosion control, and flood control. According to its website, USBR is the nation's largest wholesale water supplier, operating 338 reservoirs with a total storage capacity of 140 million ac-ft, providing western farmers with irrigation water for 10 million farmland acres that produce 60%of the nation's vegetables and one-quarter of its fresh fruit and nut crops. USBR operates 50 hydroelectric power plants that annually produce, on average, 40 billion kilowatt-hours per year. The agency manages, with partners, 249 recreation sites.

USBR constructed and operates Lakes Mead and Powell impounded by Hoover and Glen Canyon Dams on the Colorado River, which are the two largest reservoirs in the nation. The Central Valley Project of California is another famous extremely large-scale USBR accomplishment.

The Bureau of Reclamation has planned and constructed the following five reservoir projects in Texas: Travis, Twin Buttes, Texana, Choke Canyon, and Meredith. The five reservoirs have been turned over to local sponsors to own, maintain, and operate. USACE is responsible for flood control operations at the three USBR-constructed reservoirs that contain designated flood control storage capacity (Travis, Twin Buttes, and Meredith). USBR also constructed several water conveyance and distribution facilities for irrigation, municipal, and industrial water use in the Rio Grande, Colorado, and Canadian River Basins that are now operated by local sponsors.

The Reclamation Acts of 1902 and 1939 and other legislation mandate that costs allocated to irrigation in federal projects be reimbursed by the project beneficiaries. The details of repayment requirements for USBR irrigation projects have varied over the years with changes in reclamation law. Congressional acts authorizing specific USBR projects have often included repayment provisions tailored to the circumstances of the individual project. However, irrigation is treated similarly to municipal and industrial water supply for USBR reservoirs in Texas following the provisions of the Water Supply Act of 1958.

Natural Resource Conservation Service

The NRCS of the US Department of Agriculture was originally created in 1935 as the Soil Conservation Service (SCS) with the mission of working with farmers to reduce accelerated land erosion associated with farming operations. In response to a broadening of the scope of its programs and responsibilities, the name of the SCS was changed in 1994 to the NRCS. The NRCS has about 12,000 employees in 2,900 offices nationwide who work with farmers, ranchers, landowners, environmental groups, and state and local organizations in carrying out programs in managing and protecting land, water, and environmental resources.

Under the provisions of the Watershed Protection and Flood Prevention Act of 1954, the SCS/NRCS has constructed 11,850 flood-retarding dams, including 2,040 dams in Texas, to reduce flooding from small rural watersheds. Design and construction costs for the flood-retarding structures constructed usually on private land are borne by the federal government. Nonfederal sponsors pay for real estate acquisition and relocation of highways and utilities if required and, upon completion of construction, maintain and operate the projects at nonfederal expense.

The NRCS can also include water supply storage in its reservoir projects in accordance with the Water Supply Act of 1958 subject to all costs allocated to water supply being reimbursed by nonfederal sponsors. Rita Blanca Lake and Lake Clyde constructed by the SCS are now owned and operated by the cities of Dalhart and Clyde, respectively. Construction of the 12,100 ac-ft and 5,750 ac-ft reservoirs was completed in 1941 and 1969, respectively.

US Geological Survey

The mission of the Water Resources Division (WRD) of USGS of the Department of the Interior is to provide hydrologic information needed to support water management. USGS organizational structure includes 48 district offices generally located in state capitals and four regional offices. USGS WRD conducts three major types of interrelated activities: data collection and dissemination, problem-oriented water resources appraisals and interpretive studies, and research. Data collection programs provide water quality and quantity data for streamflow, reservoir storage, and groundwater. Data are stored and disseminated in the National Water Information System discussed in Chapter 4. Other divisions of USGS provide mapping products and information regarding geological and biological resources that are also used extensively in the professional water management community.

National Weather Service

The NWS is responsible for the hydrologic services programs of the National Oceanic and Atmospheric Administration (NOAA) of the Department of Commerce. NWS collects and disseminates weather and climatic data, which is widely used throughout the water management community. NWS also provides weather, hydrologic, and climate forecasts and warnings.

Thirteen NWS river forecast centers covering the nation produce and disseminate flood and drought forecast information. The Arkansas–Red Basin Forecast Center in Tulsa covers the Red River Basin portion of Texas. Most of Texas is covered by the West Gulf Coast Forecast Center in Fort Worth. The river forecast centers collaborate with other agencies discussed in this chapter. The NWS river forecast centers provide hydrologic forecasts for almost 4,000 locations across the United States that are distributed to a wide range of customers including reservoir operators.

The Advanced Hydrologic Prediction Service (AHPS) is a major component of the services provided by the NWS. The AHPS is a web-based suite of forecast products that display the magnitude and uncertainty of occurrence of floods or droughts from hours to days and months in advance. These products are useful planning and operational tools for many economic and emergency managers. The forecast products enable government agencies, private institutions, and individuals to make more informed decisions about risk-based policies and actions to mitigate the dangers posed by floods and droughts.

Zhang and Fernando (2020) report the findings of a workshop conducted by TWDB, NOAA NWS, and the University of Texas at Arlington in September 2019 in Arlington, Texas, to (1) identify obstacles to the operational use of NWS forecasts in reservoir operations and (2) motivate actions to facilitate adoption of forecast-informed reservoir operation (FIRO) paradigms for drought and flood preparedness in Texas. A key finding was the diversity of FIRO practices both in Texas and nationwide. The form of forecast, sources of forecasts, and methods of using forecasts varied greatly between reservoir operators. Recommendations were developed for NOAA consideration in improving the usefulness of forecast products.

US Fish and Wildlife Service

The USFWS created in 1939 is responsible for enforcing wildlife laws, protecting endangered species, and conserving and protecting wildlife and habitat on public and private lands nationwide. These responsibilities significantly affect the activities of other agencies in developing and managing water projects. USFWS also manages a system of 520 national wildlife refuges, thousands of wetlands and other special management areas, and many national fish hatcheries and fishery and ecological services offices.

The Fish and Wildlife Conservation Act of 1958 established the policy that fish and wildlife conservation be coordinated with other reservoir project purposes and receive equal consideration. Requirements for conservation of endangered species, pursuant to the 1973 Endangered Species Act as amended in 1978 and 1979 and other legislation, are administered by USFWS in coordination with other agencies. Endangered species are officially identified, and they and their habitat are protected from actions that could cause their destruction. Endangered species protection requirements have significantly impacted water management nationwide and in Texas.

Environmental Protection Agency

The EPA was established in 1970 to consolidate in one agency a variety of federal research, monitoring, standard-setting, and enforcement activities to ensure environmental protection (Collin 2006, National Research Council 2012). EPA's mission is to protect human health and safeguard the natural environment (air, water, and land) upon which life depends. EPA has more than 18,000 employees. Regional offices in each of 10 regions of the United States are responsible for execution of EPA's programs within the states in that region. The EPA Office of Research and Development operates 14 national laboratories.

The EPA Office of Water, working through EPA regional offices in collaboration with state environmental agencies, is responsible for implementing the Clean Water Act, Safe Drinking Water Act (SDWA), and other federal laws pertaining to protection of water quality, public health, and environmental resources. The National Pollution Discharge Elimination System (NPDES), safe drinking water programs, and an array of other regulatory activities are accomplished through issuance and enforcement of permits by state regulatory agencies that meet requirements outlined by EPA. TCEQ implements national EPA-administered programs in Texas.

A variety of federal environmental laws, policies, and programs significantly affect water management (Collin 2006, Benson et al. 2021). Several key laws are highlighted as follows.

The National Environmental Policy Act (NEPA) of 1970 articulated the policy of protecting the environment and established requirements for evaluating the environmental impacts of federal actions. NEPA requirements for diverse scientific and stakeholder input to environmental impact assessment have improved decisions, but the process can be bureaucratic, expensive, and time-consuming. Although NEPA is applicable only to federal actions, USACE Section 404 permitting of nonfederal construction activities is a federal action subject to NEPA.

The Water Pollution Control Act Amendments of 1972, as amended by the Clean Water Act of 1977 and other legislation, established NPDES and other programs for protecting water quality in the rivers and lakes of the nation. NPDES wastewater and stormwater permits are administered by state environmental agencies such as TCEQ in collaboration with the federal EPA. As noted earlier, Section 404 of the Clean Water Act authorizes the dredge and fill permit program administered by the USACE. The SDWA enacted in 1974 and amended in 1986 and 1996 established programs administered by EPA and state environmental agencies to assure safe public drinking water. Water quality standards, permit requirements, information, and other SDWA-mandated programs significantly affect cities and water utilities. The overall collection of environmental laws significantly affects river/reservoir water management as well as most other aspects of economic and natural resource development throughout the nation.

Department of Energy Agencies

The Federal Energy Regulatory Commission (FERC) is the agency in the US Department of Energy that regulates the transmission and sale of natural gas, oil, and electricity in interstate commerce and licenses and inspects private, municipal, and state hydroelectric projects. Hydroelectric power regulation includes issuing preliminary permits, project licenses, and exemptions from licensing; ensuring dam safety; performing project compliance activities; and coordinating with other agencies.

WAPA and SWPA are two of four Department of Energy agencies that market and transmit wholesale electricity generated by hydroelectric power plants at federal multipurpose reservoirs. The four electric power marketing agencies each has responsibility for an assigned service area. WAPA has a 15-state service area that includes 57 hydropower plants operated by USBR, USACE, and IBWC, which includes International Amistad and Falcon Reservoirs. SWPA is responsible for a six-state area that includes the hydroelectric power plants at USACE reservoirs in Texas. WAPA and SWPA sell and transmit the electricity generated at hydropower plants at federal multipurpose reservoirs to other agencies, cities, rural electric cooperatives, and electric utility companies that distribute the power to retail customers.

Federal Emergency Management Agency

The mission of FEMA, created in 1979, is to reduce loss of life and property and protect the nation's infrastructure from all types of hazards through a comprehensive, risk-based, emergency management program of mitigation, preparedness, response, and recovery. The two major water management programs of FEMA are (1) managing emergency response to the full spectrum of disasters including hurricanes, floods, and droughts; and (2) administration of the NFIP, which includes both floodplain management and flood insurance components. The Flood Insurance Administration of FEMA administers the flood insurance component of

NFIP, and the Mitigation Directorate of FEMA oversees the floodplain management aspect of NFIP. Knowles and Kunreuther (2014) trace the evolution of NFIP as a major component of the national strategy for dealing with flooding.

Strategies for Reducing Flood Damages

Stormwater management, floodplain management, and flood control endeavors differ institutionally, though all deal with excess runoff from storms producing high rainfall. Local communities are responsible for stormwater management including both (1) facilities for draining rainfall runoff from streets and residential, commercial, and public properties; and (2) detention facilities to mitigate the impacts of development in increasing flows to downstream properties. Municipalities are subject to NPDES requirements administered by EPA and TCEQ for dealing with water quality aspects of stormwater runoff. Local cities and towns are also responsible for regulating residential and commercial development, including development in floodplains. The federal government in collaboration with local communities has played major roles in both construction of flood control structures and encouraging floodplain management.

Prior to 1936, protection from floods was the responsibility of each individual property owner. With enactment of the Flood Control Act of 1936, the Congress tasked USACE with nationwide flood control responsibilities for major rivers. The Watershed Protection and Flood Prevention Act of 1954 authorized the SCS (later renamed the NRCS) rural stream tributary watershed flood control program. The numerous flood control dams, levees, and channel improvements constructed at federal expense are highly effective at partially controlling flood water and reducing damages. However, flood losses continued to increase despite large federal investments in flood control structures due to intensified development of floodplains. The need for floodplain regulation, flood insurance, and other nonstructural measures has been highlighted by many studies over many years before and after the 1966 House Document 484 documenting the federal task force findings and recommendations that resulted in creation of the NFIP (Task Force on Federal Flood Control Policy 1966).

Managing Susceptibility to Flood Damages

The Flood Control Act of 1960 authorized the USACE to develop information needed by local governments to guide and regulate the use of floodplain land. NFIP was established by the National Flood Insurance Act of 1968, expanded by the Flood Protection Act of 1973, and refined by the 2012 Flood Insurance Reform Act and other amending legislation.

NFIP is a collaborative endeavor of the federal government, local communities, and the private sector. FEMA administers the NFIP. The 23,000 communities that participate in NFIP are responsible for restricting construction of residential and commercial properties in floodplains following standards set by FEMA. Engineering consulting firms perform hydrology and hydraulics studies to delineate and update delineations of floodplains. A network of more than 50 insurance companies coordinated by FEMA sells flood insurance to property owners.

Floodplain regulation by a local government entity is a requirement for community participation in NFIP, which is required for property owners to have access to flood insurance and certain other federally regulated programs. Restrictions on development in the 1.0% annual exceedance frequency (100-year recurrence interval) floodplain is the minimum base standard set by FEMA. Local communities can adopt more stringent standards but typically adopt the 1.0% annual exceedance frequency flood standard for delineating regulated floodplains. Rules are established and enforced by local communities within FEMA guidelines regulating commercial and residential construction within the delineated floodplains.

Economic Feasibility of Federal Flood Control Projects

Cities, flood control and drainage districts, and other local agencies construct and maintain street culverts and bridges, stormwater drainage and detention facilities, flood-retarding structures, stream channel improvements, and levees. These local nonfederal entities also maintain channel improvements and levee projects constructed by USACE. USACE constructs, maintains, and operates federal flood control reservoirs totally at federal expense. Cost-sharing by local sponsors is required for USACE-constructed local flood protection projects, which typically consist of stream channel improvements and/or levees. Nonfederal sponsors cover real estate and utility relocation costs for construction of USACE local flood control channel and levee projects and are responsible for project ownership and maintenance upon completion of construction.

Local government entities construct and maintain drainage, stormwater management, and flood control facilities with the objective of providing needed services in a cost-effective manner but normally do not perform detailed economic benefit–cost analyses. The Flood Control Act of 1936 states that federal participation in constructing flood control projects is justified only if the economic benefits of a project exceed economic costs. USACE and other federal water agencies have developed economic benefit–cost analysis methods for navigation, flood control, agricultural irrigation, and other water development projects (Wurbs and James 2002).

USACE constructed many flood control projects throughout the nation and Texas following the Flood Control Act of 1936. Additional new flood control projects became extremely difficult to economically justify by the 1970s for multiple reasons (Wurbs 1983). Many of the most economically favorable projects had already been constructed by the 1970s. Floodplain management requirements of NFIP precluded claiming benefits for protecting future residential and commercial development in 10% annual exceedance frequency floodplains since such development should not be allowed to occur.

The discount rate used by federal water agencies in discounting benefits and costs to a common time base also significantly affects economic feasibility. The discount rate conceptually represents societal willingness to invest now to achieve benefits in the future. The discount rate used in evaluating federal water projects is updated each fiscal year and published in the Federal Register. The rate was 2.500% during 1957 to 1960 and increased steadily each fiscal year to a peak of 8.875% in 1987, 1989, and 1990. The rate decreased steadily each fiscal year to 2.250% in 2022 and increased to 2.500% and 2.750% in 2023 and 2024. At the 1960 and 2023 discount rate of 2.500%, each reduction in flood damages (benefit) of \$1.00 occurring 30 years from now justifies a present investment cost of \$0.477. At the 1987 and 1990 rate of 8.875%, a beneficial reduction in flood damages of \$1.00 30 years in the future warrants a present investment of \$0.0780 as computed below. P and F denote present and future values in the following discounting formula. N is 30 years.

$$P = F(1.0/(1.0 + i)^N) = \$1.00(1.0/(1.0 + 0.08875)^{30}) = \$0.0780$$

The steady increase in federal discount rate between 1960 and 1987 contributed to increased difficulty in developing economically feasible flood control projects. The decrease in discount rate from 1990 to 2022 increased the likelihood of additional new federal flood control dams, levees, and channel improvements being economically feasible.

Federal and State Dam Safety Programs

Risks of dams overtopping or breaching or structural failures of outlet structures are related to both flood risk mitigation and concerns nationwide and in Texas regarding rehabilitation of aging infrastructure. Security against criminal or terrorist attacks on dams and other structures is also of concern nationwide.

Past Dam Failures Nationwide

Dam safety has been an important consideration in the design, construction, and operation of dams since people first started to develop water projects. The initial creation of present federal and state dam safety programs in the United States during the 1970s was motivated by catastrophic dam failures including but not limited to the following (Wurbs 1985). Failure of Buffalo Creek Dam in West Virginia in 1972 resulted in 125 deaths. Failure of Teton Dam in Idaho upon completion of construction by USBR in 1976 resulted in 14 deaths and $1 billion in losses. Collapse of Kelly Barnes Dam in Georgia in 1977 killed 39 people and caused millions of dollars of damages.

Normand (2019) estimates that about 1,600 dam failures resulting in 3,500 casualties have occurred in the United States since the mid-1800s. Normand (2019) further states that 294 failures and 537 non-failure dam safety incidents were reported in the United States between 2000 and 2019. About 200,000 people were evacuated in 2017 owing to structural deficiencies in the spillway of the extremely high Oroville Dam in California. More than 100 dam breaches occurred in North Carolina and South Carolina during extreme storms from 2015 to 2018.

McDaniels (1964) describes the perhaps most infamous dam failure in Texas history. The City of Austin constructed Austin Dam and Lake McDonald on the Colorado River to develop a municipal water supply and hydropower. Construction of the initial reservoir project with a storage capacity of 49,300 ac-ft was completed in 1883. The dam was reportedly the largest masonry overflow-type dam in the world at that time (McDaniels 1964). The dam was breached by a flood in April 1900. Reconstruction of a 32,000 ac-ft reservoir project at the same site was initiated in 1911 and completed in 1915. Floods in September 1915 and April 1918 carried away some of the spillway gates. The flood of June 1935 destroyed the remaining gates and gate piers. The storage capacity was reduced by sedimentation resulting from the floods. Construction of Tom Miller Dam and Lake Austin at this site was completed by the LCRA in 1939. Construction of Lakes Buchanan and Travis further upstream on the Colorado River were completed in 1937 and 1940. These floods and the impacts of Lakes Buchanan and Travis on flood flows in Austin can be seen in the plot of 1900 to 2023 daily flows at a USGS gauge on the Colorado River at Austin (see Figure 5-30).

Dascher and Meitzen (2020) found 328 instances of reported dam failures or related incidents in Texas since 1900. Fifty small mostly privately owned dams in Texas were removed between 1983 and 2016. Most were older small dams removed by private owners in response to liability concerns. Most of the incidents in Texas over the last 100 years have involved small dams with relatively minimal consequences compared to catastrophic failures of large dams that have occurred elsewhere in the nation and world such as those noted earlier in this section.

The BRA contracted with a consulting engineering firm and construction contractor to perform major emergency repairs to Morris Sheppard Dam that were recognized by the National Society of Professional Engineers as one of the eight outstanding American engineering achievements of 1989. Upon completion of construction in 1941, Morris Sheppard Dam impounding Possum Kingdom Reservoir was the tallest flat-slab buttress dam in the United States. The unique dam and outlet structures were considered an engineering marvel for that time. In December 1986, during a routine inspection required by FERC, cracks were detected in spillway footings suggesting structural distress. Hydrostatic pressure through shale seams under the dam was threatening the integrity of the structure. Measurements revealed that the dam had slipped about 4.5 in. downstream in the decades since the reservoir began filling in 1941. Failure could result in catastrophic loss of life and property.

The dam and its foundation were massively reinforced during 1986 to 1991 to deal with the potentially damaging pressures. BRA launched a rigorous maintenance program in 1992 to keep the aging dam and appurtenant structures in top condition. BRA has a permanent staff crew with specialized skills and equipment dedicated solely to maintaining the large project.

Issues encountered in 2016 and 2019 when spillway gates broke loose at Lakes Dunlap and Wood on the Guadalupe River are discussed in Chapter 7. The small GBRA hydropower dams constructed during the 1920s also illustrate problems in maintenance and rehabilitation of aging infrastructure.

Dam Safety Programs

Responsibilities for dam safety are shared among federal agencies, nonfederal agencies, and private dam owners. FEMA administers a National Dam Safety Program (NDSP) that facilitates collaboration between federal and state dam safety activities that include dam inspection programs, risk reduction strategies, rehabilitation measures, funding, and information sharing (Normand 2019). The federal government is responsible for maintaining the safety of federally owned dams. The USACE and USBR own and operate most of the large federal reservoirs in the United States and are responsible for dam safety activities related to their projects. Several other federal agencies own and are responsible for other federal dams. Most of the other federal reservoir projects are much smaller than USACE and USBR projects.

The states are responsible for administering dam safety programs for nonfederal dams that include regular inspections and regulatory actions following federally approved guidelines. Nonfederal dam owners are subject to dam safety regulations administered by a state regulatory agency, which in Texas is the TCEQ. State dam safety programs are subject to federal oversight and support.

The NID maintained by USACE in collaboration with FEMA and the state dam safety programs is a component of the NDSP. Congress authorized an NDSP in 1972 that included creation and maintenance by the USACE of an NID in collaboration with other federal and state agencies along with a National Dam Inspection Program and other related activities. NID is a database maintained by USACE that is accessible through a website included in Table 4-1. The database consists of conveniently accessible information regarding each dam including location, size, type of dam and appurtenant structures, project purposes, hazard potential, and safety inspection history.

For purposes of inclusion in the NID, a dam is defined as any artificial barrier that has the ability to impound water, wastewater, or any liquid-borne material for the purpose of storage or control of water that (1) is at least 25 ft in height with a storage capacity of more than 15 ac-ft, (2) is greater than 6 ft in height with a storage capacity of at least 50 ft, or (3) poses a significant threat to human life or property should it fail. Numerous dams located throughout the nation do not meet these criteria for inclusion in the NID.

The NID includes 91,843 dams as of January 2024. The states have regulatory authority for over 69% of the dams in the NID. Many of the largest dams are owned by federal agencies that have authority directly for safety regulation of their own dams. More dams in the NID are in Texas than in any other state. Texas has 7,384 dams in the NID.

The Texas Dam Safety Program is included in the multiple regulatory programs conducted by the TCEQ. A wealth of information regarding the Dam Safety Program available through the TCEQ website includes regulations, manuals, and guidelines for periodic dam inspections and technical publications regarding design, construction, operation, and rehabilitation of dams and outlet structures. The TCEQ Dam Safety Program is responsible for inspection and administrative regulation of 4,048 nonfederal dams that include 1,507 dams with high hazard potential, 304 with significant hazard potential, and 2,237 classified as low hazard potential. TCEQ dam safety staff inspect dams at 5-year intervals and provide technical information and assistance to dam owners. Dam safety regulatory policies are outlined in the Texas Administrative Code, Part 1 TCEQ, Chapter 299 "Dams and Reservoirs," which is accessible online. Safety and rehabilitation concerns grow as dams and appurtenant structures age and watersheds and floodplains urbanize.

As previously noted, NRCS has constructed 11,850 flood-retarding dams in rural watersheds nationwide, including 2,040 in Texas. According to the Texas State Soil and Water Conservation Board (TSSWCB) website, 622 of NRCS dams in Texas are classified as high hazard, meaning potential exists for loss of life in case of dam failure. Many of these dams do not meet recommended safety criteria. NRCS administers a watershed rehabilitation program to assist local sponsors in the rehabilitation of aging NRCS dams that are reaching their design life and/or no longer meet federal or state safety criteria or performance standards.

ASCE Texas Section has published an infrastructure report card periodically since 1998 that uses the old A, B, C, D, and F grading scales to describe the condition of 12 categories of infrastructure in Texas (ASCE 2021). The overall 2021 grade of C was a slight improvement over the overall grade of C– on the preceding 2017 report card. The 2021 grades for water-related categories are dams (D+), levees (D), flood mitigation (C–), drinking water (C–), and wastewater (D). The rationale presented in the report for a low D+ for dams highlights concern regarding rehabilitation needs and inadequate safety inspections for many smaller old nonfederal dams (ASCE 2021).

Programs and Responsibilities of State Water Agencies

Information describing the programs and responsibilities of TWDB, TCEQ, and TPWD is found on their websites listed in Table 2-1. These three agencies play interconnected roles in statewide water management in Texas that are important to river and reservoir system water management. TWDB has both planning and financing responsibilities. TCEQ has water allocation, public health and safety, and environmental protection responsibilities. TPWD's responsibilities for both recreation and conservation of natural resources intersect with TWDB and TCEQ endeavors.

Water Resources Planning

TWDB has a governing board of three members appointed by the governor and about 400 employees. The agency is responsible for statewide planning and administering grant and loan programs to assist public water agencies and communities in financing water management endeavors. The State Water Implementation Fund for Texas (SWIFT), established in 2014 and administered by TWDB, provides low-interest loans with favorable terms for implementation of projects included in the most recently adopted state water plan.

The infamous devastating 1950 to 1957 drought ended by extreme flooding in April–May 1957 motivated the creation of TWDB by legislative act in 1957. A $200,000,000 water development fund administered by TWDB to help local communities develop water supplies was created by a voter-approved constitutional amendment in 1957.

In 1977, legislation combined three state water agencies (TWDB, Water Rights Commission, and Water Quality Board) creating the Texas Department of Water Resources (TDWR). TDWR was disbanded in 1985, and TWDB again became a separate agency. A 1984 version of the Texas Water Plan developed by TDWR includes a section summarizing state water resources planning and development from 1900 through 1983 focusing on planning studies and legislation enacted by the Texas legislature (TDWR 1984).

TWDB staff in 1966 completed and released for public review a preliminary version of the first state water plan, which contained proposed strategies for meeting the state's water needs through 2020. The preliminary plan proposed 53 new reservoirs and a 980 mi long water transport project beginning in northeast Texas and ending in the Lower Rio Grande Valley. A 1968 revised plan with significant revisions recommended development of 67 major new reservoirs with storage capacities greater than 5,000 ac-ft (TWDB 1968). The 1968 Texas Water Plan was formally adopted by TWDB governing board and legislature in 1969.

Between adoption of the initial statewide water plan in 1969 and the 1984 update, construction of 43 major reservoirs and three reservoir enlargements added almost 10 million ac-ft of storage capacity. Twenty-four of the new reservoirs were for water supply, 18 were off-channel cooling ponds for steam-electric power generation, and one was for natural salt pollution control (TDWR 1984). TWDB released water plan update reports in 1990 and 1992, which were relatively brief additions to the much more voluminous 1984 report.

Motivated by severe drought conditions during 1995 to 1996, the 1997 Senate Bill 1 (SB1) created the expanded regional and statewide water planning process introduced in Chapter 1. Local stakeholder-guided consensus-based planning was integrated with statewide TWDB planning. The 1997 SB1 also authorized development of the WAM system discussed in Chapter 4 to support both the planning process and TCEQ-administered water rights system.

The 1997 SB1 established a process for developing 16 regional plans and a statewide plan at 5-year planning cycles, with a 50-year future planning horizon. Updated plans were completed in 2002, 2007, 2012, 2017, and 2022. The current cycle is scheduled for completion in 2027. Updates of the 16 regional plans and statewide plan are documented by voluminous reports available on the TWDB website along with rules governing the planning process and other information. The plans focus on water supply resources and future water needs and use.

The 16 planning regions were defined considering river basins, aquifers, and municipal and agricultural water use areas. The 16 planning groups have a total of more than 400 members representing diverse interests. TWDB provides administrative leadership and technical and funding support. Consulting firms provide services as needed. TWDB coordinates with the public in developing each regional plan and integrating the regional plans into a statewide plan.

Senate Bill 8 enacted by the Texas legislature in 2019 authorized the creation of the first State Flood Plan under the leadership of TWDB. The new flood control planning process is patterned after the regional and statewide water supply planning process established pursuant to the 1997 SB1. TWDB has divided the state into 15 flood control planning regions based on river basin boundaries. A regional flood planning group for each region is responsible for developing a regional plan in accordance with requirements outlined in the Texas Water Code and TWDB administrative rules and technical guidelines. TWDB will combine 15 regional plans developed during 2021 to 2023 to create the first official statewide flood plan in 2024.

Water Allocation in the United States

The institutional framework for water resources management includes a hierarchy of water allocation systems (Wurbs 2013). Water resources are allocated between nations by treaties and other agreements. In the United States, water is allocated among states through interstate river compacts. Within individual states, water is shared by regional water authorities, municipal utility districts, cities, irrigation districts, farmers, and private companies through water right systems. Water supply entities service their customers in accordance with contracts and other commitments.

Water allocation systems (1) equitably apportion water resources among users, (2) protect existing water users from having their supplies diminished by new users, (3) govern the sharing of limited water resources during droughts when supplies are inadequate to meet all needs, (4) facilitate efficient use of water resources, and (5) protect against wasteful overexploitation. Effective water allocation becomes particularly important as demands exceed reliable supplies.

Benson et al. (2021) and Getches et al. (2015) describe the law of water rights as administered in the United States. Each state has developed its own rules and practices, which have evolved historically and continue to change. Western and eastern states have adopted different approaches to water rights due to the western states having much drier climates. Water allocation and accounting systems tend to be more rigorous in regions where demands approach or exceed supplies. Groundwater and surface water are treated as separate resources with different water right systems. Texas has a rich heritage of developing water allocation strategies as a central thrust of its water resources management efforts (Wurbs 1995, Porter 2014, Sahs and Heinrich 2022).

Legal rights to the use of streamflow are based on two alternative doctrines: riparian and prior appropriation. The riparian doctrine is based on the concept that water rights are incidental to the ownership of land adjacent to a stream. The prior appropriation doctrine is based on protecting senior water users from having their supplies diminished by newcomers developing water supplies later in time. In a prior appropriation system, water rights are not inherent in land ownership, and priorities are based on dates that water is appropriated.

Variations of the riparian doctrine apply in 29 states in the eastern and central United States (Getches et al. 2015). The prior appropriation doctrine governs water rights in nine western states. Ten states, including Texas, originally recognized riparian rights but later converted to a system of appropriation while preserving existing riparian rights. Hawaii and Louisiana have their own unique water right systems. Most of the western states have established permit systems in which a state agency issues permits to water right holders specifying amounts and conditions of water use. Riparian and/or appropriative rights may be incorporated into the original development of the permit system, with additional new permits being issued based on prior appropriation. With growing demands on limited water resources nationwide, permit systems will likely continue to be developed in the eastern states like those already in place in the drier western states.

Texas Water Rights System

Water rights in Texas evolved over several centuries into an unmanageable assortment of poorly recorded and often conflicting rights (Wurbs

1995, 2004; Porter 2014). The severe drought during 1950 to 1957 motivated a massive lawsuit in the Lower Rio Grande Valley that demonstrated the impracticality of a purely judicial adjudication of water rights statewide. Thus, the Water Rights Adjudication Act was enacted by the Texas legislature in 1967 to require a recording of all claims for water rights, limit the exercise of those claims to actual use, and provide for the adjudication and administration of water rights. The water rights adjudication process required for transition to the permit system was initiated in 1968 and completed by about 1990.

Riparian water rights were merged into the prior appropriation permit system, creating two variations. One water allocation approach is designed for managing use of the Texas share of waters of the Rio Grande stored in International Amistad and Falcon Reservoirs, and the other water rights system is applicable to the remainder of Texas. The differences between water rights in the Lower Rio Grande Valley and the rest of Texas are outlined in Chapter 3.

Some type of water rights system has been administered statewide since 1913 by a centralized agency, but that agency has changed over time (Wurbs 2004). The Board of Water Engineers was established in 1913; reorganized as the Texas Water Commission (TWC) in 1962; and renamed the Texas Water Rights Commission in 1965 with non-water rights functions being transferred to the TWDB. In 1977, the Texas Department of Water Resources (TDWR) was created by combining the Water Rights Commission, TWDB, and Water Quality Board. In 1985, the TDWR was dissolved, and TWC and TWDB became separate agencies. The Texas Natural Resource Conservation Commission (TNRCC) was created in 1993 by merging the TWC and Texas Air Quality Board. TNRCC was renamed the Texas Commission on Environmental Quality (TCEQ) effective September 2002.

TCEQ consists of three full-time commissioners appointed by the governor and a professional and administrative staff of about 2,800 employees. Water rights for use of streamflow and reservoir storage are one of many regulatory responsibilities of TCEQ.

Surface water rights in Texas are in the form of certificates of adjudication or permanent or term water use permits. Water rights resulting from adjudication of preexisting rights pursuant to the Water Rights Adjudication Act of 1967 are recorded as certificates of adjudication. Rights created after completion of the adjudication process are documented as water use permits. Modifications to water rights are recorded by permit amendments. Most water rights are permanently in effect as long as the holder continues to use water as specified by the water right. TCEQ also issues term permits valid for only a specified period, typically from one to several years, to deal with emergencies or special situations. Some water rights include water management plans outlined in auxiliary documents that are periodically updated as water right amendments. Different forms of water supply contracts are also incorporated into the water rights administrative system.

Water rights are granted by a state license, or permit, which authorizes the license holder to divert a specified amount of water annually at a specific location, for a specific purpose, and to store water in reservoirs of specified capacity. A water right permit holder has no actual title of ownership of the water but only a right to store water in reservoirs and withdraw the water for beneficial use. However, water rights can be sold, leased, or transferred, subject to TCEQ approval.

Any organization or person may apply to TCEQ for a new water right or to change an existing water right at any time. TCEQ will approve the application if unappropriated water is available, the proposed beneficial need for water will be supplied at an acceptable level of reliability, existing water rights are not impaired, efficient water conservation will be practiced, and proposed actions are consistent with relevant SB1 statewide and regional water plans. During the 1968 to 1990 adjudication process, priority dates were established based on historical water use. Since then, priorities are based on the dates that the permits are administratively approved. As discussed in Chapter 3, the very different priority system for the Lower Rio Grande Valley is based on classifying water use as municipal, agricultural type A, or agricultural type B.

Water rights are subject to water availability. Streamflow amounts authorized for storage or diversion may not be available during drought conditions. Water supply reliabilities are dependent on reservoir storage capabilities and water right seniorities.

Most surface water rights are held by cities, river authorities, municipal water districts, and irrigation districts. Some water rights are held by electric utilities or other large private companies. Most of the 30 million residents of Texas and most private businesses are provided water by water utilities of the cities within which they reside. The cities and/or water districts or river authorities that supply water to the cities hold water rights. The 6,235 currently active surface water rights are discussed further in Chapters 3, 4, and 6.

As discussed in Chapter 1, groundwater rights in Texas have historically been based on the common law rule allowing landowners to pump as much water as they wish from under their land (Porter 2014). However, increased regulation of groundwater has evolved over time primarily through the establishment of local groundwater conservation districts.

Diffuse surface runoff from precipitation belongs to the landowner until the runoff reaches a stream. Landowners can capture surface water in reservoirs with storage capacities of less than 200 ac-ft without obtaining a water right permit. The Texas Water Code also allows a person to divert water from a stream for domestic and livestock purposes on land adjacent to the stream owned by the person without obtaining a permit. Water used for wildlife management on private land is also exempt from water right permit requirements.

Several western states have watermaster operations, but other states do not. The TCEQ Lower Rio Grande Water Master Office, established during the 1970s, administers a relatively precise accounting of water use, working closely with irrigators, cities, and IBWC. Watermaster operations and associated monitoring and accounting procedures have been established more recently for several South Texas River Basins and the Brazos River Basin. For the remainder of the state for which watermaster offices have not yet been created, TCEQ administers curtailment actions during drought and takes enforcement action anytime to stop reported unauthorized water use but does not otherwise closely monitor water use other than through periodic water use reports. Water right holders throughout Texas are required to submit annual reports recording their water use.

Effective water management requires an understanding of the reliabilities that water needs can be supplied and the effects of actions of each water user on other water users. Pursuant to the 1997 SB1, the WAM system introduced in Chapter 4 was developed by TCEQ as lead agency in collaboration with TWDB and TPWD and contractors consisting of consulting firms and university researchers. Water right permit applicants, or their consultants, and TCEQ staff employ the WAM system to assess water supply reliabilities for proposed new or revised water right permits and effects on other water rights. TWDB and regional planning groups apply the modeling system to assess water supply capabilities and consequences of proposed management plans in the regional and statewide planning studies.

Environmental Flow Standards

The importance of policies and practices to protect instream flows for fish, riverine ecosystems, wetlands, and freshwater inflows to bays and estuaries has been recognized in Texas since the 1980s and earlier. Efforts to formulate and implement environmental flow standards in Texas intensified pursuant to legislation enacted in 2001 as SB2 and in 2007 as SB3. SB2 created the TIFP jointly administered by TPWD, TWDB, and TCEQ to improve capabilities for understanding and protecting aquatic ecosystems (National Research Council 2005). Recognizing that many years will be required to perform detailed studies for all reaches of all streams of the state under the 2001 SB2 TIFP, the legislature enacted SB3 in 2007, creating an accelerated process for establishing environmental instream flow standards for selected priority river systems using the best available information and science.

The SB3 process results in environmental standards that are incorporated in the WAM system maintained by TCEQ (Wurbs 2017a). Flow standards consist of metrics and rules that vary with location, season, and hydrologic conditions that govern curtailment of diversion and/or storage of streamflows by junior water rights. Environmental flow standards are defined in terms of regimes with subsistence, base, and high pulse flow components required to maintain a sound ecology. The SB3 process also includes periodically reevaluating and updating the flow standards.

TCEQ, TWDB, and TPWD provide administrative leadership and technical and funding support for the SB3 process. A stakeholder committee reflecting an equitable balance of interest groups is appointed for each priority river system. An expert science team is also appointed for each priority river system to develop a recommended flow regime considering only environmental needs. The stakeholder committee reviews the science team's recommendations and develops environmental flow requirements based on a comprehensive consideration of all water needs. TCEQ adopts flow standards based on recommendations of the science teams and stakeholder committees, along with public review and comment.

The environmental flow standards are incorporated into the WAM system with a priority based on the date that TCEQ receives the environmental flow regime recommendations from the applicable science team. Thereafter, TCEQ may not issue a permit for a new appropriation or amendment to an existing water right permit that increases the amount of water authorized to be stored or diverted if any environmental flow standard requirement would be impaired. However, holders of existing senior water rights are not required to curtail appropriations of water to maintain junior environmental flow standards.

Environmental policy has been established primarily by federal laws, which provide some degree of consistency nationwide. However, the states have major responsibilities in implementing national environmental policies. TCEQ implements NPDES and other national environmental regulatory programs in Texas under the administrative oversight of USEPA. Establishment of environmental flow requirements is an example of state-level instigation of environmental programs. The states have primary responsibility for water allocation, and environmental flow standards are a component of water allocation. TCEQ is responsible for water allocation. TCEQ, TWDB, and TPWD share responsibility for determining environmental instream flow needs.

Water Districts and River Authorities

Texas has more than a thousand water management jurisdictions of various forms that include river authorities, municipal water districts, irrigation districts, groundwater conservation districts, soil and water conservation districts, and flood control districts. River authorities are jurisdictions created by the Texas legislature to manage the water resources of all or a major portion of a river basin. Municipal water districts are created to develop and operate regional water supply and wastewater management facilities for multiple member cities. The regional approach is often more efficient than each city owning and operating its own individual facilities. Irrigation districts supply water to multiple farmers. Groundwater conservation districts regulate or support in various ways the development and use of groundwater in their specified counties.

Soil and water conservation districts work to mitigate land erosion, flooding, and non-point source pollution and manage land and water resources. The Texas State Soil and Water Conservation Board coordinates conservation and nonpoint source water pollution abatement programs and provides technical assistance to the 216 soil and water conservation districts. Levee districts allow farmers or other landowners to pool their resources to protect their land from flooding. Local districts can also sponsor NRCS projects.

Urban drainage and flood control districts encompass areas larger than a single city. The Harris County Flood Control District (HCFCD) encompasses Houston and surrounding areas. HCFCD is the nonfederal sponsor for USACE Addicks and Barker Reservoirs. Other primary responsibilities of HCFCD include regulation of commercial and residential development in floodplains and watersheds throughout Harris County and operation of a flood warning system.

The approximately 20 river authorities in Texas are special political subdivisions of state government created by the Texas legislature to manage the water resources of all or a major portion of a river basin. Some water management districts may be categorized as either regular water districts or smaller river authorities. River authorities have no taxing authority or regular legislative appropriations. They are funded through the sale of water and other services. The different river authorities vary in size and types of activities.

Major Texas river authorities (RAs) are listed as follows in alphabetical order: Angelina and Neches RA, Bandera County RA and Groundwater District, Brazos RA, Central Colorado RA, Guadalupe-Blanco RA, Lavaca-Navidad RA, Lower Colorado RA, Lower Neches RA, Nueces River RA, Palo Duro RA, Red RA, Sabine RA, San Antonio RA, San Jacinto RA, Sulphur RA, Trinity RA, Upper Colorado RA, and Upper Guadalupe RA. The LCRA, Brazos River Authority (BRA), and Trinity River Authority (TRA) are the largest.

LCRA is a major provider of both electricity and water, operating four thermal-electric power plants and six water supply reservoirs while also generating hydropower. LCRA operates the six Highland Lakes on the Colorado River to supply water for Austin, other cities, and farmers in the lower basin. These six reservoirs also generate hydropower and are extremely popular for recreation. LCRA also owns and operates cooling water lakes for its thermal-electric power plants.

The Brazos River Authority, created in 1930, manages the water resources of the 45,600 sq. mi Brazos River Basin, which is larger than many states and nations. BRA constructed and now operates three water supply reservoirs, contracts for the conservation storage capacity of nine USACE reservoirs, and operates a regional water supply system and three regional wastewater treatment systems.

TRA is the water supply sponsor for four USACE reservoirs in the upper Trinity River Basin and owns and operates Lake Livingston on the lower Trinity River in partnership with the city of Houston. Water is conveyed by pipeline from Lake Livingston to Houston located in the San Jacinto River Basin. TRA also owns and operates regional water and wastewater treatment facilities serving Dallas, Fort Worth, Houston, and smaller cities.

The river authorities contract for conservation storage capacity in USACE reservoirs. USACE provides access to a volume of reservoir storage space under these contracts but is not responsible for providing water. The river authorities sell water to cities and other customers under various agreements that commit to deliver water rather than only provide reservoir storage space. The river authorities or in some cases their municipal or industrial customers obtain and hold water right permits. USACE is not responsible for water rights permits for storage capacity contracted to river authorities or other nonfederal sponsors such as water districts and cities.

Cities and Towns

Municipalities provide water supply, wastewater collection and treatment, and stormwater management services for their communities. Cities and towns provide recreation facilities and electrical energy services that also involve water management. Cities have constructed and now directly own 47 of the 195 major reservoirs and contract with other reservoir owners in Table 2-2 for water supply. Grigg (2011, 2012), Debo and Reese (2003), and the American Water Works Association (2001) describe the planning, design, financing, construction, maintenance, operation, and administration of municipal water, wastewater, and stormwater infrastructure.

Water Supply and Sanitation

For regulatory oversight purposes, TCEQ (2022) defines a public water system (PWS) as providing water via piping or other constructed conveyances for human consumption to at least 15 service connections or at least 25 people for at least 60 days each year. PWSs are categorized as (1) community PWSs such as towns or other communities where people live, (2) non-transient noncommunity PWSs such as schools or factories where people work but do not live, and (3) transient noncommunity systems such as rest stops, parks, and restaurants where people frequently come and go. As of July 2022, TCEQ regulated 7,108 PWSs, providing drinking water to approximately 29,989,617 customers (TCEQ 2022). Approximately 29,199,096 people receive drinking water from 4,636 community water systems. About 484,653 people receive drinking water from 889 non-transient noncommunity water systems. Approximately 305,868 people receive drinking water from 1,583 transient noncommunity water systems.

Cities provide water supply and wastewater management services for their citizens and commercial and institutional customers. Cities own, maintain, and operate groundwater wells, storage facilities, conveyance pipelines, water treatment facilities, water distribution networks, and wastewater collection and treatment facilities. Many cities own and manage their own sources of water supply, which may include multiple surface and/or groundwater sources. Some cities sell water to other neighboring cities. Many cities contract with river authorities, water districts, or USACE for water supply diversions or storage in reservoirs.

Of all the water use sectors in Texas and the United States, water for producing and condensing (cooling) steam for thermal-electric power plants accounts for the largest volume diverted from supply sources, but most of the diverted water is returned to streams (Stillwell et al. 2011, US Department of Energy 2014). Private and public electric utilities own reservoirs supplying cooling water. Agricultural irrigation accounts for the greatest consumptive use of water. Municipal use involves the greatest number of people directly consuming water. Municipal use has also been the most rapidly growing type of water use in Texas (TWDB 2022).

Stormwater Management and Floodplain Management

Stormwater management, floodplain management, and flood control are also functions of municipal government. Local communities are responsible for street bridges and culverts; draining rainfall runoff from streets and residential, commercial, and public properties and detention facilities to mitigate the impacts of development in increasing flows to downstream properties; and providing flood protection from overflowing streams.

Federal responsibility through USACE for the reduction of flood risk and damages on major rivers and NRCS on rural tributary streams has been well established for many decades. USACE and NRCS provide capabilities for partially controlling flood waters using dams and other constructed facilities, working in collaboration with local nonfederal sponsors. Floodplain management is a major objective of the NFIP administered by FEMA. However, the federal government legally cannot regulate private land use in local communities. City governments can legally regulate land use through building permits and other regulatory authorities. Cities regulate residential, commercial, and public development in floodplains as a condition for participation in NFIP.

Consulting Firms and Construction Contractors

Water management agencies often hire engineering consulting firms to perform feasibility and design studies, prepare water right permit applications, update floodplain delineations, administer and supervise contracts with construction companies, and perform other professional services. About 73,000 professional engineers, 2,700 surveyors, and more than 11,400 engineering firms practicing in Texas were registered with the Texas Board of Professional Engineers and Land Surveyors as of September 2023. Many civil engineering firms with offices in Texas include water-related disciplines in their repertoire of expertise. Many consulting companies specialize in water resources engineering and/or other professional fields relevant to water management. The firms range in size from one or several professionals to firms with more than 100,000 employees housed in many offices located throughout the nation and world. Many engineering companies performing water-related work have multiple offices located in different cities in Texas.

Private sector construction companies contract with water management agencies to construct and rehabilitate water projects. All major construction is performed by private construction companies. Equipment suppliers manufacture and sell pumps, pipes, irrigation equipment, and various other types of equipment required for water management.

Universities and Professional Organizations

Universities provide undergraduate and graduation education, research, extension, and service in the various water-related disciplines. Extension connects the products of university research with the practicing water management community. University faculty service often involves voluntary work with professional society organizations as well as diverse types of service to the professional community and the public.

The author has served since 1980 on the faculty of the Zachry Department of Civil and Environmental Engineering at Texas A&M University (TAMU) and has also been a member since its creation in 2005 of the TAMU interdepartmental graduate program in Water Management and Hydrologic Science (WMHS). Faculty from several colleges and multiple departments within the colleges participate in the WMHS graduate program. Graduate students in the WMHS program come from a diverse variety of academic disciplines. At TAMU, like many other universities, water resources engineering is one of several specialty areas within the Department of Civil and Environmental Engineering for which civil engineering undergraduate students can organize their technical electives and graduate students can focus their studies. Environmental engineering is both a specialty area within civil engineering and a separate degree-granting program. Many other academic departments also offer courses and specialty tracks in water-related studies.

The Water Resources Research Act of 1964 authorized establishment of a water institute at a land-grant university in each state to facilitate federal and state partnerships in university research and extension. The 54 institutes that comprise the National Institutes for Water Resources (NIWR) include a university-based institute in each state, District of Columbia, and three US territories. USGS manages the NIWR at the federal level. The Texas Water Resources Institute at TAMU is the NIWR member for Texas. In addition to NIWR institutes, many universities have centers to support research in hydrologic science, environmental studies, water engineering, and water resources management.

Professional societies and associations facilitate sharing of information, foster professional growth of members, influence political processes, and provide leadership in improving the effectiveness of the professional community in serving the public. These organizations are often identified with their conferences, training opportunities, journals, magazines, and manuals.

Several of the numerous national and international professional societies and associations that deal with water-related disciplines are listed as follows: American Water Resources Association (AWRA), International Water Resources Association (IWRA), International Water Association (IWA), American Water Works Association (AWWA), Water Environment Federation (WEF), Environmental and Water

Resources Institute (EWRI) of ASCE, US Society on Dams (USSD), Association of State Dam Safety Officials (ASDSO), International Commission on Large Dams (ICOLD), Floodplain Management Association (FMA), National Groundwater Association (NGWA), and Universities Council on Water Resources (UCOWR). These organizations all maintain websites that explain their missions, programs, and membership structures.

Many of the national organizations such as ASCE also have Texas chapters or sections. Many, including ASCE, have student chapters at universities. Other organizations such as the Texas Water Conservation Association and Texas Groundwater Association focus specifically on Texas.

The American Academy of Water Resources Engineers (AAWRE) is one of nine specialty field-advanced certification institutes affiliated with ASCE. State registration boards grant professional engineer licenses upon completion of specified education, experience, and other requirements. AAWRE certifies professional engineers in the specialty area of water resources engineering upon completion of additional experience as a licensed professional engineer, advanced graduate education, and other qualifications.

Other Contributors to Water Management

The responsibilities of elected officials in the US Congress, Texas legislature, and city government include dealing with water problems and needs and the professional water management community. Elected officials and appointed boards and commissions play important roles in directing water policy, authorizing water projects, and funding water management programs.

Certain organizations lobby for political support for water development projects. Environmental and natural resource conservation organizations such as the Nature Conservancy and Sierra Club encourage wise stewardship of water resources. A myriad of local, regional, and statewide citizen advisory groups guide water management programs and endeavors.

Water is essential for everyone from various perspectives. Water users, which include everyone, are foundational participants in water management.

Interstate Reservoirs

The portion of the border between the United States and Mexico defined by the Rio Grande extends through International Amistad and Falcon Reservoirs. State boundaries defined by the Red and Sabine Rivers extend through Lakes Texoma and Toledo Bend. These reservoirs are impounded by dams constructed on rivers that serve as boundaries. A segment of the border between Texas and Louisiana extends through Caddo Lake on Cypress Creek, but Cypress Creek is not a state border.

Referring to Tables 1-4 and 1-5 and Figure 1-9, International Amistad and Falcon Reservoirs on the Rio Grande operated by IBWC are the second and fifth largest reservoirs in Texas. Management of the waters of the Rio Grande is discussed later in the last section of this chapter. Texoma and Toledo Bend are the first and third largest reservoirs in Texas. Caddo Lake, with a storage capacity of 129,000 ac-ft, is the only natural lake in Texas with a storage capacity exceeding 5,000 ac-ft. Caddo Lake is an ecologically unique resource.

Lakes Texoma and Toledo Bend on State Borders

Lake Texoma impounded by Denison Dam on the Red River is operated by USACE Tulsa District primarily for flood control, hydropower, and recreation. Lake Texoma has by far the largest flood control pool of any reservoir in Texas. Flood control operations do not differentiate between Texas and Oklahoma flood protection. The hydroelectric power generated at Lake Texoma and seven other USACE reservoirs in Oklahoma is marketed by the SWPA to about a hundred electric utilities in a six-state region. The conservation pool was originally used only for hydroelectric power and recreation. High natural salinity constrains the use of Lake Texoma for water supply. However, the City of Sherman, Texas, uses desalinated water from Lake Texoma combined with groundwater for municipal supply. A pipeline from Lake Texoma to Lake Lavon in the Trinity River Basin constructed many years ago as a backup supply for cities in the Dallas area during drought has been used only infrequently.

Toledo Bend Dam and Reservoir on the Sabine River is owned and operated jointly by the Sabine River Authorities of Texas and Louisiana. The Sabine River Authority of Texas was created by the Texas legislature in 1949. The Sabine River Authority of Louisiana was created by the Louisiana legislature in 1950. The primary purpose for creating the two river authorities was to plan, design, construct, and operate the Toledo Bend Dam and Reservoir project to supply the municipal and industrial water use, hydroelectric energy generation, and recreation needs of the region. Because the project was implemented without federal funding, unlike USACE and IBWC reservoirs, Toledo Bend Reservoir contains no designated flood control storage. Project costs were shared equally by Texas and Louisiana. The two states arranged for revenue bond financing in 1959. Land acquisition began in 1963. Impoundment began in October 1966. The hydroelectric power plant began operation in 1969.

Caddo Lake on the Texas and Louisiana Border

The 129,000 ac-ft Caddo Lake is the only natural lake in Texas with a capacity greater than 5,000 ac-ft. A dam was constructed in 1914 to prevent erosion of the downstream channel from draining the lake. In 1971 for safety reasons, the Corps of Engineers completed construction of a replacement dam. The border between Texas and Louisiana extends through the lake.

Prior to 1976, the dam was owned and maintained by the Caddo Lake Levee District. The Water Resources Development Act of 1976 transferred maintenance responsibility to USACE. The cities of Marshall, Texas, and Oil City and Mooringsport, Louisiana, withdraw municipal supplies from the lake. Cooling water is also withdrawn from the lake by a steam-electric power plant located near Mooringsport. The city of Shreveport, Louisiana, operates a pumping and conveyance system that delivers water from Caddo Lake to Cross Lake which supplies the city.

The Caddo Lake Compact augments and amplifies the Red River Compact discussed later in this chapter. The stated purpose of the Caddo Lake Compact is to preserve and protect Caddo Lake while allowing its use for supplying water needs of adjacent portions of Louisiana and Texas. Rules are established for apportioning withdrawals between water users in Texas and Louisiana while preserving the environmental attributes of the lake.

Caddo Lake and adjoining wetlands include one of the largest flooded cypress forests in the United States (Figure 1-7) as well as seasonally flooded bottomland hardwood forests and mixed pine–hardwood upland forests. The many flora and fauna species at Caddo Lake include a large number of species classified as endangered, threatened, or rare.

The Nature Conservancy purchased 7,000 ac of Caddo Lake and adjacent area in 1992 that were merged with Caddo Lake State Park to become the Caddo Lake State Park and Wildlife Management Area managed by the TPWD. In 1993, Caddo Lake was selected as "a wetland of international importance, especially as waterfowl habitat" under the Ramsar Convention. The Ramsar designation includes state, federal, and private lands surrounding Caddo Lake. The Ramser Convention on Wetlands signed in 1971 is an international treaty for the conservation and sustainable use of wetlands at more than 2,300 locations throughout the world.

Interstate River Compacts

Texas participates in five interstate river compacts that have been executed by member states and approved by the US Congress and administered by compact commissions. Many similarly administered interstate river compacts have been established throughout the United States, especially in the western states. The rivers and the dates that compact between Texas and neighboring states became effective are Rio Grande, 1939; Pecos, 1948; Canadian, 1952; Sabine, 1954; and Red, 1980. The compacts for each of the five rivers are published as Chapters 41, 42, 43, 44, and 45, respectively, of "Title 3 River Compacts" of the Texas Water Code (TCEQ 2023). The compacts represent both state and federal laws. The purposes of the compacts are to provide for equitable allocation of water between the states and to facilitate cooperative planning and management.

The water apportionment rules are structured differently between the five compacts though general administrative procedures are the same. Each compact is administered by a commission with representatives from each member state and a representative of the federal government appointed by the president of the United States. The commissioners rely on state water agencies for technical support. The Texas compact commissioners rely on TCEQ staff for technical and administrative support.

A website maintained by TCEQ (2023) provides convenient access to copies of the compacts and related documents and contact information for Texas commissioners and TCEQ technical and legal advisory staff. The website also provides a brief history of each of the five compacts.

Disputes are settled by the US Supreme Court if compact commissions cannot work out disagreements between the states through negotiation. A major objective of compact commissions is to avoid lawsuits, but disputes in the western states have often reached the litigation stage. Various states throughout the western United States have experienced lengthy and costly water allocation disputes settled through the US Supreme Court (Longo 2002).

Rio Grande Compact

The Rio Grande Compact approved by the affected states and US Congress in 1939 and amended through a resolution in 1948 apportions the waters of the Rio Grande and its tributaries above Fort Quitman between Colorado, New Mexico, and Texas. The Compact obligates Colorado to deliver specified quantities of water each year to New Mexico at the Colorado–New Mexico state boundary. Likewise, New Mexico is required to deliver specified quantities of water into Elephant Butte Reservoir located on the Rio Grande in New Mexico. Compact provisions include a system of annual and accrued debits and credits and limit the maximum accrued debits and credits under various scenarios.

Water in Elephant Butte Reservoir is allocated between New Mexico and Texas. The water shares of New Mexico and Texas are managed by Elephant Butte Irrigation District and El Paso County Water Improvement District No. 1, respectively. The original compact allocated 57% of the water to New Mexico and 43% to Texas. The apportionment was later revised such that Texas receives 376,000 ac-ft annually.

Elephant Butte Dam and Reservoir on the Rio Grande is part of the USBR Rio Grande Project comprised of several reservoirs and other facilities constructed to provide irrigation, hydroelectric power, and flood control in the upper Rio Grande Basin. Elephant Butte owned and operated by the USBR is the largest reservoir in New Mexico with a total storage capacity of 2,065,000 ac-ft. Construction of the Elephant Butte Dam and Reservoir project was completed by USBR in 1916.

Pecos River Compact

Water allocation each year under the Pecos River Compact is based on the principle that New Mexico must not deplete by human activities the annual volume of water available to Texas at the New Mexico–Texas border under 1947 conditions with a 43% Texas and 57% New Mexico apportionment. Unappropriated flood flows originating in New Mexico are shared equally between the two states. Flows of the Pecos River originating from the Texas portion of the watershed belong to Texas. Red Bluff Reservoir is in Texas on the Pecos River just below the state border. Water delivered from New Mexico flows into Red Bluff Reservoir. Gauge measurements and calculations are employed each year to determine annual volumes to be delivered from New Mexico to Texas. Deficits in one year can be balanced by extra flows the next year.

Texas filed a lawsuit against New Mexico in 1974, claiming that Texas did not receive its full share of water allocated by the 1948 Pecos River Compact due to overuse in New Mexico (Hall 2002). In 1988, based on the recommendations of a court-appointed special master, the US Supreme Court awarded Texas a cash settlement for past damages and established a Pecos River Master to administer future water allocations under the compact.

Canadian River Compact

The Canadian River Commission administers the Canadian River Compact, which includes the states of New Mexico, Oklahoma, and Texas. Signed in 1950 by the member states, the compact was subsequently ratified by the respective state legislatures, approved by Congress, and signed into law by the president in 1952. The Canadian River Commission includes one state commissioner appointed by the governor of each member state and one federal commissioner appointed by the president.

The compact gives New Mexico free and unrestricted use of all waters of the drainage basin of the Canadian River above Conchas Dam. New Mexico is also given free and unrestricted use of all waters of the drainage basin in New Mexico below Conchas Dam, but a maximum limit of 200,000 ac-ft of reservoir storage is placed on the use of this water. The compact gives Texas free and unrestricted use of all waters of the Canadian River in Texas subject to limitations on the volume of water that can be stored. Oklahoma has free and unrestricted use of all waters of the Canadian River in Oklahoma.

In response to litigation filed by Oklahoma and Texas against New Mexico, the US Supreme Court ruled in December 1993 that New Mexico had been in violation of the compact from 1987 to 1993. New Mexico was required to make specified releases from reservoirs in New Mexico for the benefit of the two downstream states. New Mexico was also required to pay attorney fees incurred by Texas and Oklahoma.

Red River Compact

The Red River Compact allocates the waters of the Red River and its tributaries between Texas, Oklahoma, Arkansas, and Louisiana and establishes an interstate compact commission to administer the compact. Texas is represented on the interstate Red River Compact Commission by the executive director of TCEQ and one commissioner who is appointed by the governor for a term of six years.

The compact divides the Red River and its tributaries into five reaches: (1) above Denison Dam, (2) from Denison Dam to Arkansas–Louisiana border, (3) tributaries that cross the Texas–Louisiana and Arkansas–Louisiana borders, (4) tributaries east of the Red River in Arkansas, and (5) Red River and tributaries in Louisiana not included in other reaches. The compact further divides the first four reaches to define water allocations at subbasin levels. Rules for apportionment of water between relevant states are specified for each subbasin of each reach.

Sabine River Compact

About 135 mi of the more than 300-mi-long Sabine River form the majority portion of the boundary between Texas and Louisiana. Texas and Louisiana contain 77.6% and 22.4% of the 9,756 sq. mi Sabine River Basin.

The Sabine River Compact defines the term *Stateline* as "the point on the Sabine River where its waters in downstream flow first touch the States of both Louisiana and Texas." Stateline flow is measured at a gauge on the Sabine River at US Highway 84 near Logansport. The gauge is about 4 mi downstream of Stateline. The compact term *Stateline reach* refers to the Sabine River from Stateline downstream to its outlet at Sabine Lake.

The compact allows Texas free and unrestricted use of all waters of the Sabine River above the Stateline site. All free water at Stateline is divided equally between Texas and Louisiana with provisions for maintaining a minimum flow of 36 ft^3/s at Stateline. Reservoirs constructed before January 1953 are exempted from obligations to maintain the 36 ft^3/s minimum flow at Stateline. Domestic and stock water reservoirs are also exempt from maintaining the minimum flow requirement. All free water in the Stateline reach is divided equally between the two states.

The compact prohibits either state from constructing a dam on the shared state border without the other state's consent. The two states can jointly construct and operate reservoirs on the Stateline reach of the Sabine River, with water in such reservoirs allocated proportional to the contribution to the cost of developing and maintaining the reservoir storage.

International Rio Grande

The Rio Grande Basin shown in Figure 2-1 is the fifth largest river basin in the United States. The Mississippi, Yukon, Columbia, and Colorado (western states, not the Texas Colorado) River Basins are larger. The waters of the Rio Grande are shared by Mexico and the United States, and the river ("rio" in Spanish) serves as an international boundary. The Rio Grande is also called the Rio Bravo in Mexico. International Amistad and Falcon Dams and Reservoirs are operated jointly by the two nations. Within the United States, the waters of the Rio Grande are allocated between Texas, New Mexico, and Colorado by interstate compacts. The Rio Grande is different from the other rivers in Texas both institutionally as discussed in here in Chapter 2 and continuing in Chapter 3 and hydrologically as discussed in Chapter 5.

Water Allocation

The share of the waters of the Rio Grande allocated to Texas is further allocated to water right holders through a water allocation system administered by TCEQ. As discussed earlier in this chapter, TCEQ administers a surface water rights permit system for all of Texas. However, water rights in the Rio Grande Basin below Fort Quitman are significantly different than surface water rights everywhere else in Texas as discussed in Chapter 3.

Fort Quitman shown in Figure 2-1 is a key location in both the international and the state water allocation systems. Fort Quitman was

Figure 2-1. Rio Grande Basin.

established in 1858 as a US Army installation but is now an abandoned historical site. The site is located about 90 river mi downstream of the city of El Paso and 1,150 river mi upstream of the Gulf of Mexico.

The international boundary between the United States and Mexico follows the middle of the Rio Grande from its mouth at the Gulf of Mexico 1,255 mi to a point just upstream of El Paso, Texas. From there, the boundary follows an alignment westward overland for 533 mi to the Colorado River, follows the middle of that river for 24 mi and then extends overland for 140 mi to the Pacific Ocean. The governments of the two countries through the Convention of 1889 established the International Boundary Commission to settle questions arising regarding the location of the boundary when the two meandering rivers changed their course.

A 1944 treaty allocating the waters of the Rio Grande and Colorado River between the two nations also changed the name of the International Boundary Commission to International Boundary and Water

Commission (IBWC). The Mexico section and US section of IBWC are headquartered in Ciudad Juarez, Chihuahua, Mexico, and El Paso, Texas, which are separated by the Rio Grande. IBWC administers the allocation of the waters of the Rio Grande and Colorado River between the two nations and operates a multipurpose reservoir system on the Rio Grande for water supply, flood control, hydroelectric power, and recreation. IBWC also conducts planning studies and implements projects for border water supply and sanitation, salinity mitigation, local flood control, and stream bank stabilization.

A 1906 convention between the United States and Mexico provides for delivery of 60,000 ac-ft/year of Rio Grande water to Mexico in the El Paso-Juarez Valley above Fort Quitman. Elephant Butte Reservoir in New Mexico, operated by USBR, and the American and International diversion dams near El Paso, operated by IBWC, are used to implement the water allocation provisions of the 1906 convention.

The Water Treaty of 1944 provided for the distribution of waters of the Rio Grande from Fort Quitman to the Gulf of Mexico between the two nations and authorized construction of the International Amistad and Falcon Reservoir projects.

The 1944 treaty also allocates the waters of the Colorado River between the upstream and downstream countries. The Colorado River flows through Lakes Powell and Mead, the two largest reservoirs in the United States, impounded by Glen Canyon and Hoover Dams, before flowing from the United States through Mexico to the Gulf of California. These two reservoir projects are operated by the USBR. The Colorado River and its tributaries drain the fourth largest river basin in the United States encompassing portions of seven western states, which jointly administer interstate river basin compacts. The Colorado River has been called the most institutionally regulated river in the world.

The Rio Grande Compact approved by the legislatures of Colorado, New Mexico, and Texas in 1939 allocates the remaining waters of the Rio Grande above Fort Quitman that are not committed to Mexico by the 1906 convention. The Pecos River Compact adopted in 1949 allocates the waters of that tributary of the Rio Grande between Texas and New Mexico. The 307,000 ac-ft Red Bluff Reservoir is located on the Pecos River in Texas just downstream of the border with New Mexico. Construction of the project was completed in 1936. The Red Bluff Water Control District operates the reservoir for agricultural irrigation and hydroelectric power.

International Amistad and Falcon Reservoirs

The International Amistad and Falcon Reservoirs shown in Figures 2-2 and 2-3 are operated by the IBWC primarily for water supply and flood control for the Lower Rio Grande Valley below Falcon Dam. The reservoir projects also provide hydroelectric power and recreation. As indicated by Table 1-5, these are the second and fifth largest reservoirs located partially or wholly in Texas. The locations of the dams are shown in Figures 1-9 and 2-1.

In accordance with the 1944 treaty, the United States has 56.2% and 58.6% of the conservation storage capacity of Amistad and Falcon, respectively, with Mexico owning the remainder. IBWC operates Anzaldus and Retamal Dams on the lower reach of the Rio Grande to facilitate water supply diversions. The travel time for releases from Falcon Reservoir to reach the most downstream diversion locations is about a week.

Streamflows into Falcon and Amistad Reservoirs are allocated between the two countries. Flows on several major tributaries named in the 1944 treaty are gauged and allocated as specified by the treaty. All other flows not otherwise allocated are divided equally between the two countries. Computations are performed routinely by IBWC staff to allocate the reservoir inflow and evaporation volumes which are

Figure 2-2. International Amistad Dam and Reservoir on the Rio Grande. *Source*: Courtesy of TWDB.

Figure 2-3. International Falcon Dam and Reservoir on the Rio Grande.
Source: Courtesy of TWDB.

combined with recorded releases to determine the amount of water that each country has in storage.

IBWC is responsible for flood control and hydropower operations. Hydroelectric power generation is usually limited to using water released anyway for other purposes.

The US share of the water supply storage in Amistad and Falcon Reservoirs is used to meet demands in the lower basin pursuant to the state water rights system administered by TCEQ. Irrigation districts, individual farmers, and cities communicate their water needs directly to the TCEQ watermaster office, which in turn schedules releases from Falcon and Amistad Reservoirs with IBWC. The watermaster office routinely maintains an accounting of the amount of water used and the amount of water in reservoir storage allocated to each of about 1,600 water rights accounts. Most of the water is used in the very productive agricultural region below Falcon Reservoir.

References

Arnold, J. L. 1988. *The evolution of the 1936 Flood Control Act*. EP 870-1-29. Washington, DC: Government Printing Office, USACE Office of History, Fort Belvoir.

American Water Works Association. 2001. *Water resources planning*. Manual of water supply practice M50. Denver: AWWA.

ASCE. 2021. *Texas infrastructure report card*. Austin, TX: ASCE Texas Section.

Benson, R. D., B. W. Griggs, and A. D. Tarlock. 2021. *Water resource management: A casebook in law and public policy*. 8th ed. New York: Foundation Press.

Boes, R. M., ed. 2011. *Dams and reservoirs under changing challenges*. Boca Raton, FL: CRC Press.

Collin, R. W. 2006. *The Environmental Protection Agency: Cleaning up America's mess*. Westport, CN: Greenwood Press.

Council on Environmental Quality. 2014. "Economic and environmental principles and guidelines for planning water and related land resources implementation studies." *Fed. Reg.* 79 (247): 77460–77461.

Dascher, E. D., and K. Meitzen. 2020. "Dams are coming down, but not always by choice: The geography of Texas dams, dam failures, and dam removals." *Tex. Water J.* 11 (1): 89–129.

Debo, T. N., and A. J. Reese. 2003. *Municipal stormwater management*. Boca Raton, FL: Lewis Publishers.

Dworsky, L. B., D. J. Allee, and R. M. North. 2020. "Water resources planning and management in the United States federal system: Long term assessment and intergovernmental issues." *Nat. Res. J.* 31 (3): 475–547.

Executive Office of the President. 2020. *Modernizing America's water resource management and water infrastructure*. Executive Order 13956. Fed. Reg. 85 FR 65647.

Getches, D. H., S. Zellmer, and A. Amos. 2015. *Water law in a nutshell*. 5th ed. Eagan, MN: West Academic Publishing.

Godfrey, F. A., and C. L. Dowell. 1968. *Major hydroelectric powerplants in Texas*. Rep. 81. Austin, TX: Texas Water Development Board.

Grigg, N. S. 1985. *Water resources planning*. New York: McGraw-Hill.

Grigg, N. S. 2011. *Water finance: Public responsibilities and private opportunities*. New York: Wiley.

Grigg, N. S. 2012. *Water, wastewater, and stormwater infrastructure management*. 2nd ed. Boca Raton, FL: Lewis Publishers.

Hall, G. E. 2002. *High and dry: The Texas-New Mexico Struggle for the Pecos River*. Albuquerque, NM: University of New Mexico Press.

Haynes, W. P. and M. C. Barnes, ed. 2009. *Dams: Impacts, stability, and design*. New York: Nova Science Publishers.

Jackson, D. C. 1997. *Dams, studies in the history of civil engineering*. Vol. 4. Farnham, UK: Ashgate Publishing.

Jansen, R. B., ed. 1988. *Advanced dam engineering for design, construction, and rehabilitation*. New York: Van Nostrand Reinhold.

Knowles, S. G. and H. C. Kunreuther. 2014. Troubled waters: The National Flood Insurance Program in historical perspective. *J. Policy Res.* 26 (3): 327–353.

Longo, P. J. 2002. "Water across borders: Judicial realities." In *Water on the Great Plains: Issues and policies*, edited by P. J. Longo and D. W. Yoskowitz. Lubbock, TX: Texas Tech Press.

McDaniels, L. L. 1964. *Conservation storage reservoirs in Texas, some aspects and chronology of surface-water development*. Bulletin 6404. Austin, TX: Texas Water Commission.

National Research Council. 2005. *The science of instream flows: A review of the Texas instream flow program*. Washington, DC: The National Academies Press.

National Research Council. 2012. *Science for environmental protection*. Washington, DC: The National Academies Press.

Normand, A. E. 2019. *Dam safety overview and the federal role*. R45981. Washington, DC: Congressional Research Service.

NWS 2024. "Modernizing probable maximum precipitation estimation." Accessed March 5, 2024. https://www.weather.gov/owp/hdsc_pmp.

Patterson, L. A. and M. W. Doyle. 2018. A nationwide analysis of US Army Corps of Engineers reservoir performance in meeting operational targets. *J. Am. Water Resour. Assoc.* 54 (2): 543–564.

Porter, C. R. 2014. *Sharing the common pool: Water rights in the everyday lives of Texans*. College Station, TX: Texas A&M University Press.

Reuss, M., ed. 1993. *Water resources administration in the United States: policy, practice, and emerging issues*. East Lansing, MI: American Water Resources Association and Michigan State University Press.

Sahs, M. K. and H. Heinrich, ed. 2022. *Essentials of Texas water resources*. 7th ed. Austin, TX: State Bar of Texas.

Schnitter, N. J. 1994. *A history of dams: The useful pyramids*. Brookfield, VT: Balkema.

Stillwell, A. S., C. W. King, M. E. Webber, I. J. Duncan, et al. 2011. "The energy-water nexus in Texas." *Ecol. Soc.* 16 (1): 2.

Task Force on Federal Flood Control Policy. 1966. *A unified national program for managing flood losses*. House Document 465. 89th Congress, 2nd Session. Washington, DC: Government Printing Office.

TCEQ (Texas Commission on Environmental Quality). 2022. *State of Texas public drinking water program 2021 annual report*. Austin, TX: Office of Water, Water Supply Division, TCEQ.

TCEQ. 2023. "Interstate river compact commissions." Accessed November 27, 2023. https://www.tceq.texas.gov/permitting/compacts/interstate.html.

TCEQ. 2024. "Adopted environmental flow rules." Accessed March 5, 2024. https://www.tceq.texas.gov/permitting/water_rights/wr_technical-resources/eflows/rulemaking.

TDWR (Texas Department of Water Resources). 1984. *Water for Texas: A comprehensive plan for the future*. Vol. 1 and 2. GP-4-1. Austin, TX: Texas Department of Water Resources.

TPWD (Texas Parks and Wildlife Department). 2018. *Texas outdoor recreation plan*. Austin, TX: State Parks Division, Texas Parks and Wildlife.

TWDB (Texas Water Development Board). 1968. *The Texas water plan*. Austin, TX: Texas Water Development Board.

TWDB. 2022. *Water for Texas: 2022 state water plan*. Austin, TX: Texas Water Development Board.

USACE (US Army Corps of Engineers). 2016. "Use of US Army Corps of Engineers reservoir projects for domestic, municipal & industrial water supply." *Fed. Reg.* 81 (242): 91556–91590. Washington, DC.

US Department of Energy. 2014. *The water-energy nexus: challenges and opportunities*. Washington, DC: Water-Energy Technology Team, Office of Energy Policy and Systems Analysis.

Viessman, W. and C. Welty. 1985. *Water management technology and institutions.* Cambridge, MA: Harper & Row.

Water Resources Council. 1973. "Principles and standards for planning water and related land resources." *Fed. Reg.* 38 (24784): 248222–248223.

Wurbs, R. A. 1983. "Economic feasibility of flood control improvements." *J. Water Resour. Plan. Manag.* 109 (1): 29–47.

Wurbs, R. A. 1985. *State-of-the-art review and annotated bibliography of dam-breach flood forecasting.* Military Hydrology Report 9 and Miscellaneous Paper EL-79-6. Vicksburg, MS: USACE Waterways Experiment Station.

Wurbs, R. A. 1994. "Supply storage in federal reservoirs." *J. Am. Water Work. Assoc.* 86 (4): 188–197.

Wurbs, R. A. 1995. "Water rights in Texas." *J. Water Resour. Plan. Manag.* 121 (6): 447–454.

Wurbs, R. A. 2004. "Water allocation systems in Texas." *Int. J. Water Resour. Dev.* 20 (2): 229–242.

Wurbs, R. A. 2013. "Water allocation systems." In Chap. 8, *Water resources planning, development, and management*, edited by R. A. Wurbs. London: InTechOpen.

Wurbs, R. A. 2017a. "Incorporation of environmental flows in water allocation in Texas." *Water Int.* 42 (1): 18–33.

Wurbs, R. A. 2017b. "Institutional framework for water management." In Chap. 137, *Handbook of applied hydrology*, 2nd ed., edited by V. P. Singh. New York: McGraw-Hill.

Wurbs, R. A. and W. P. James. 2002. *Water resources engineering.* Upper Saddle River, NJ: Prentice Hall Division, Pearson Education Publishing.

Zhang, Y. and N. Fernando. 2020. *Forecast-informed reservoir operations (FIRO) and water resources management in Texas.* Austin, TX: Texas Water Development Board, University of Texas at Arlington, and National Integrated Drought Information System.

Chapter 3

Reservoir and River System Operations

Operating plans and practices consist of policies, guidelines, and procedures for determining the quantities of water to maintain in rivers as instream flows, to store and release from reservoirs, and to divert for beneficial use under various conditions. The terms *operating* or *regulation* or *water control procedures, rules, schedules, policies, practices,* or *plans* are used here interchangeably. Operating decisions involve allocation of river flows, reservoir storage capacity, and water releases or withdrawals between multiple reservoirs, between project purposes, between water users, and between time periods. Regulation plans include sets of quantitative criteria within which significant flexibility exists for qualitative judgment. Operating plans combine fixed requirements, flexible guidance, and decision-support information and tools. River/reservoir system operations include

- Operations during normal hydrologic conditions from the perspective of optimizing the present day-to-day, seasonal, or year-to-year use of available infrastructure and water resources;
- Operations during normal hydrologic conditions from the perspective of maintaining capabilities for responding to infrequent floods and droughts expected to occur at unknown times in the future;
- Operations during flood events; and
- Operations during low flow or drought conditions.

For many water supply reservoirs, operations are based simply on making withdrawals or releases as necessary to meet water demands subject to water rights and contractual agreements. Flood flows pass through uncontrolled spillways if the reservoir is full. Predeveloped water conservation plans are in place for responding to supply depletion during infrequent severe droughts. Conversely, complex regulation plans guide operations of many reservoirs including large multiple-purpose, multiple-reservoir systems. Operating practices combine a framework of quantitative rules with experience-based operator judgment. Day-to-day operating decisions may be influenced by a complex array of factors and often are based on professional judgment and experience. Operating procedures may change over time with experience and changing conditions.

Dams and Associated Facilities

Reservoir projects include dams, outlet structures, pumping plants, pipelines, canals, river channel improvements, hydroelectric power plants, electric transmission facilities, recreation resources, roads, buildings, and various other components. The design, construction, maintenance, and physical operation of these facilities are outside the scope of this book. However, water management strategies addressed in this book are dependent on the capabilities provided by constructed infrastructure to store and regulate river flows and beneficially use water resources.

Golze (1977), Jansen (1988), Siddiqui (2009), Tanchev (2014), and Lewis (2014) address methods applied in the design, construction, and maintenance of dam and reservoir projects throughout the world. USACE and USBR have developed many engineering manuals and technical documents to guide the design and construction of federal projects. TCEQ (2006, 2009) disseminates guidelines for design, construction, operation, and maintenance of dams in Texas in conjunction with the statewide dam safety program.

Although many concrete dams are found in Texas, most of the major reservoirs are impounded by earth-fill embankment dams. Earth embankments may have a concrete spillway section. Reservoir releases are made through overflow spillways and outlet works conduits. Outlet works gates control releases through conduits through the dam from both conservation and flood control pools. Spillways provide additional capability to release high flows without overtopping or damaging the dam and appurtenant structures. Spillways are required to allow flood inflows to safely flow through the reservoir regardless of whether the project includes designated flood control storage. Spillways may be controlled (gated) or uncontrolled (no gates). A spillway with crest gates allows the outflow rate to be adjusted. For uncontrolled spillways, the outflow rate is simply a function of the height of the water surface above the spillway crest. Some reservoirs have more than one spillway. Service spillways convey smaller, more frequently occurring release rates, with flows over an emergency spillway occurring only during infrequent extreme floods.

The major portion of the storage volume in most reservoirs is located below the spillway crest. Outlet works may be used for releases from storage both below and above the spillway crest. Outlet works are used to release water for water supply, maintenance of instream flows, or other purposes. Components of an outlet works include an intake structure in the reservoir, one or more conduits or sluices through the dam, gates located on either the intake structure or conduits, and a stilling basin or other energy dissipation structure at the downstream end.

Water supply diversions may be either lakeside or downstream. Lakeside withdrawals require intake structures, pumps, and pipeline or canal conveyance facilities. Downstream releases through the outlet works of one or multiple dams may be diverted from the river at downstream locations that may be great distances below the dam or multiple dams. Downstream releases may be made through hydroelectric power penstocks as well as outlet conduits and overflow spillways.

Most of the major reservoirs listed in Appendixes A and B are impounded by earthen or concrete dams constructed across a river that flows into the reservoir. However, several of the reservoirs are off-channel storage facilities. The reservoir may be located some distance from the river from which most or all the water is supplied. Water is diverted from the river through a pumping plant and conveyed by pipeline or canal to the off-channel reservoir for storage. Water supply withdrawals from the off-channel reservoir are made as needed to supply water demands.

Many of the water supply reservoir projects in Texas include pipeline and pumping stations that transport water long distances, in some cases exceeding a hundred miles, from a diversion structure in the lake or river to water treatment plants, water distribution systems, or other reservoirs in the city or other location near where the water is used. Water users and the reservoirs supplying their water use are usually located in the same river basin. However, transport of water between river basins is common in Texas. Several of the water supply systems discussed in Chapter 7 include major interbasin transfers of water.

Descriptive information for USACE reservoirs is found at USACE websites. A TWDB website in Table 4-1 of Chapter 4 describes about 200 individual dams and appurtenant structures. The three-volume TWDB Report 126 (Dowell and Petty 1971, 1973, 1974) comprises drawings, dimensions, storage and outlet capacities, and other information describing the dam and outlet structures of 161 large reservoir projects in operation in Texas as of the early 1970s. The USACE NID website in Table 4-1 includes a link labeled "Learn More about Dams" providing a basic tutorial describing the components of dam and reservoir projects. The NID photos webpage comprises photographs of many dams and appurtenant structures at reservoir projects located throughout the United States including several in Texas.

Both flat broad-crested weir and ogee weir spillways are common. Many dams have concrete ogee weir spillways controlled by Tainter gates. The curved shape of an ogee weir is designed to achieve hydraulic efficiency by replicating the curvature of the lower side of the nappe of a sharp-crested weir (Wurbs and James 2002). The Tainter gate invented by Jeremiah Tainter in 1886 is a type of radial arm floodgate employed at numerous reservoirs nationwide and worldwide that may range in size from small to very large.

Several photographs of dam and reservoir projects are included in various chapters of this book. The photographs in the figures referenced in the following paragraphs include dams, gated and ungated spillways, outlet works, and other facilities. Hydropower plants located at Whitney, Toledo Bend, Amistad, Falcon, and Denison dams can also be seen in the photographs.

The spillway at Whitney Dam and Reservoir shown on the front cover of this book consists of a 680 ft long concrete overflow ogee weir controlled with seventeen 40 × 38 ft Tainter gates. Water from Whitney Reservoir is also released through sixteen 5 × 9 ft rectangular conduits through the dam. The concrete gravity and earth-fill Whitney Dam is 1,674 ft long and has a maximum height of 159 ft.

The photo of Toledo Bend Dam and Reservoir in Figure 1-10 shows the 9,070 ft long earth-fill embankment dam and 440 ft long concrete ogee weir spillway with eleven 40 × 28 ft Tainter gates. Pertinent elevations in feet above mean sea level (msl) are as follows: Top of dam (185 ft), maximum design water surface (175.3 ft), top of spillway gates (173 ft), top of conservation pool (172 ft), spillway weir crest (145 ft), low flow conduit outlet invert (100 ft), and streambed just below dam (73 ft).

The 800 ft long ogee spillway at International Amistad Dam shown in Figure 2-2 is controlled with sixteen 50 × 54 ft Tainter gates. The dam consists of a concrete section containing the spillway, which has a maximum dam height of 254 ft above the riverbed, flanked by earth embankments on each side of the concrete section. The 6.07 mi long dam includes 4.25 mi in Mexico and 1.82 mi in the United States.

The 2,000 ft long uncontrolled (ungated) concrete ogee weir spillway at Denison Dam shown in Figure 3-1 has a crest elevation of 640 ft above msl, which is the top of the flood control pool. Controlled releases are made through eight 20 ft diameter, 800 ft long conduits through the dam controlled by vertical lift gates with an intake invert elevation of 523 ft above msl. The 15,200 ft long dam has a maximum height of 165 ft.

The outlet intake structure at Benbrook Dam is shown in Figure 3-2. The bridge from the dam to gate tower is seen in the photo. Flow through a gated 13 ft diameter concrete conduit and two 2.5 ft diameter steel

Figure 3-1. Spillway and hydropower house at Denison Dam on the Red River. *Source*: Courtesy of USACE Tulsa District.

Figure 3-2. Intake structure for the outlet works of Benbrook Dam. *Source*: Courtesy of TWDB.

pipes through the dam is controlled in this intake structure. Benbrook Dam also has an ungated 500 ft long ogee weir emergency spillway. The dam is on the Clear Fork of the Trinity River near the city of Benbrook upstream of Fort Worth.

The outlet structure for the two gate-controlled 10 ft diameter conduits through the dam at Lake O' the Pines on Big Cypress Creek in northeast Texas is shown in Figure 3-3. The 10,600 ft long dam has a maximum height of 97 ft. In addition to the outlet works in Figure 3-3, the dam has a 200 ft long uncontrolled ogee spillway with the crest at top of flood control pool.

Reservoir Pools

Operating rules include dividing the total storage capacity of a reservoir into pools defined by designated water surface elevations. A typical reservoir consists of one or more of the vertical zones, or pools, illustrated by Figure 3-4. As discussed later in this chapter, portions of the storage capacity may be viewed as being reserved for estimated volumes of future accumulated sediment deposits. Sediment accumulation represents a loss of storage capacity.

Conservation storage purposes, such as municipal and industrial water supply, agricultural irrigation, and hydroelectric power generation

Figure 3-3. Outlet structure for Lake O' the Pines outlet works.
Source: Courtesy of USACE.

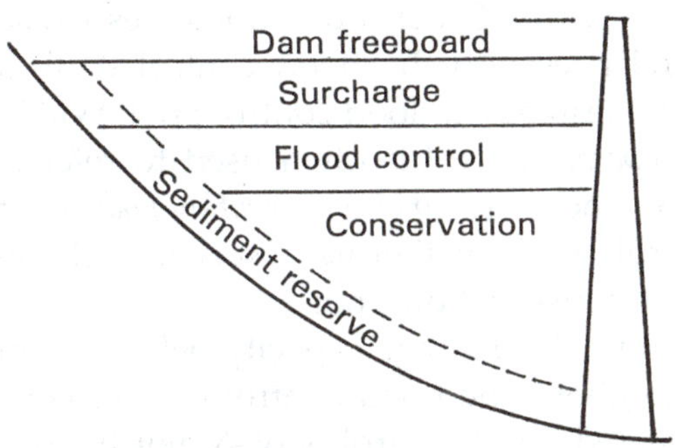

Figure 3-4. Reservoir pools.

involve storing water during periods of high streamflow and/or low water demand for later beneficial use as needed. Conservation storage also provides head for hydropower and opportunities for recreation. The reservoir water surface is maintained at or as near the designated top of conservation pool elevation as streamflows and water demands allow.

Drawdowns are made as required to meet the various needs for water. *Normal operating level* is another expression commonly used for the top of conservation pool elevation.

The conservation pool may include active and/or inactive storage capacity. Water is not withdrawn from the inactive portion of the conservation pool, except through the natural processes of evaporation and seepage. The top of inactive pool elevation may be fixed by the invert of the lowest outlet or, in the case of hydroelectric power, by conditions of operating efficiency for the turbines. An inactive pool also may be contractually set to facilitate withdrawals from outlet structures that are significantly higher than the invert of the lowest outlet structure at the project. The inactive pool is sometimes called *dead storage*. It may provide a portion of the sediment reserve, head for hydroelectric power, and a pool for recreation and fish habitat.

The flood control pool remains empty except during and immediately following flood events. The bottom of the flood control pool is the top of the conservation pool. The top of flood control elevation is often set by the crest of an uncontrolled emergency spillway, with releases being made through other outlet structures. Gated spillways allow the top of flood control pool elevation to exceed the spillway crest elevation.

Multipurpose Corps of Engineers reservoirs are divided into a flood control pool and a conservation pool, allowing the two pools to be operated separately. Nonfederal water supply sponsors control the conservation pool, with USACE operational responsibilities activated as the water level rises into the flood control pool. Most nonfederal reservoirs have no designated flood control pool. Without flood control, the term *normal operating level* is often used to refer to the top of the conservation pool. The term *top of conservation pool elevation* is employed in this book regardless of whether the reservoir includes a flood control pool on top of the conservation pool.

The surcharge pool is storage capacity below the maximum design water surface and above the flood control pool or conservation pool if there is no designated flood control pool. Major flood events exceeding the capacity of the flood control pool encroach into surcharge storage. Without a flood control pool, surcharge storage occurs whenever inflow to a full reservoir exceeds outflow.

Reservoir storage capacities are defined as volumes below a specified elevation or between two specified elevations. Likewise, pool delineations are defined as specified elevations in feet above mean sea level. Although pool elevation designations imply a flat, horizontal water surface, the water surface along the length of a reservoir slopes in the downstream direction as inflows flow through the reservoir and outlet structures. At an instant in time, the water surface elevation in the upstream reaches of a reservoir can be expected to be higher than the water surface elevation near the dam. Conceptually, the reservoir surface is precisely horizontal and flat only when there is no inflow or outflow. Accepted practice is to define pool designations and operating rules as a function of water surface elevations in the main reservoir near the dam but outside of drawdowns near the outlet structures.

The top of conservation pool elevation or normal target operating level may be precisely fixed or alternatively some degree of flexibility may be allowed to enhance operations. Toledo Bend Reservoir has the largest conservation storage capacity (water supply, hydropower, recreation) in Texas but no designated flood control pool. The lake is more than 100 miles long. The storage volume and surface area are 4,477,000 ac-ft and 181,600 ac (284 sq. mi) at top of conservation pool elevation 172.0 ft above mean sea level. The overflow spillway crest is 1 ft above the normal operating pool (top of conservation pool) adding an additional 184,000 ac-ft of storage volume. One foot of additional operating flexibility in Toledo Bend represents a large storage volume for attenuating flood flows or generating hydropower.

Water supply storage capacities have been increased significantly without structural modifications in several reservoirs by permanent or temporary reallocation of flood control storage capacity to water supply by raising the designated top of conservation pool (bottom of flood control pool) elevation. Dredging to increase storage capacity has been employed in a few cases but is not typically economically feasible for widespread use in preserving or increasing storage capacity.

Maximum Design Water Surface

The maximum design water surface, or top of the surcharge storage in Figure 3-4, is set during project design from the perspective of dam safety. Reservoir design and operation are based on assuring that the reservoir water surface will never exceed the designated maximum design water surface elevation under any conditions. For most dams, particularly earth-fill embankments, the top of dam elevation includes a freeboard allowance above the maximum design water surface elevation to account for wave action and provide an additional safety factor against overtopping.

Water management capabilities provided by conservation and flood control storage are assessed from a likelihood, probability, frequency, or reliability perspective employing computer-based modeling tools discussed in Chapter 4. Relationships between water supply diversion targets or hydroelectric energy generation targets and the reliabilities with which these targeted demands can be supplied are assessed. Probabilities of failures in supplying demands are estimated. Flood control storage capabilities are measured in terms of the probability of the flood control pool storage capacity being equaled or exceeded in any future year. However, protection against overtopping and the resulting failure of a dam is based on extremely high levels of safety evaluated deterministically rather than probabilistically in the design process for new projects or in dam safety assessments for existing dams (Wurbs and James 2002).

Estimates of the extreme upper limits of precipitation and runoff are required for design situations in which failure could result in catastrophic consequences. A flood overtopping a dam could breach the dam and result in severe downstream flooding that is much worse than if the dam did not exist. Design of a dam and emergency spillway is based on a probable maximum flood (PMF) computed following certain criteria. Likewise, evaluation of capabilities of existing dams for handling PMF is an important aspect of federal and state dam safety programs.

The probable maximum precipitation (PMP) is the estimated greatest depth of precipitation for a given duration that is physically possible over a given size storm area at a particular geographical location for a certain time of the year. The probable maximum storm (PMS) is PMP with appropriate temporal and spatial distribution. PMS is provided as input to a watershed model to develop PMF stream discharge hydrographs.

For a reservoir with conservation storage but no flood control storage capacity, the spillway and outlet works are designed with sufficient outflow discharge capacity to pass PMF without overtopping the maximum design water surface. The reservoir is assumed to be full to capacity at the beginning of PMS and PMF. For a reservoir with flood control storage capacity, the maximum design water surface is a key consideration in establishing the flood control operating rules discussed later in this chapter as well as design of the dam and outlet structures.

The NWS has developed PMP depth–duration–area relationships by increasing observed precipitation to physical limits. The process includes (1) storm transposition, (2) moisture maximization, and (3) envelopment of the adjusted transposed storms. Transposition enhances observed precipitation records by considering not only storms that have occurred at the location of interest but also those that actually occurred elsewhere but could have occurred there. Moisture maximization consists of increasing precipitation observed for an extreme past storm event by a factor that reflects the maximum amount of moisture that physically could have existed in the atmosphere for the storm location and season of the year. Envelopment involves construction of smooth curves that envelope precipitation maxima for various durations and area sizes. Geographic smoothing is performed to ensure regional consistency.

Generalized estimates of PMP for the entire United States are reported by NWS in its Hydrometeorological Report (HMR) series publicly available at its PMP website. HMR 51 (NWS 1978) provides PMP information for the United States east of the 105th meridian. Other reports cover the western states, which have rainstorm characteristics significantly influenced by mountains. PMP at a location is a function of rainfall duration, areal extent, and season of the year. The NWS PMP atlases provide maps for various combinations of these variables. The World Meteorological Organization (2009) addresses PMP estimates for regions throughout the world.

HMR 52 (NWS 1982) provides criteria and a procedure for developing a PMS using PMP estimates from HMR 51. The USACE Hydrologic Engineering Center (HEC) has developed computer software that performs the computations outlined in NWS HMR 52. The following characteristics of the synthetic design storm are reflected in the procedure for developing a PMS: isohyet pattern, location of the storm center, orientation of the storm, storm area size, spatial variability, duration, temporal distribution, and antecedent storms.

Watershed rainfall-runoff modeling methods are applied to compute the PMF hydrograph to result from PMS. Ely and Peters (1984a, b) outline the process of developing PMF hydrographs employing USACE HEC computer software discussed in Chapter 4. PMF design flood hydrographs are computationally routed through a reservoir to determine the maximum design water surface.

Except for several hundred square miles in the western tip of Texas that includes the city of El Paso, the entire state is east of the 105th meridian, which means NWS HMR Reports 51 and 52 are applicable for PMP and PMS development. The all-season PMP from HMR 51

(NWS 1978) for 24-hour duration and 10 sq. mi areal coverage for different regions of Texas increases from 32.0 in. in northwest Texas to 47.1 in. in southeast Texas near the Gulf of Mexico. For comparison, the greatest amount of rainfall occurring over 24 hours in Texas is 43.0 in., measured at an unofficial gauge in the city of Alvin, south of Houston, during Tropical Storm Claudette on July 25–26, 1979. The greatest 24-hour rainfall recorded at an official NWS gauge in Texas was 29.05 in. in the city of Albany in Shackelford County on August 4, 1978.

The NWS NOAA has provided PMP information and guidance since the 1940s at the request of USACE, USBR, FEMA, and other federal agencies with funding from these agencies. The funding diminished and gradually ceased in recent decades. The US Congress in its 2022 Bipartisan Infrastructure Law provided authorization and funding for NOAA NWS to update PMP estimates for the entire United States. NWS initiated a study in 2023 in collaboration with other entities called *Modernizing Probable Maximum Precipitation Estimation* to develop a framework for this endeavor (NWS 2024). The effects of climate change on extreme precipitation events will be considered. NWS is maintaining information regarding this PMP update program at its PMP website along with its many relevant reports developed in the past. The updated and refined PMP methods and rainfall data will be important for the design of new reservoir projects and federal and state dam safety program evaluations of existing dams.

Modifying Reservoir Operations

Construction of several reservoir projects in Texas inundated other smaller reservoirs that had been constructed at the site many years earlier. Dam heights of several reservoir projects have been raised many years after the original construction to increase storage capacity. However, the following discussion focuses on modifying operations of existing reservoirs without major structural modifications to the dam or outlet structures. The modifications discussed here are implemented by changing operating strategies and/or the designated storage elevations and release or diversion rates used to define operating rules. Reservoir operating procedures and revisions thereto are discussed in this section.

Public needs and numerous factors affecting reservoir operation change over time. An increasing necessity to use limited storage capacity as effectively as possible warrants periodic reevaluations of operating practices. Water right permits for some more complex systems require periodic reviews and updates of water management plans. Operations at many reservoirs in Texas have changed since their construction. Additional modifications are likely to be proposed and in many cases implemented in the future as demands on limited resources increase.

Storage Reallocations

Wurbs (1990), Patterson and Doyle (2018), and Doyle and Patterson (2019) review USACE policies and experience in modifying reservoir operations nationwide. The Office of the Chief of Engineers has discretionary authority to approve reallocations of up to the lesser of 50,000 ac-ft or 15% of the total storage capacity for federal purposes in a USACE reservoir. Congressional authorization is normally required for larger reallocations. Storage has been reallocated between purposes in reservoirs in Texas in some cases with and other cases without Congressional approval. The cost allocation provisions of the Water Supply Act of 1958 apply to operational modifications of existing federal reservoirs as well as construction of new projects.

Water supply diversions and storage authorized by TCEQ-administered water rights are in most cases based on initial reservoir site topography prior to sediment accumulation. If a storage reallocation can be shown to replace capacity loss due to sedimentation, the water right previously granted for the original project could possibly still cover the reallocation. Otherwise, water rights can be amended or new rights obtained through the normal TCEQ-administered process.

In reviewing USACE reservoirs nationwide, Doyle and Patterson (2019) found that of 36 district offices operating multipurpose reservoirs, 93% of the reallocated storage volume was in the five districts of Fort Worth, Kansas City, Little Rock, Tulsa, and Vicksburg. The FWD had reallocated the most storage volume, changing the authorized purposes for 7.2% of the total storage capacity in reservoirs owned by the district and proposing other changes totaling an additional 5.1% of the storage.

Wurbs and Carriere (1988) investigated potentialities for improving federal and nonfederal reservoir operations in Texas through storage reallocations and other reservoir system operating strategies. The SB1 regional and statewide water plans (TWDB 2022) explore potential storage reallocation projects and other related types of operational modifications. The two most common storage reallocations implemented or proposed to date in Texas are reallocating flood control storage capacity to water supply by raising the designated top of conservation pool elevation and reallocation of hydropower storage to water supply within the conservation pool. In some cases, reallocations in an existing reservoir occurred in coordination of completion of construction of another reservoir. Reallocations may be permanent, seasonal, or temporary. Examples of permanent reallocations are noted in the following paragraphs.

Construction of Lakes Belton and Proctor on the Leon River was completed in 1954 and 1963. The top of conservation pool at Lake Belton was raised 25 ft in 1972. The loss of flood control capacity in Lake Belton was mitigated by addition of Lake Proctor upstream.

Ray Roberts Reservoir is an unusual Corps of Engineers project in that its construction provided only additional water supply and recreation. The total two-reservoir system flood control storage capacity remained unchanged. USACE completed construction of Lewisville and Ray Roberts Reservoirs on the Elm Fork of the Trinity River in 1954 and 1987. The flood control pool of Lake Lewisville extends upstream to the dam at Lake Ray Roberts. Lake Ray Roberts was constructed with flood control, conservation, and sediment reserve capacities of 260,800 ac-ft, 749,200 ac-ft, and 54,600 ac-ft, respectively. Lake Lewisville originally had flood control, conservation, and sediment reserve capacities of 525,200 ac-ft, 436,000 ac-ft, and 20,500 ac-ft, respectively. Upon completion of Ray Roberts Reservoir, the top of conservation pool elevation of Lewisville Reservoir was raised, relocating flood control storage capacity to water supply in the amount of the flood control added by the new reservoir.

Construction of Lake Waco on the Bosque River was completed in 1965 with flood control, conservation, and sediment reserve capacities of 553,300 ac-ft, 104,100 ac-ft, and 69,000 ac-ft, respectively. In 1979, the BRA in cooperation with the city of Waco requested USACE FWD to investigate the feasibility of increasing the conservation storage capacity in Lake Waco to provide a greater dependable water supply yield. A proposed reallocation was approved by the USACE Office of the Chief of Engineers in 1983. A water supply contract and water right permit were developed and approved. The top of conservation pool elevation was raised 7.0 ft in 2003, converting 47,500 ac-ft (8.6%) of the flood control pool to water supply. Relocation of park roads, boat ramps, and other recreation facilities around the lake was necessitated by the pool raise. However, major modifications to the dam and outlet structures were not required.

The top of conservation (bottom of flood control) pool may be varied seasonally, particularly in regions with distinct flood seasons. Seasonally varying the allocation of storage capacity between flood control and conservation purposes is called *seasonal rule curve operations*, where the seasonally varying "rule curve" is the designated top of conservation pool elevation. Seasonal rule curve operations are common in some regions outside of Texas where summer months are characterized by high water demands, low streamflows, and a low probability of floods. Use of large permanent seasonal conversions between flood control and conservation pools has been limited in Texas due to the possibility of floods occurring during any season of the year. However, several reservoirs in Texas including Lakes Texoma, Wright Patman, Lake 'O the Pines, Toledo Bend, Amistad, and Falcon have now or in the past had variable designated top of conservation pool elevations either permanently or temporarily over several years facilitating an extra degree of operating flexibility.

The conservation pool of Lake Texoma impounded by Denison Dam on the Red River has been used primarily for hydroelectric energy generation and recreation. Water supply capacities are constrained by upstream natural salt pollution. The Water Resources Development Act of 1986 (Public Law 99-662) authorized the USACE Tulsa District to reallocate conservation storage from hydropower to water supply in increments as needed to supply growing water demands in the region. Several reallocations have been implemented as more nonfederal water supply sponsors contracted for conservation storage capacity in the federal reservoir.

The conservation pool of Lake Whitney on the Brazos River is also used primarily for hydroelectric energy generation and recreation. Water supply use has been hindered by salinity. The following data for Lake Whitney are available from the USACE FWD website in Table 4-1. The elevations in feet above mean sea level for the top of dam, maximum design water surface, top of flood control pool, and top of conservation pool are 580.0, 573.0, 571.0, and 533.0. The corresponding storage volumes below these elevations are 2,664,000 ac-ft, 2,118,000 ac-ft, 1,999,500 ac-ft, and 554,203 ac-ft. The top of conservation pool at spillway crest elevation of 533.0 ft is also called the top of power pool.

The storage volume below the lowest outlet structure intake invert elevation of 448.8 ft is 854 ac-ft. The streambed elevation at the dam is about 425 ft. Impoundment began in 1951.

The top of conservation pool of Lake Whitney was raised from 520 to 522 ft in March 1968, from 522 to 523 ft in March 1969, and from 523 to 533 ft in May 1972. The top of inactive pool is set at 520 ft to provide a minimum head for hydroelectric power generation. BRA holds a water supply storage contract and water right permit for use of 22% of the conservation storage capacity provided by a reallocation in 1982 from hydropower to water supply.

Storage Triggers Activating Operating Rules

Operating rules include varying release or withdrawal rates as a function of current or forecasted reservoir storage contents. Reservoir storage levels trigger activation of different components of operating strategies or alterations in rates at which water is released or withdrawn from a reservoir. The term *storage trigger level* is adopted to refer to specified levels of reservoir storage at which certain aspects of operations are initiated, terminated, or altered.

The top of conservation (bottom of flood control) pool elevation provides a switch or trigger to activate and deactivate flood control operations. Either the flood control or conservation pool can be subdivided into any number of vertical zones to facilitate specifying reservoir releases as a function of amount of water in storage as monitored by water surface elevations. Modifications to reservoir operations can be implemented by changing designated storage elevation triggers.

Flood control operations explained in the next section are based on allowable nondamaging flow rates at downstream gauges. Flood control pools are emptied as expeditiously as possible without releases contributing to exceeding specified nondamaging downstream flow rates. The designated allowable flow rates for some or all the downstream gauges vary as a function of reservoir storage levels within the flood control pools for many of the USACE reservoirs. The specified allowable downstream flow rates controlling reservoir releases are increased, allowing some level of downstream damage, as remaining flood control pool storage capacity is depleted.

Conservation pool operating plans can be modified to supply different water users at different levels of reliability. Volume versus reliability preferences vary with different types of water use and alternative backup supply capabilities. Municipal water use requires a high level of reliability. Agricultural water users may be willing to accept lower reliabilities to increase the volume of water available to them most of the time. Alternative water supply strategies may be implemented during droughts that are not normally employed. Implementation of drought contingency plans may be triggered by reservoir storage levels falling below specified levels. More severe demand management options may be implemented as storage contents fall below various prespecified levels. Conjunctive management of groundwater and surface water sources may involve shifting to greater use of groundwater whenever reservoir storage falls below designated levels.

A WAM system described in Chapter 4 and employed in Chapters 5, 6, and 7 provides capabilities for quantifying water supply reliabilities for alternative river and reservoir system operating plans. Studies have demonstrated that the annual volume of water committed to water users can be increased greatly by accepting relatively small additional risks of supply shortages or failures. For example, an annual water supply diversion volume supplied with a reliability of 99.0% might be increased by 40% with a corresponding decrease in reliability to 96.2%.

Lower Colorado River Authority and BRA system operations discussed in Chapter 7 employ the concept of combining firm (high reliability) and interruptible (lower reliability) water supply commitments. Water supply deliveries for certain water customers are curtailed whenever storage in specified reservoirs falls below specified trigger levels. Water is allocated between users by water right permits and associated water management plans and storage contracts in terms of both quantity and reliability. Modifications in reservoir system operations implemented by storage triggers modify water supply volume commitments and associated reliabilities.

Flood Control Operations

Flood control pool operations are based on minimizing the risk and consequences of making releases that contribute to downstream flooding, subject to the constraint of assuring that the maximum design water surface is never exceeded. Flood control pools are emptied as quickly as downstream flooding conditions allow to reduce the risk of future highly damaging releases being necessitated by filling of the available storage capacity. Minimizing the risks and consequences of storage backwater effects contributing to flooding upstream of the dam is also an important trade-off consideration at some reservoir projects.

One type of reservoir system operation task consists of developing an operating plan, often called a *regulation schedule*. Another related but distinctly different reservoir system operation endeavor involves making release decisions during real-time flood control operations, within the framework of the regulation schedule. The operation plan provides guidance for real-time release decisions but typically still leaves a significant degree of flexibility. Information regarding current storage levels and streamflows is used, in combination with the regulation schedule, to make release decisions. Real-time operations often use current precipitation and streamflow data to forecast flows to be expected at pertinent locations during the next several hours or days, to enable more effective release decisions. During normal non-flooding conditions, flood control operations consist simply of passing inflows to maintain empty flood control pools.

The Corps of Engineers is responsible for operating most of the major flood control reservoir systems throughout the nation including throughout Texas. The most notable exception is International Amistad and Falcon Reservoirs on the Rio Grande operated by the IBWC. Flood control regulation plans are developed to address conditions associated with each individual reservoir and multiple-reservoir system. However, flood control operating rules for most reservoirs follow the same general strategy outlined in this chapter.

Flood control pool release decisions for USACE reservoirs occur at centralized control centers located in district offices. USACE reservoir control centers have direct communications with NWS river forecast centers. Gate operations are performed by USACE personnel working at the individual reservoir project offices located at the dams as directed by the reservoir control center.

Addicks and Barker Reservoirs in Houston are owned and operated by the USACE Galveston District. Lake Texoma on the Red River is owned and operated by the USACE Tulsa District. The Tulsa District is also responsible for flood control operations at the multipurpose Lake Kemp, which is owned and otherwise operated by the city of Wichita Falls and Wichita County Water Improvement District. The Fort Worth District Reservoir Control Center is directly responsible for flood control operations of 25 USACE FWD multipurpose reservoirs and has oversight responsibility for flood control operations at three multipurpose reservoirs constructed by the USBR and now owned by nonfederal water management agencies.

The regulation criteria controlling flood control operations of USACE FWD projects are publicly accessible information available at the USACE FWD website listed in Table 4-1 of the next chapter. These criteria established by USACE include allowable flow rates at specified USGS gauge sites. The Galveston and Tulsa Districts similarly provide public access to flood control operating plans and other relevant information for reservoirs in their districts.

Release decisions depend on whether the flood control storage capacity is expected to be exceeded. Water levels encroach into flood control pools often, but the flood control pools are seldom filled to their full storage capacity. Although flood control storage capacities at some projects are exceeded more frequently, federal reservoirs are typically sized to contain at least a 2% annual exceedance probability (50-year recurrence interval) flood and, for many projects, design floods greater than the 1% annual exceedance probability (100-year recurrence interval) flood. Therefore, flood control storage capacities in USACE and IBWC reservoirs are not exceeded often. As discussed in Chapter 6, the flood control pools of some of the reservoirs constructed in Texas several decades ago have not yet been filled completely to capacity even once. Other reservoirs have had their flood control pools overtopped one or multiple times since their construction.

A specified set of rules, based on downstream flow rates, is followed if sufficient storage capacity is available to handle a particular flood event without having to deal with the water surface rising above the top of flood control pool. Operation is switched over to an alternative approach, based on reservoir inflows and storage levels, during extreme flood conditions when the anticipated inflows are expected to deplete the remaining empty flood control pool storage space. The reservoir release rates necessitated by the flood control pool storage capacity being exceeded will contribute to downstream flooding. The objective is to assure that reservoir releases do not contribute to downstream damages if the storage capacity is not exceeded. However, for extreme flood events that would exceed the reservoir flood control pool storage capacity, moderately high damaging discharge rates beginning before the flood control pool is full are preferable to waiting until a full reservoir necessitates much higher release rates.

Operations Based on Downstream Flow Rates

Flood control operations are based on minimizing the risk and consequences of making releases that contribute to downstream flooding. Maximum allowable stages and flow rates at downstream control points are set based on bank-full stream capacities, stages at which significant damages occur, environmental considerations, and/or constraints such as inundation of road crossings or other facilities. USGS stream gauge stations are located at the operational control points. Releases are made to empty flood control pools as quickly as possible without contributing to streamflows exceeding specified maximum allowable flow levels at downstream gauges. Maximum allowable flow rates just below the dams are also set.

When a flood occurs, the spillway and outlet structure gates are closed. The gates remain closed until a determination is made that the flood has crested and flows are below the target levels specified for each of the gauge sites. The gates are then operated to empty the flood control pool as quickly as practical without exceeding the allowable flows at the downstream locations.

As a result of flood inflow entering the river downstream of the dams, flooding may still occur in downstream floodplains when the gates are fully closed with no outflows from upstream reservoirs. However, the concept is to make no reservoir releases that will worsen the flooding downstream.

In many cases, the allowable flow rate associated with a given location is constant regardless of the volume of water in storage in one or multiple upstream reservoirs. At other projects, the allowable flow rates at one or more gauge sites vary depending upon the volume of water currently stored in the upstream flood control pools. This allows stringently low flow levels to be maintained at certain locations if only a relatively small portion of the flood control pool capacity is occupied, with the flows increased to a higher level, at which minor damages occur, as the reservoirs fill.

Flood control reservoirs are operated based on maintaining flow rates at several gauges located various distances below one or multiple dams as illustrated by Figure 3-5. The most downstream gauge sites may be many miles below the dams. Lateral inflows from uncontrolled watershed areas below the dams increase with distance downstream. The impact of reservoirs on flood flows decreases with distance

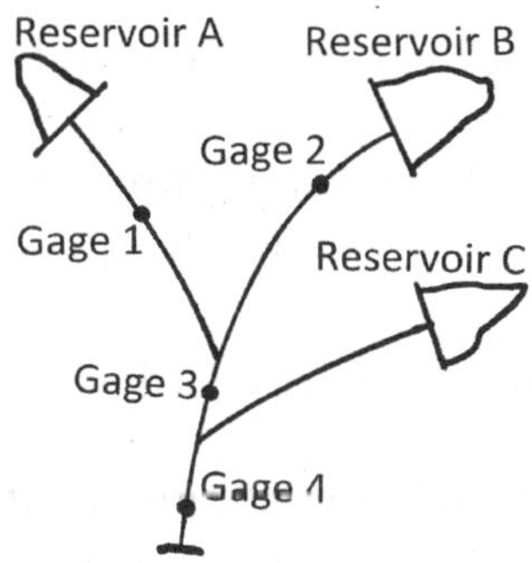

Figure 3-5. Flood control operations based on flows at downstream gauge sites.

downstream. Operations benefit from forecasts of streamflows. Flood attenuation and travel time from the dams to the gauge sites and inflows from watershed areas below the dams may be estimated as an integral part of the reservoir operating procedure.

Most flood control reservoirs are components of multiple-reservoir systems. The two largest USACE reservoir systems in Texas are the Brazos River Basin system with nine reservoirs and the Trinity River Basin system with eight reservoirs.

Two or more reservoirs located in the same river basin will have common control gauges. A reservoir may have one or more downstream control gauge locations affected only by that reservoir and several other control gauge sites with flows also dependent on other reservoirs. Multiple-reservoir release decisions may be based on maintaining some specified relative balance between the percentages of flood control storage capacity utilized in each reservoir. For example, if unregulated flows are below the maximum allowable flow rates at all the control gauge sites, the reservoir with the greatest amount of water in storage, expressed as a percentage of flood control storage capacity, might be selected to release water. Various balancing criteria may be adopted. Releases from all reservoirs, as well as runoff from uncontrolled watershed areas, must be considered in forecasting flows at downstream control gauge locations.

Maximum allowable rate of change of release rates is also specified. Abrupt gate openings causing a flood wave with rapid changes in stage are dangerous from the perspective of downstream hazards to public safety. Rapid variations in flow rates also contribute to streambank erosion.

The flood control reservoir operating strategy outlined in the preceding paragraphs is commonly called *regular flood control operations*. The additional component of the operating procedures outlined in the next section is often referred to as emergency operating rules.

Operations Based on Reservoir Inflows and Storage Levels

Emergency operations during an extreme flood are based on assuring that the maximum design water surface is never exceeded while fully using flood control pool storage to reduce downstream flooding. If releases are based on downstream non-damaging or allowable flow criteria until the flood control pool fills, later spills at high flow rates could result that would be more damaging than a lower release rate with a longer duration beginning before the flood control pool is full. On the other hand, an operator would not want to make excess releases during a storm and then later learn that the flood control pool never filled and the releases unnecessarily contributed to downstream damages. Although forecasts of streamflows several hours or days into the future are often made during real-time operations, future flows are still highly uncertain.

The overall strategy for operating the outlet works and spillway gates of a flood control reservoir typically consists of two component types of regulation procedures. The type of procedure requiring the largest release rate controls for given flooding and storage conditions. The regulation approach discussed previously, based on downstream allowable flow rates, is followed until such time during a flood that the release rate indicated by the schedule outlined next is higher than that indicated by the downstream allowable flow rates. The regulation procedure outlined next is based on reservoir inflows and storage levels without consideration of downstream streamflows.

The emergency schedule controls reservoir releases during an extreme flood that would otherwise exceed the capacity of the flood control pool. The controlling objective is to assure that the maximum design water surface is not overtopped. Releases are determined as a function of current reservoir water surface elevation and inflow rate. An alternative version of the schedule specifies releases as a function of the current water surface elevation and rate of rise of water surface. The two forms of the schedule are intended to result in the same releases. Release rates are typically determined at a reservoir control center which has access to real-time streamflow measurements or forecasts and can base release rates on inflow rates. If communications between the control center and the operator at the project are interrupted during a flood, the operator can determine gate releases based on the rate of rise of the water surface elevation measurements without needing to know inflow rates.

The operating plan is prepared during preconstruction planning of the project. The emergency regulation component specifying releases as a function of storage levels and inflow rates is developed based on estimating the minimum volume of inflow that can be expected in a flood, given the current inflow rate and reservoir elevation. The corresponding outflow required to limit storage to the available capacity is determined by mass balance computations. For a given current inflow rate, the minimum inflow volume for the remainder of the storm is obtained by assuming the inflow hydrograph has just crested and computing the volume under the recession side of the hydrograph. For conservatively low inflow volume estimates, the assumed recession curve is steeper than the average observed recession. The complete regulation schedule which allows the outflow to be adjusted based on the current inflow and empty storage space remaining in the reservoir is developed by making a series of computations with various assumed values of inflows and amounts of remaining storage available.

Conservation Storage Operations

An array of considerations may be important in the operation of specific reservoir systems for water supply, hydroelectric power, recreation, and other conservation purposes. Each reservoir and multiple-reservoir system has unique aspects. A variety of mechanisms are used to define operating rules. There is no standard format for specifying operating rules that is applicable to all situations. However, several basic concepts pertinent to a wide range of practices are noted as follows.

Although seasonality is a significant factor in operating reservoirs in Texas, threat of severe multiple-year drought at some unknown time in the future is the dominant controlling consideration. In general, conservation operations can be categorized as being primarily influenced by either seasonal fluctuations in streamflow and/or water use or long-term threat of drought. In some regions of the United States and the world, a reservoir will be filled during a distinct season of high rainfall or snowmelt and emptied during a dry season with high water demands.

Thus, the reservoir level fluctuates each year in a predictable seasonal cycle. In other cases, surface water management is predominantly influenced by a long-term threat of severe drought. Water is stored through many wet years to be available during drought conditions. Although reservoir storage may be significantly depleted within several months, severe drought conditions in Texas are characterized as a series of several dry years rather than the dry season of a single year. Long-term drought is a dominant factor in Texas.

Reservoir operation for municipal, industrial, and agricultural water supply is based on meeting demands subject to institutional constraints related to project ownership, contractual agreements, and water rights. Either a single or multiple nonfederal sponsors may contract for conservation storage in a particular USACE reservoir. Water supply contracts may allocate the storage capacity in federal reservoirs between two or more nonfederal water management agencies. As discussed later in this chapter, the conservation storage capacity in International Amistad and Falcon Reservoirs is allocated among more than 1,000 water right accounts.

Conservation storage serves a variety of project purposes. Hydroelectric power plants are typically components of complex electric energy systems, which include thermal-electric and hydroelectric generation, as well as multiple-purpose reservoir operations. Reservoir operations are based on maintaining a high reliability of meeting both water supply and electrical energy commitments while minimizing the total costs of both thermal-electric and hydroelectric generation.

Operating considerations for reservoir recreation typically involve maintenance of desirable storage levels and minimizing fluctuations in storage levels and may also include maintenance of desirable instream flow levels. Instream flow needs also include maintenance of streamflow for water quality, fish and wildlife habitat, livestock water, river recreation, and aesthetics.

Agricultural irrigation involves consumptive withdrawals and significant fluctuations in reservoir storage levels. Conversely, in thermal-electric power plant cooling water reservoirs, most of the water withdrawn is returned to the reservoir and water surface levels fluctuate very little.

Reservoir operations also address requirements other than the primary project purposes. For example, due to water right considerations, releases may be required to pass inflows through the reservoir to other more senior water users and water management entities located downstream, which are not directly served by the reservoir. Such requirements may be specified in terms of maintaining minimum flow rates at specified downstream locations, subject to the stipulation that reservoir releases in excess of inflows are not required.

Operation during floods is an important consideration for conservation-only projects without flood control storage capacity. Another consideration involves restricting the rate of change in release rates to prevent public safety hazards. Rapid increases in stage and velocity can be dangerous for people recreating in the river downstream of a reservoir. Rapid changes in release rates are also undesirable from the perspective of riverbank erosion. Storage level fluctuations are sometimes employed to help control vectors such as mosquitos.

Water quality storage may be included in reservoirs, as a primary project purpose, to provide releases for low flow augmentation. Water quality may be an important incidental consideration in operations for other purposes. The quality of downstream flows and water supply diversions is sometimes improved by selection of the vertical storage levels from which to make the releases.

Multiple-Purpose and Multiple-User Operations

Multiple-purpose reservoir operation involves various interactions and trade-offs between purposes, which may be either complimentary or competing. Reservoir operation may be based on the conflicting objectives of maximizing the amount of water available for conservation purposes and maximizing the amount of empty space available for storing future flood waters to reduce downstream damages. Conservation pools are shared by various purposes that involve both consumptive withdrawals and in-reservoir and in-stream uses.

As discussed in the preceding sections of this chapter, reservoirs may be operated for water conservation storage only, flood control only, or a combination of flood control and conservation with separate pools designated for each. Interactions between flood control and conservation purposes in a multiple-purpose reservoir involve allocation of storage capacity as represented by the designated top of conservation pool elevation. Modifications to the operations of completed projects may involve either permanent long-term reallocations of storage capacity or establishing or refining seasonally varying rule curves for joint use storage.

Normally, no flood control releases are made if the storage level is at or below the top of conservation pool. However, in some cases, if flood forecasts indicate that the inflow volume will exceed the available conservation storage, flood control releases from conservation storage may be made if downstream conditions permit. The idea is to release some water before the stream rises downstream, if practical, to maximize storage capacity available for regulating the forecasted flood.

Operation during floods of nonfederal water supply reservoirs that have no flood control pool is a widespread concern. Without a flood control pool, pre-releases from a conservation pool prior to a forecasted flood may be a relevant consideration under appropriate circumstances. The effectiveness of this strategy is limited by imperfect flood forecasting capabilities, limited reservoir release and downstream channel capacities, and long periods of time required to drawdown storage enough to significantly affect flooding. Risks of adversely affecting water supply reliabilities is a major concern.

Hydroelectric power operations may also involve flood control pool release rates. For example, in some cases like the Highland Lakes on the Colorado River, flood control pool releases may be passed through hydroelectric power plants and limited to the maximum discharge that can be used to generate power. Releases from conservation storage may be made to generate hydroelectric energy along with partially drawing down pool levels in anticipation of forecasted flood inflows.

Conservation pools typically serve multiple purposes with at least some complimentary characteristics. Water stored for water supply and hydroelectric power provides opportunities for recreation and reservoir fisheries. Hydroelectric power releases contribute to other instream flow uses and can be diverted at downstream locations for water supply. Sharing of reservoir storage capacity and limited water resources by multiple users also involves conflicting demands.

Certain water users require a high degree of reliability. For other water users, obtaining a larger quantity of water with some risk of shortage may be of more value than a supply of greater reliability but smaller quantity. Municipal and industrial water supply operations are typically based on assuring a high degree of reliability in meeting demands during anticipated infrequent but severe droughts. Supplying water for agricultural irrigation often involves acceptance of greater risks of shortages than municipal and industrial supply and is based more on maximizing economic benefits.

Storage triggering designations may provide a mechanism for reflecting relative priorities or trade-offs between water users. A reservoir operating plan may involve assuring a high degree of reliability for municipal use and lesser reliability for agricultural irrigators. All demands are met if storage is above a specified level, but irrigation demands are curtailed whenever storage falls below the specified level. Alternative drought contingency plans are implemented when reservoir storage levels fall below specified trigger levels.

Supplies from groundwater aquifers may be preserved by relying upon surface water reservoirs to supply water most of the time. However, groundwater used only during infrequent drought conditions can greatly increase the water supply capabilities of a reservoir system. The volume of water from the reservoir system that can be committed to water users is greatly increased by accepting a higher risk of shortage facilitated by having access to groundwater as a backup supply.

Multiple-Reservoir System Operations

Multiple-reservoir release decisions occur in situations in which water needs can be met by releases from two or more reservoirs. In Figure 3-6, diversions 1 and 3 are from specific reservoirs, but diversion 4 can be met by releases from either of the three reservoirs. Instream flow as well as diversion requirements at location 4 can be met by releases from the reservoirs.

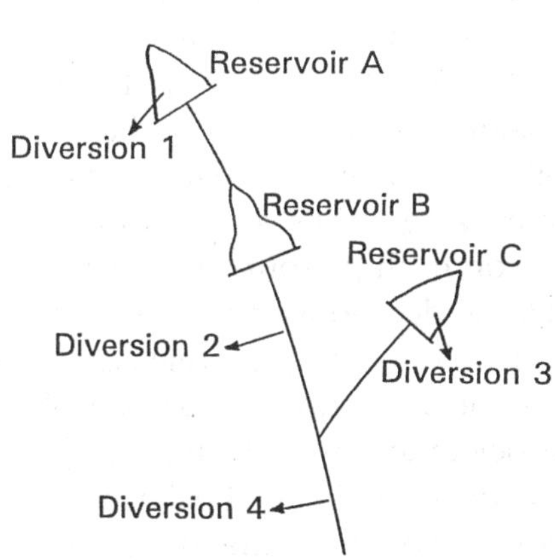

Figure 3-6. Multiple-reservoir water supply system.

Water supply reliabilities are increased by operating multiple reservoirs as a system rather than independent individual projects. Risks of water supply shortages are reduced by balancing storage between multiple reservoirs. Various release criteria may be adopted for balancing storage contents.

One criterion for deciding from which reservoir to release is to minimize spills, which represent water loss from the system. The term *spill* refers to discharges through an uncontrolled spillway or controlled releases made simply to prevent the reservoir surface from rising above the designated top of conservation pool. Spills from an upstream reservoir (such as Reservoir A in Figure 3-6) may still be stored in a downstream reservoir (Reservoir B) and thus are not lost to the system. For reservoirs in series, such as Reservoirs A and B in Figure 3-6, the downstream reservoir would be depleted before using upstream reservoir water to meet downstream demands. In addition to minimizing spills from the downstream reservoir, this procedure maximizes the amount of water in storage above and thus accessible by gravity flow to each diversion location. For example, water stored in Reservoir A can be used to meet diversions 1, 2, and 4, but water stored in Reservoir B can be used to meet only diversions 2 and 4.

For reservoirs in parallel, such as Reservoirs B and C in Figure 3-6, minimizing spills involves balancing storage depletions in the different reservoirs. The simplest approach may be to release from the reservoir with the largest ratio of conservation pool storage content to storage capacity. Thus, release decisions would be based on balancing the percent depletion of the conservation pools. Other more precise and more complex approaches can be adopted to select the reservoir with the highest likelihood of incurring future spills.

Numerous other considerations may be reflected in multiple-reservoir release decisions. If the reservoirs have significantly different evaporation potential, minimization of evaporation may be an objective. The criteria of minimizing spills or evaporation are pertinent to either single-purpose or multiple-purpose systems. Multiple-purpose, multiple-reservoir release decisions can involve a wide variety of interactions and trade-offs. For example, releases to meet downstream municipal, industrial, or irrigation water supply demands may be passed through hydroelectric power turbines. Thus, multiple-reservoir water supply release decisions may be based on optimizing power generation. Recreational aspects of the system could motivate release decisions that minimize storage level fluctuations in certain reservoirs.

Water Rights System Allocation of Storage and Diversions

The Texas water rights system administered by TCEQ is described in Chapter 2. Water flowing in major streams and stored in major reservoirs is owned by the state of Texas. Water rights allow water to be stored and withdrawn from river and reservoir systems for beneficial use. Water right holders are granted the right to store, divert, and use water not exceeding specified maximum quantities following rules and provisions outlined in the permits or certificates of adjudication that define their water rights.

Surface water rights in Texas are comprised of two significantly different water allocation systems, both administered by TCEQ. One system controls the use of water in the Rio Grande below Fort Quitman supplied by International Amistad and Falcon Reservoirs shown in Figure 2-1. The other water rights system is applied in the management of the water resources of all other rivers and reservoirs in Texas.

International Amistad and Falcon Reservoirs on the Rio Grande

IBWC operates International Amistad and Falcon Reservoirs primarily for water supply and flood control for the Lower Rio Grande Valley below Falcon Dam. IBWC operates Anzaldus and Retamal Dams on the lower reach of the Rio Grande to facilitate water supply diversions. The reservoirs also provide hydroelectric power and recreation.

In accordance with the 1944 treaty discussed in Chapter 2, the United States has 56.2% and 58.6% of the conservation storage capacity of Amistad and Falcon, respectively, with Mexico owning the remainder. IBWC maintains a continuous accounting of the volume of water currently stored in the two reservoirs that belong to each of the two countries. Streamflows into Falcon and Amistad Reservoirs from several major tributaries named in the 1944 treaty are gauged and allocated as specified by the treaty. All other streamflows not otherwise allocated are divided equally between the two countries. Computations are performed by IBWC staff to allocate the reservoir inflow and evaporation volumes which are combined with recorded releases to determine the amount of water that each country has in storage.

The water rights system administered by TCEQ is employed to allocate the US share of the water supply storage in Amistad and Falcon Reservoirs to water right holders in the Lower Rio Grande Valley. Most of the water is used for irrigation in the very productive agricultural region below Falcon Reservoir. Several cities in the region are also supplied by diversion of water from the Rio Grande released from Amistad and Falcon Reservoirs.

Irrigation districts, individual farmers, and cities owning water rights communicate their water needs directly to the TCEQ watermaster office, which in turn schedules releases from Falcon and Amistad Reservoirs with IBWC. The watermaster office maintains an accounting of the amount of water diverted and the amount of water in reservoir storage allocated to each of about 1,600 water right accounts. The travel time for releases from Falcon Reservoir to reach the most downstream water supply diversion locations is about one week.

The Texas, share of the waters of the Rio Grande below Fort Quitman was allocated among numerous water rights holders in conjunction with a massive lawsuit, commonly called the Lower Rio Grande Valley Water Case (Wurbs 1995). The lawsuit was filed in 1956, the trial was held in 1964 to 1966, and the final judgment of the appellate court was filed in 1969. In 1971, the Texas Water Rights Commission (TCEQ predecessor) adopted rules and regulations implementing the court decision. Assorted versions of riparian and appropriative rights were combined into a permit system. The litigants in the Rio Grande lawsuit included 42 water districts and 2,500 individuals. More than 90 lawyers appeared before the court. The expense and effort involved demonstrated the impracticality of a purely judicial determination of water rights for the entire state and led to enactment of the Water Rights Adjudication Act of 1967 which authorized establishment of the current water rights system employed for the remainder of Texas.

The lawsuit resulted in water rights being divided into three categories. Municipal rights have the highest priority. Irrigation rights are divided into Class A and Class B rights, with Class A rights receiving more storage in Falcon and Amistad Reservoirs storage accounts in the allocation procedure. Although this weighted priority system for irrigation rights has little significance during years of plentiful water, its effect in water-short years is to distribute the shortage among all users, with the greater shortages occurring on lands with Class B water rights.

The allocation rules administered by TCEQ watermaster include first providing a specified volume in reserve in Amistad and Falcon

Reservoirs for domestic, municipal, and industrial uses. Next, available water is allocated to an operating reserve that provides for seepage and evaporation losses, adjustments required as IBWC computations of Mexico and US allocations are updated periodically, and emergency requirements. The remaining water in storage is allocated among all the irrigation permit holders. The storage is basically allocated in proportion to annual diversion rights, except the Class A rights are multiplied by a factor of 1.7 to allow them a greater storage allocation than Class B rights. Other provisions include limiting each storage allotment to not exceed more than 1.41 times its authorized diversion right. If an irrigation right does not use water for two consecutive years, its storage account is reduced to zero.

All Other Major Reservoirs and Rivers in Texas

All of Texas except the Rio Grande below Fort Quitman has a consistent surface water rights permit system that is different than the system described for the lower Rio Grande. TCEQ administers both systems. Water management in the Rio Grande Basin is significantly different than the other river basins of Texas from various perspectives as noted throughout this book. The differences listed focus specifically on water rights:

- The water rights system in the Rio Grande was established judicially during 1956 to 1971 in response to litigation. The water rights system for the rest of Texas was established administratively pursuant to the Water Rights Adjudication Act of 1967.
- Priorities for the Rio Grande are based on three categories: municipal, irrigation Class A, and irrigation Class B. Priorities for the rest of Texas are based on the dates that water right holders initially appropriate water.
- Storage accounting by the IBWC tracks the storage volume in Amistad and Falcon Reservoirs allocated to each of the two nations. The Rio Grande is the only international river in Texas.
- Storage accounting by TCEQ watermaster tracks the US storage contents in Amistad and Falcon Reservoirs allocated to each of well over a thousand individual Texas water right accounts. One or at most two or three cities, river authorities, water districts, or other

entities hold water rights for storing water in each of the other individual reservoirs in Texas.

- TCEQ watermaster operations have been well established for the Rio Grande for several decades. Establishment of watermaster operations for the other river basins has been slowly progressing over the last several decades. The water accounting system is much more detailed for the Rio Grande than the other river systems of Texas.

The water rights system established pursuant to the Water Rights Adjudication Act of 1967 is based on the prior appropriation doctrine. Priorities are based on the dates that each water right holder was first using water historically before the adjudication process or, since the 1990s, had a permit application administratively accepted by TCEQ. Reservoir operators may be required to curtail storage and diversions as necessary to pass inflows through their reservoirs for more senior water right holders downstream.

Implementation of water right permit requirements in actual river/reservoir systems operations is not necessarily always precise. For example, reservoir storage and water supply diversions are typically assigned the same priority in a water right permit. Assigning refilling of drawn-down reservoir storage capacity the same priority as diversions to supply agricultural, municipal, or industrial water needs is a complex issue. Curtailment of storage or diversions of a junior right to protect a more senior water diversion right may have a certain urgency. Curtailment of junior water supply diversions to allow refilling of a senior storage right for a large almost full reservoir may not be the optimal use of limited water resources.

Most of the water supply reservoirs in Texas and conservation storage pools of federal multipurpose reservoirs are controlled by a river authority, city, water district, or electric utility company that holds all water rights associated with their one or multiple reservoirs. In some cases, two or three nonfederal water supply sponsors may share the conservation pool of a USACE multipurpose reservoir. The reservoir owner/operator supplies water to numerous water users who are not burdened with having to deal with water rights. The storage accounting is thus much simpler than dealing with many hundreds of water right accounts in International Amistad and Falcon Reservoirs.

Water rights are an important driver of reservoir operations for water supply. Water rights specify allowable storage capacities and annual diversion volumes. Various other specifications affecting reservoir operations may also be incorporated in water right permits. As discussed in Chapter 2, with the notable exception of Lake Texoma, hydroelectric power operations in Texas tend to be incidental to water supply. Spills from full conservation pools and releases for downstream water supply diversions are passed through hydropower plants. Most water right permits or certificates of adjudication for hydropower are in the same format as water supply based on storage volume and diversion amounts. Flood control storage capacity is not included in water rights.

Environmental Flow Standards

Instream flow requirements have been included in the Texas water rights system dating back several decades to protect ecosystems and serve other purposes. Instream flow requirements established before the 2007 SB3 are primarily in the form of maintaining specified minimum instream flow rates. More recently, increased concern with maintaining environmental instream flows is reflected in the 2001 SB2 and 2007 SB3. Improved capabilities for determining and implementing instream flow requirements are being developed pursuant to the 2001 SB2. Environment flow standards are being established following a process created by the 2007 SB3 (National Research Council 2005, Wurbs 2017).

SB2 enacted by the Texas legislature in 2001 created the TIFP jointly administered by TWDB, TCEQ, and TPWD. The goal is to establish appropriate flow regimes for an ecologically sound environment, conserving fish and wildlife resources while also providing sustained benefits for human uses of water resources. SB2 directed the agencies to establish data collection and assessment programs and to conduct studies and analyses to develop appropriate methodologies for determining environmental flow needs considering a broad range of factors such as habitat, hydrology, biology, water quality, geomorphology, and stream system connectivity. TWDB maintains a TIFP website at which relevant reports and information are compiled for public access. Results of continuing TIFP studies are expected to facilitate future refinements of environmental flow standards (EFS) established through a process created by the 2007 SB3.

Process for Establishing Environmental Flow Standards

The Texas legislature enacted SB3 in 2007 creating a process for establishing EFS for selected priority river systems using existing information and the best available science (Wurbs 2017). The 2007 SB3 introduced a new regulatory strategy for protecting environmental flows that generates a sequence of recommendations regarding flow regimes developed by expert science teams and stakeholder committees that lead to the adoption by TCEQ of standards and set-asides to satisfy the standards. *Set-asides* refer to unappropriated water reserved by TCEQ in administration of the water rights permit process to meet specified EFS. EFS are incorporated in the TCEQ WAM system introduced in Chapter 4 and employed in Chapters 5 through 7 of this book.

Only new water right permits or new amendments to existing water rights are subject to EFS established through the process created by the 2007 SB3. Little unappropriated streamflow is available for additional new water rights. Practically no unappropriated flow is available in many river reaches. Developing and implementing management strategies involving existing water right holders and stakeholders that further increase the protection of environmental flows while respecting existing water rights is an important current challenge. Rubinstein et al. (2022) explore future strategies for addressing this challenge.

The legislatively mandated set-asides for the SB3 EFS are assigned a priority corresponding to the date TCEQ receives the appropriate environmental flow regime recommendations from an expert science team appointed through the SB3 process. After official establishment of EFS, TCEQ may not issue a permit for a new appropriation or amendment to an existing water right permit that increases the amount of water authorized to be stored or diverted if any EFS would be violated. In amending existing permits, this restriction applies only to the increase in amount of water to be stored or diverted.

The SB3 regulatory process is based on public participation, statewide agency oversight, and technical support by the science and engineering community. Local stakeholders and science experts develop recommendations regarding the appropriate environmental flow regime for assigned river systems. TCEQ, TPWD, and TWDB provide administrative oversight and technical support. EFS established through the SB3 process are periodically reevaluated and modified as warranted. Studies

Table 3-1. River Systems with Environmental Flow Standards.

River Basin and Bay System	Effective Date	Priority Date	Number of Sites
Trinity and San Jacinto Rivers and Galveston Bay	May 2011	Nov 2009	6
Sabine and Neches Rivers and Sabine Lake Bay	May 2011	Nov 2009	10
Colorado and Lavaca Rivers and Matagorda and Lavaca Bays	Aug 2012	Mar 2011	20
Guadalupe, San Antonio, Mission, and Aransas Rivers, and Mission, Aransas, and San Antonio Bays	Aug 2012	Mar 2011	15
Nueces River and Corpus Christi and Baffin Bays	Mar 2014	Oct 2011	18
Brazos River Basin and Bay and Estuary System	Mar 2014	Mar 2012	19
Rio Grande, Rio Grande Estuary, and Laguna Madre	Mar 2014	Jul 2012	3

conducted through TIFP are designed to improve capabilities for improving the SB3 EFS and more effectively implementing the SB3 EFS.

The process will require many years to fully implement statewide, but flow standards were established by TCEQ based on recommendations of science teams and stakeholder committees for several selected priority river systems during 2009 to 2014. TCEQ has published EFS listed in Table 3-1 as Subchapters B through H of Texas Administrative Code Chapter 298 titled, "Environmental Flow Standards for Surface Water" (TCEQ 2024). Subchapter A of Chapter 298 titled, "General Provisions" provides general information relevant to all EFS. EFS, supporting science team and stakeholder committee recommendation reports, and other related information are conveniently accessible from the TCEQ website.

Structure of Environmental Flow Standards

The river systems with established EFS are listed in the first column of Table 3-1. The effective date in the second column reflects the official approval and publication of EFS. The water right priority date is established as the date of TCEQ administrative acceptance of the science team recommendations report. The number of sites with EFS shown in the fourth column includes 89 USGS gauges and two IBWC gauges for a total of 91 locations.

EFS adopted by TCEQ pursuant to the 2007 SB3 comprise metrics and rules that vary seasonally and with hydrologic condition and by location that govern decisions to curtail diversion or storage of stream flows by junior water right holders. EFS flow regimes prescribe the magnitude, frequency, duration, timing, and rate of change of flows required to maintain a sound ecology. Prior to the 2007 SB3, environmental flow requirements at a location typically were specified as a minimum instream flow target that may vary seasonally. However, the SB3 process has adopted a framework recommended by studies performed pursuant to TIFP that defines a flow regime that includes subsistence flows, base flows, and high flow pulse components that vary seasonally and with hydrologic conditions.

Subsistence flows occur during drought or very dry conditions. The primary objective of subsistence flow standards is to maintain water quality. Other objectives include ensuring that species populations recolonize the river system after normal base flow rates return.

Base flows represent the range of average or normal flow conditions without the effects of recent rainfall. A primary objective of base flow standards is to ensure adequate habitat conditions, including variability, to support the natural biological community of the river system.

High flow pulses are short-duration, high-magnitude flow events that occur during and immediately following rainfall storms. The high pulse components of EFS are designed to maintain physical habitat features and longitudinal connectivity along the river channel. High flow pulses include both within-bank and overbank flows. Overbank flows are infrequent, high-magnitude flow events that exceed channel banks resulting in water entering the floodplain. The primary objectives of overbank flow requirements are to maintain riparian areas and provide lateral connectivity to the floodplain. Other objectives may include transporting organic matter to the main channel, providing life cycle cues for various species, and maintaining the balance of species in aquatic and riparian communities.

The general strategy adopted in Texas for defining environmental flow requirements is consistent with methods investigated and/or applied in other regions of the world. Publications by Tharme (2003), Mathews and Richter (2007), Poff and Zimmerman (2010), and O'Keefe (2012)

Table 3-2. Metrics for EFS at the USGS Gauge on the Trinity River Near Romayor.

Season	Subsistence Flow	Base Flow	High Pulse Flows (Two per Season)		
			Trigger	Volume	Duration
	(ft^3/s)	(ft^3/s)	(ft^3/s)	(ac-ft)	(days)
Winter	405	875	8,000	80,000	7
Spring	700	1,150	10,000	150,000	9
Summer	200	575	4,000	60,000	5
Fall	230	625	4,000	60,000	5

are representative of the extensive literature on environmental instream flow requirements.

The general format of the SB3 EFS is illustrated by the metrics in Table 3-2 for EFS at the USGS gauge on the Trinity River near the city of Romayor (Wurbs 2019). This gauge site is located about 20 mi below Livingston Dam and 50 mi above the Trinity River outlet at Galveston Bay shown in the basin map of Figure 7-4 in Chapter 7. The watershed area above the gauge site is 17,200 sq. mi. Observed daily flows at this gauge are plotted in Figure 5-26 of Chapter 5.

For junior water right holders to which EFS apply, streamflow, storage, and diversions that diminish flows at the Romayor gauge are not allowed unless the streamflow at the Romayor gauge is above the subsistence flow limit shown in Table 3-2. If the flow at the gauge location is above the subsistence flow limit but below the base flow limit, junior water right holders may divert or store water to the extent that the flow at the gauge does not fall below the subsistence flow limit. If the flow is above the base flow limit, the water right holder may store or divert streamflow until the flow falls below the base flow standard.

EFS include preservation of two high flow pulses during each season if the specified pulse flow events occur. Quantities used to define high flow pulse events are tabulated in the last three columns of Table 3-2. A qualifying pulse event is initiated when the flow exceeds the prescribed trigger flow shown in Table 3-2 in cubic feet per second. A pulse flow event is terminated when either the volume limit in acre-feet in Table 3-2 or the duration limit in days is reached. Pulse flow events initiated in a particular season continue into the following season if and as necessary to meet the volume and/or duration termination criteria. Junior water rights are not allowed to store or divert streamflow in a manner that would adversely affect preservation of the high flow pulses.

Seasons are defined as follows for EFS for the Trinity River system: Winter (December, January, February), Spring (March, April, May), Summer (June, July, August), and Fall (September, October, November). The EFS for the different river systems in Table 3-1 differ a little in the selection of which months to define seasons.

Unlike EFS established for other river systems, hydrologic conditions are not specified for the Trinity River system EFS. For several of the other river systems in Table 3-1, the subsistence, base flow, and high pulse criteria in Table 3-2 vary with hydrologic conditions as well as season of the year. Hydrologic conditions are defined in most EFS as a function of either storage content of one or more specified reservoirs or cumulative streamflow at a specified gauge over a specified preceding period. EFS for the Brazos River system use the Palmer hydrologic drought index to define hydrologic conditions (Wurbs 2019).

EFS for the Trinity River system specify that two high flow pulses be preserved in each of the four seasons of the year. EFS for other river systems vary the number of high flow pulses to be protected in each of the seasons of the year.

Erosion and Sedimentation

Stream erosion and deposition processes are altered by the construction and operation of reservoir projects. Impacts of individual reservoirs vary significantly depending on flow and sediment characteristics of the stream and storage and operating characteristics of the reservoir project. Interruption of the natural sediment transport processes by construction of a dam and reservoir project results in deposition of sediment in the upstream reservoir area and erosion and degradation of the streambed and banks downstream from the dam.

Reservoir Sedimentation

Storage capacity is lost over time due to sediment deposits (Morris and Fan 1998, Annandale et al. 2016). Soil transported by streamflow is deposited with decreases in flow velocity as inflows are stored in a reservoir. As streamflow velocities decrease in the upper reaches of a reservoir, sediments are deposited forming deltas. Smaller particles will move further into the reservoir before depositing. Differences between sediment loads of reservoir inflows and outflows result in sediment accumulation. The rate of sediment deposition varies between reservoirs, depending on stream inflow rates, sediment loads from watershed and stream bank erosion, and reservoir sediment trap efficiencies. Because sediment transport increases greatly during high flows, the accumulation of sediments in reservoirs is highly dependent on the frequency and timing of major floods.

No attempt is made to estimate the volume and location of past or projected future sediment deposits for many smaller reservoirs. For most federal and other large reservoirs, sediment reserve storage volume is designated for sedimentation estimated to occur over a period of typically 50 to 100 years. The volume and location of the sediment deposition are predicted using methods outlined by USBR (1987) and USACE (1995). Storage capacity reserved for future sediment accumulation is reflected in water supply contracts and planning.

Some reservoirs have existed for decades without sediment surveys being performed. Reservoir sedimentation surveys are performed occasionally for many major reservoirs. Larger reservoir owners such as USACE can perform their own sediment surveys. TWDB has operated a nonprofit cost-reimbursable hydrographic survey program since 1991, which is described at the TWDB website. Reservoir owners contract with TWDB to perform surveys to determine storage capacity, sedimentation rates, updated elevation-area tables, and bathymetric contour maps. TWDB completed 197 hydrographic surveys on 114 reservoirs between 1991 and August 2021.

Streambank and Reservoir Shoreline Erosion

Large reservoirs may trap and retain almost all the inflowing suspended sediment and bed material load, thus releasing sediment-free water. These releases may erode the bed and banks of the river downstream of the outlet structures. The extent of this erosion is related to the composition of the bed and bank material, the volume of water released on an annual or seasonal basis, release rates, and flow velocities. Fluctuating releases often result in an initial loss of the banks. This loss is closely related to the magnitude of the stage fluctuation. The recession of banks due to fluctuating releases usually stabilizes during the first few years of operation, as the underwater slope reaches a quasi-equilibrium state. Once this equilibrium slope has been achieved, the bank erosion process behaves as in the natural channel. Periodic wetting and drying of the banks through fluctuating releases accelerate this process. Reservoir releases also result in lowering the streambed, with the maximum amount of lowering occurring immediately downstream from the outlet structures and decreasing in the downstream direction. This degradation process continues until the slope is reduced to its equilibrium value and/or the bed becomes naturally armored by removal of the fines, which exposes coarser, nonerodible bed materials. After the bed becomes naturally armored, future lowering of the streambed is usually insignificant (USACE 1995).

Channels downstream from small- and medium-sized reservoir projects often exhibit different characteristics than described above for large reservoirs. Channel capacity below smaller reservoirs tends to be lost over time. Reservoir projects that make only limited releases may result in extensive deposition and subsequent vegetative encroachment in the downstream channel. With construction of a reservoir, the preconstruction periodic flushing flows that remove deposits near the mouth of tributaries are often replaced by low nonerosive reservoir releases. This contributes to the loss of channel capacity and reservoir operating flexibility (USACE 1995).

Reservoir shorelines are subject to various forces contributing to their instability and frequently undergo major changes during the life of a project. Fluctuating pool levels saturate previously unsaturated material, resulting in soil material slides when the pool is drawn down to lower levels. This material accumulates at the base of the slope, and often forms an underwater bench, leaving steep unstable slopes above the waterline. Reservoir banks are also subject to attacks by both wind and waves, which tend to remove this material and undercut the banks.

Reservoir Water Quality

Water quality encompasses the physical, chemical, and biological characteristics of water. Both natural water quality and human-induced changes in quality are important considerations in river and reservoir system management (Jorgensen et al. 2005). Water quality requirements for reservoir releases may involve both flow rates and quality parameters. Low flow augmentation, or maintenance of minimum streamflow rates at downstream locations, may be a water quality operating objective.

Common reservoir water quality problems include turbidity, suspended solids, and algae. Pollution from watershed activities such as oil field operations, agricultural activities, leaking septic tanks, and municipal and industrial wastewater effluents may be problems in some reservoirs.

Eutrophication

Eutrophication is the process of excessive addition of organic matter, plant nutrients, and silt to reservoirs at rates sufficient to cause increased production of algae and rooted plants. Symptoms of eutrophication include algae blooms, weed-choked shallow areas, low dissolved oxygen, and accumulation of bottom sediments. Resulting problems include elimination of reservoir fisheries, adverse impacts on downstream ecosystems, degradation of water supplies, and reduced storage capacity. Nutrients promote ecosystem production and healthy growth of algae, larger plants, and fish and other aquatic organisms. However, excess nutrients and algae in reservoirs can have a deleterious effect on water quality.

The trophic state of a reservoir refers to its nutritional status indicated by measurements of nutrients and algae. The Section 314 Clean Lakes Program of the US Clean Water Act of 1972 requires all states to classify lakes and reservoirs according to trophic state. TCEQ reports every two years on assessments for reservoirs in Texas. Reservoirs are classified based on a scheme that groups reservoirs into discrete quality (trophic) states along a continuum from oligotrophic (poorly nourished) to hypereutrophic (overnourished).

The 2022 TCEQ assessment report (TCEQ 2022) documents water quality investigations that resulted in 139 reservoirs being classified as follows: oligotrophic (six reservoirs), mesotrophic (24 reservoirs), eutrophic (104 reservoirs), and hypereutrophic (five reservoirs). Reservoirs with the clearest water are listed in descending order of clarity as follows: Canyon, Amistad, Travis, Medina, and Tyler State Park Lake. Reservoirs with the poorest light transparency are in ascending order as follows: Rita Blanca, Cox, Palo Duro, White River, and Lake Wichita. Thirty-one reservoirs share the lowest mean total phosphorus concentration of 0.02 mg/L. Reservoirs with the highest mean total phosphorus concentrations, listed in descending order of concentration, are as follows: Rita Blanca, O. C. Fisher, Tanglewood, Woodlands, Palo Duro, and Lake Corpus Christi. The 2022 TCEQ assessment report lists 15 reservoirs with ongoing water quality restoration efforts.

Reservoir Stratification

Reservoir water quality problems may also be related to seasonal stratification. In a stratified lake, the well-mixed surface layer, called the *epilimnion*, and the colder bottom layer, called the *hypolimnion*, are separated by a layer of sharp temperature gradient, called the *metalimnion*. Most impoundments exhibit some degree of temperature stratification. In general, deeper lakes are more likely to become highly stratified each summer and are not as likely to become mixed by wind or short-term temperature changes. When the surface of the lake begins to receive a greater amount of heat from the sun and air than is lost, it becomes warmer and less dense, while the colder, more dense water remains on the bottom. In the layer of colder water near the bottom, little if any oxygen is transferred from the air to replace that depleted by oxidation of organic substances, and eventually anoxia may develop. Under this condition, a reducing environment is created resulting in elevated levels of parameters such as iron, manganese, ammonia, and hydrogen sulfide. Changes such as these may result in water that is degraded and toxic to aquatic life.

A primary means of managing the water quality of reservoir releases is to control the vertical levels at which water is withdrawn from the reservoir. Reservoir projects may include outlet works intake structures providing multilevel withdrawal capabilities. The reservoir operating strategy consists of establishing desired temperature, dissolved oxygen, and other water quality criteria and selecting the elevations at which to make releases to meet the criteria. Water from different levels may be mixed to meet water quality criteria. Management of water quality in the reservoir pool may also be a consideration in selective withdrawals from multilevel intake structures. Good-quality and poor-quality water can be blended to meet the release criteria with a minimum of good and a maximum of poor-quality water. This type of release policy will help to prevent a deterioration of quality in the reservoir which could lead to an eventual inability to meet the release criteria.

Turner (1986) assesses selective withdrawal capabilities of the 25 reservoirs owned and operated by USACE Fort Worth District. Spillways with higher crest elevations are employed for releasing high flows during floods in addition to the conservation pool outlet structures discussed here. The depth of the conservation pool surface above the lowest outlet structure intake invert at the 25 USACE reservoirs ranges from 27 ft at Wright Patman Lake to 110 ft at Belton Lake and 134 ft at Canyon Lake. Downstream releases from the conservation pool occur through a single outlet intake at 15 projects. Three of the 25 lakes have outlet structure intakes at two or three different elevations. Seven of the lakes have outlet structures allowing releases from four different levels. The general conclusion was that the selective withdrawal capabilities at 10 projects appear to justify their additional outlet structure construction costs. However, no water quality problems had been identified at the other lakes that could be addressed by adding selective withdrawal capabilities (Turner 1986).

Canyon Lake and the Guadalupe River are extremely popular for recreation. A recreational aspect of this area that is unusual for Texas is the trout fisheries maintained by TPWD in the Guadalupe River below Canyon Dam. The waters of the rivers and reservoirs of Texas are typically too warm for trout. However, releases from the lower level of the deep Canyon Lake provide flows in the Guadalupe River with temperatures cold enough to maintain trout fisheries.

Salinity

Salinity plays a key role in water resources development and management throughout the world. Salts or inorganic solutes enter the lower coastal reaches of rivers and coastal interfaces of groundwater aquifers through seawater intrusion. Salts also enter rivers throughout their watersheds from various other sources, particularly in arid regions. Salinity in natural waters occurs because of weathering of rocks and soils. Total dissolved solids or salinity concentrations increase as waters move over the land surface and through soils and aquifers. Evaporation and transpiration increase concentrations. Human activities such as irrigated agriculture and construction of reservoirs increase evaporation and the salinity of land and water resources. Groundwater pumping, oil field operations, and municipal water use and wastewater disposal activities may also increase salinity.

Both seawater intrusion in coastal reaches and addition of salts to river flows from upper watershed sources are important in Texas. Freshwater flows in the lower reaches of rivers of Texas mix with seawater as the rivers flow into the Gulf of Mexico. Intrusion of seawater moves further upstream in the rivers during periods of low flows. The following discussion focuses on the other salinity problem of salt pollution entering the rivers from sources in their upper watersheds. Natural salt pollution in upper river basins severely constrains water supply capabilities of many major reservoirs in Texas including Lakes Texoma, Whitney, Possum Kingdom, Meredith, and other large reservoirs that contain much of the storage capacity of the central and western regions of the state.

Geologic formations underlying the Permian Basin Region in the upper watersheds of the Rio Grande, Pecos, Colorado, Brazos, Red, Canadian, and Arkansas Rivers in Texas, New Mexico, Oklahoma, Kansas, and Arkansas contribute large salt loads to these rivers that severely constrain the use of water supplied by several large reservoirs (Wurbs 2002). This semiarid region consists of gypsum and salt-encrusted rolling plains containing numerous salt springs and seeps. The mineral pollutants consist of sodium chloride with moderate amounts of calcium sulfate and other dissolved solids. The primary salt source sub-watersheds of these major river basins have streams with extremely high salinity concentrations that at some locations exceed salinity concentrations of seawater. Salt concentrations in the downstream reaches of the rivers decrease with dilution from low-salinity tributary inflows. Dilution is affected by reservoir releases. Natural salt pollution in the Permian Basin Region is discussed further in Chapters 5 and 7.

References

Annandale, G. W., G. L. Morris, and P. Karki. 2016. *Extending the life of reservoirs: Sustainable sediment management for dams and run-of-river hydropower*. Washington, DC: International Bank for Reconstruction and Development.

Dowell, C. L., and R. G. Petty. 1971. *Engineering data on dams and reservoirs in Texas, Part III*. Report 126. Austin, TX: Texas Water Development Board.

Dowell, C. L., and R. G. Petty. 1973. *Engineering data on dams and reservoirs in Texas, Part II*. Report 126. Austin, TX: Texas Water Development Board.

Dowell, C. L., and R. G. Petty. 1974. *Engineering data on dams and reservoirs in Texas, Part I*. Report 126. Austin, TX: Texas Water Development Board.

Doyle, M. W., and L. A. Patterson. 2019. "Federal decentralization and adaptive management of water resources: Reservoir reallocation by the US Army Corps of Engineers." *AWRA J.* 55 (5): 1248–1267.

Ely, P. B., and J. C. Peters. 1984a. *Probable maximum flood determination: Eastern United States*. Tech. Paper 100. Davis, CA: USACE Hydrologic Engineering Center.

Ely, P. B., and J. C. Peters. 1984b. "Probable maximum flood determination: Eastern United States." *Water Resour. Bull.* 20 (3): 391–395.

Golze, A. R., ed., 1977. *Handbook of dam engineering*. New York: Van Nostrand Reinhold.

Jansen, R. B., ed., 1988. *Advanced dam engineering for design, construction, and rehabilitation*. New York: Van Nostrand Reinhold.

Jorgensen, S. E., H. Loffler, W. Rast, and M. Straskraba. 2005. *Lake and reservoir management*. Cambridge, MA: Elsevier.

Lewis, B. 2014. *Small dams: Planning, construction, and maintenance*. Boca Raton, FL: CRC Press.

Mathews, R., and B. D. Richter. 2007. "Application of the indicators of hydrologic alteration software in environmental flow setting." *AWRA J.* 43 (6): 1400–1413.

Morris, G. L., and J. Fan. 1998. *Reservoir sedimentation handbook: Design and management of dams, reservoirs, and watersheds for sustainable use*. New York: McGraw-Hill.

National Research Council. 2005. *The science of instream flows: A review of the Texas instream flow program*. Washington, DC: National Academies Press.

NWS (National Weather Service). 1978. *Probable maximum precipitation, United States east of the 105th meridian*. Hydrometeorological Report No. 51. Silver Spring, MD: NOAA National Weather Service.

NWS. 1982. *Application of probable maximum precipitation estimates, United States east of the 105th Meridian*. Hydrometeorological Report No. 52. Silver Spring, MD: NOAA National Weather Service.

O'Keefe, J. 2012. "Chapter 4: Environmental flow allocation as a practical aspect of IWRM." In *River conservation and management*, edited by P. J. Boon, and P. J. Raven. New York: Wiley.

Patterson, L. A., and M. W. Doyle. 2018. "A nationwide analysis of US Army Corps of Engineers reservoir performance in meeting operational targets." *AWRA J.* 54 (2): 543–564.

Poff, N. L., and J. K. H. Zimmerman. 2010. "Ecological responses to altered flow regimes: A literature review to inform the science and management of environmental flows." *Freshw. Biol.* 55 (1), 194–205.

Rubinstein, C., C. Seaton, and R. E. Mace. 2022. "Beyond Senate Bill 3: How to achieve environmental flows in Texas under prior appropriation." *Tex. Water J.* 13 (1): 13–26.

Siddiqui, I. H. 2009. *Dams and reservoirs: Planning and engineering*. Oxford, UK: Oxford University Press.

Tanchev, L. 2014. *Dams and appurtenant hydraulic structures*. 2nd ed. Boca Raton, FL: CRC Press.

TCEQ (Texas Commission on Environmental Quality). 2006. *Guidelines for operation and maintenance of dams in Texas*. GI-357. Austin, TX: TCEQ.

TCEQ. 2009. *Design and construction guidelines for dams in Texas*. RG-473. Austin, TX: TCEQ.

TCEQ. 2022. *Trophic classification of Texas reservoirs: 2022 Texas integrated report for Clean Water Act Sections 305(b) and 303(d)*. Austin, TX: Water Quality Division, TCEQ.

Tharme, R. E. 2003. "A global perspective on environmental flow assessment: Emerging trends in the development and application of environmental flow methodologies for rivers." *River Res. Appl* 19 (5–6) 397–441.

Turner, R. L. 1986. "A review of selected withdrawal performance in the Fort Worth District." In *Proc., CE Workshop on Design and Operation of Selective Withdrawal Intake Structures. MP HL-86-3*. Vicksburg, MS: USACE Waterways Experiment Station.

TWDB (Texas Water Development Board). 2022. *Water for Texas: 2022 state water plan*. Austin, TX: TWDB.

USACE (US Army Corps of Engineers). 1995. *Sedimentation investigations of rivers and reservoirs*. Engineering Manual 1110-2-4000. Washington, DC: US Army Corps of Engineers.

USBR (US Bureau of Reclamation). 1987. *Design of small dams*. 3rd ed. Denver: US Bureau of Reclamation, US Department of the Interior.

World Meteorological Organization. 2009. *Manual for estimation of probable maximum precipitation*. WMO 1045. 3rd ed. Geneva: WMO.

Wurbs, R. A. 1990. *Modifying reservoir operations to improve capabilities for meeting water supply needs during drought*. RD-31. Davis, CA: USACE Hydrologic Engineering Center.

Wurbs, R. A. 1995. "Water rights in Texas." *J. Water Resour. Plan. Manag.* 121 (6): 447–454.

Wurbs, R. A. 2002. "Natural salt pollution control in the Southwest." *J. Am. Water Work. Assoc.* 94 (12): 58–57.

Wurbs, R. A. 2017. "Incorporation of environmental flows in water allocation in Texas." *Water Int.* 42 (1): 18–33.

Wurbs, R. A. 2019. *Daily water availability model for the Brazos River Basin and San Jacinto-Brazos Coastal Basin*. TR-513. College Station, TX: TWRI.

Wurbs, R. A., and P. A. Carriere. 1988. *Evaluation of storage reallocation and related strategies for optimizing reservoir system operations*. TR-145. College Station, TX: TWRI.

Wurbs, R. A., and W. P. James. 2002. *Water resources engineering*. Upper Saddle River, NJ: Prentice Hall Division, Pearson Education.

Chapter 4

Computer Databases and Modeling Systems

Online databases and computer-based decision-support modeling systems play important roles in planning, development, and management of river and reservoir systems. The information sources and modeling tools discussed in this chapter also provide quantitative information presented in the tables and figures of several chapters of this book. Table 4-1 provides an outline for this chapter, as well as a catalog of several key sources of information and software tools.

Water agencies in the United States are unique, compared to other countries, in their development and maintenance of extensive collections of useful information and decision-support tools publicly accessible online free of charge. The 10 websites selected for inclusion in Table 4-1 are particularly relevant to the topics covered in this book. Reports, graphs, tabulations, computer software tools, and databases accessible at these websites are fundamental to water management throughout Texas. A wealth of information and decision-support tools can be downloaded free of charge at these selected websites. Many other agencies maintain numerous other websites that also provide additional relevant information and decision-support tools.

National Water Information System

The Water Resources Division of USGS is responsible for the National Water Information System (NWIS). The NWIS website listed in Table 4-1 provides access to water resources data collected at about 1.9 million

Table 4-1. Websites for Accessing Databases, Decision-Support Software, and Other Information.

Database or Decision-Support Tools	Responsible Agency
National Water Information System (NWIS) https://waterdata.usgs.gov/nwis	US Geological Survey (USGS)
National Inventory of Dams (NID) https://nid.sec.usace.army.mil/#/	US Army Corps of Engineers (USACE)
Water Management Software https://www.hec.usace.army.mil	USACE Hydrologic Engineering Center (HEC)
Information on USACE FWD Reservoirs https://www.swf-wc.usace.army.mil/	USACE Fort Worth District (FWD)
Information on Texas Reservoirs https://www.twdb.texas.gov/surfacewater/rivers/reservoirs/index.asp	Texas Water Development Board (TWDB)
Precipitation and Reservoir Evaporation https://waterdatafortexas.org/lake-evaporation-rainfall	Texas Water Development Board (TWDB)
Regional and Statewide Water Plans https://www.twdb.texas.gov/waterplanning/index.asp	Texas Water Development Board (TWDB)
Water Rights Viewer https://www.tceq.texas.gov/gis/water-rights-viewer	Texas Commission on Environmental Quality (TCEQ)
Water Rights Analysis Package (WRAP) https://wrap.engr.tamu.edu/	Texas A&M University (TAMU)
Water Availability Models (WAMs) https://www.tceq.texas.gov/permitting/water_rights/wr_technical-resources/wam.html	Texas Commission on Environmental Quality (TCEQ)

sites in all 50 states, the District of Columbia, and 5 US territories. The data are divided into the categories of surface water (streamflow and lake storage levels), groundwater, water quality, and water use. Although periods of record vary between gauge sites, historical measurements date back to the late 1800s at some sites. Many USGS gauge sites include current real-time as well as past recorded observations.

USGS state offices in each of the 50 states manage data collection and dissemination for that state. The first step in downloading data from the NWIS website for gauges on rivers and lakes located in Texas is to select the Texas component of the website. As of January 2024, water quantity and/or quality observations recorded daily for flows in streams and storage in lakes in the NWIS database include 29,577 gauge sites that include 1,145 gauges in Texas. Annual maximum streamflow rates are published for 30,078 sites, including 1,272 sites in Texas. Other types of observed data are provided for many other locations.

The NWIS website includes tutorials explaining the data and procedures for downloading datasets. Descriptive map information is provided for each gauge location. Various options are provided for accessing data online in tabular and graphical formats and for downloading data.

A later section of this chapter includes discussion of a computer program called HEC-DSSVue available from the USACE HEC that provides an array of capabilities including downloading and managing datasets from NWIS and other online databases. The daily flows in Figure 1-5 were downloaded from the NWIS website into a data storage system (DSS) file using HEC-DSSVue and plotted using HEC-DSSVue. The annual flow volumes of Figure 1-6 were computed from the daily mean flows of Figure 1-5 within HEC-DSSVue and plotted. HEC-DSSVue was also employed with daily streamflow datasets from NWIS to prepare graphs presented in Chapter 5.

National Inventory of Dams

The NID introduced in the Dam Safety Program section of Chapter 2 is the second website referenced in Table 4-1. NID is maintained by USACE in collaboration with FEMA and federal and state dam safety regulatory agencies to provide a central source of information on 91,843 dams located throughout the United States and its territories as of January 2024. Support of dam safety regulatory programs is a main motivation for NID. The database contains information about each dam's location, owner, type of dam and appurtenant structures, project purpose, uses and benefits, hazard potential, emergency action plans, date of last safety inspection, and other information. NID includes interactive maps and search features that facilitate a focus on dams in Texas or any other selected state.

More of the 91,843 dams in NID are in Texas than in any other state. Entering Texas in the website search box accesses a map showing locations of 7,384 dams in Texas and a list of the dams that provides access to information regarding each dam and reservoir.

Congress first authorized USACE to inventory dams in the United States with the National Dam Inspection Act of 1972. NID has been reauthorized and expanded periodically since its 1972 initial authorization. The online database has been improved with advances in information technology. Criteria for selection of dams for inclusion in the NID database are based on dam height and the potential human and economic consequences of downstream flooding in the event of dam failure. These criteria are described in Chapter 2.

Generalized Simulation Modeling Systems

Computer simulation models, statistical analysis software, and other computer-based decision-support tools are extensively employed in a broad spectrum of hydrologic science, water resources engineering, and water management applications (Wurbs 1994, 1995). State-of-the-art reviews of generalized modeling systems for simulating and evaluating river and reservoir system water resources management are provided by Wurbs (1996, 2005a, 2011). *Generalized* means the computer software can be applied by professionals other than the original software developers to river/reservoir systems located anywhere. Generalized modeling systems should be well documented and convenient to obtain and apply. A significant level of expertise is typically required to understand the hydrologic and hydraulic processes and water management facilities and practices being modeled.

The generalized computer modeling systems developed and disseminated by USACE HEC discussed in the next section include the most extensively applied hydrologic, hydraulic, and water management computer programs employed throughout the United States, including Texas. The HEC software is also used in other countries. Most applications of HEC generalized simulation modeling systems involve flood control, floodplain management, stormwater management, hydraulic structures, and multiple-purpose reservoirs.

The Water Rights Analysis Package (WRAP) is introduced in the last section of this chapter and employed in Chapters 5, 6, and 7. The generalized WRAP modeling system is applicable anywhere. The modeling system is applied routinely within the water management community of Texas as discussed later in this chapter. WRAP simulates comprehensive river and reservoir system water management. WRAP focuses on providing particularly detailed modeling capacities for analyzing water allocation, water supply, and environmental flow requirements.

TWDB developed a series of generalized river and reservoir system water management simulation models during the 1960s to 1980s used by TWDB staff in statewide planning in Texas (TWDB 1974, Martin 1987, Wurbs 1996). TWDB replaced these early computer simulation models during the early 2000s with adoption of the WRAP modeling system.

Early reservoir system analysis models developed during the 1950s to 1960s were typically for specific reservoir systems. Site-specific models were later replaced with generalized models with broader applicability (Wurbs 1996).

The published literature on computer modeling of reservoir operations is massive. Many hundreds of mathematical models for optimizing reservoir operations developed by university researchers and published in journal papers over the last several decades have been applied only by the original model developers in research studies at their home universities. The number of journal papers reporting academic faculty and graduate student research endeavors continues to grow. Although many generalized modeling systems are routinely applied by the professional community in actual reservoir system water management, the number of such models is small relative to the massive number of models found in the published literature.

USACE Hydrologic Engineering Center Modeling Systems

The USACE Civil Works Program is discussed in Chapter 2. The Hydrologic Engineering Center in Davis, California, was created within USACE in 1964 to develop computer software for use throughout all

USACE offices. HEC products are also provided free of charge to any interested persons outside of the Corps of Engineers. The latest versions of software, manuals, and relevant technical reports can be downloaded from the HEC website shown in Table 4-1. The software is continually expanded and improved with periodic releases of the latest new versions.

HEC modeling systems are routinely applied within local and federal agencies, civil engineering consulting firms, and universities as well as in USACE offices throughout the United States. Several of the most extensively employed HEC software packages are listed as follows.

- HEC-HMS Hydrologic Modeling System
- HEC-RAS River Analysis System
- HEC-ResSim Reservoir System Simulation
- HEC-SSP Statistical Software Package
- HEC-FDA Flood Damage Analysis
- HEC-DSSVue Data Storage System Visual Utility Engine

HEC models were originally employed primarily by civil engineers in USACE district offices in the planning, design, and operation of dams, spillways, flood control projects, and river navigation facilities. Over time, a growing number of engineers and scientists in agencies, consulting firms, and universities have adopted HEC models for a variety of applications. For example, HEC-HMS and HEC-RAS have been used by civil engineering consulting firms and city engineering staff to delineate floodplains and periodically update the delineations in most of the more than 20,000 communities participating in the National Flood Insurance Program as well as in the planning and design of numerous stormwater management facilities and hydraulic structures in local communities throughout the United States and abroad.

HEC-HMS, HEC-RAS, and other HEC modeling systems are covered in civil engineering courses in hydrology, hydraulics, and water resources engineering at various universities including TAMU. Examples of textbooks that incorporate HEC-HMS and HEC-RAS include Wurbs and James (2002), Bedient et al. (2019), and Akan and Iyer (2021). Universities and other entities offer continuing education short courses for practicing professionals in applying these modeling systems as well as employing the software in regular undergraduate and graduate courses.

The Corps Water Management System (CWMS) is applied by USACE district and division offices to support real-time flood control operations of reservoir systems and operations of other water control facilities. The LCRA of Texas was the first non-USACE agency to employ the integrated CWMS (Ickert and Luna 2004) and has used CWMS to model flood control operations of the Colorado River Highland Lakes System.

CWMS combines the generalized simulation models listed earlier, other data management and visualization software, and automated systems for collecting observed rainfall and streamflow data. The NWS real-time observed data from rain gauges and in some cases radar measurements are accessed. HEC-HMS is applied in real-time reservoir system operations to predict stream flows given real-time measurements of rainfall. Reservoir system operations are simulated using HEC-ResSim with observed and forecasted stream inflows from HEC-HMS provided as input. HEC-RAS computes river stages for discharge hydrographs. The information developed with these simulation models is combined within CWMS to support reservoir operating decisions.

HEC-DSS incorporated in the HEC simulation models reads time series input from DSS files and stores time series simulation results in DSS files. Data are conveniently transported between modeling subsystems.

HEC-DSSVue is an interface that allows users to access DSS files. HEC-DSSVue is employed to organize, manage, and display data, perform statistical analysis and mathematical operations, and access various online databases. As noted earlier, HEC-DSSVue is used to download data from NWIS. HEC-DSS and HEC-DSSVue are incorporated in the WRAP modeling system described in the last section of this chapter as well as in other non-HEC modeling systems. HEC-DSSVue was used to develop the time series plots of precipitation, evaporation, streamflow, and reservoir storage found in Chapters 1, 5, 6, and 7.

USACE Fort Worth District Reservoir Information

The fourth website link referenced in Table 4-1 accesses a water information system maintained by the FWD of the USACE. This website provides an array of information about activities related to the 25 reservoirs owned and operated by USACE FWD. The "Water Management Menu" at the website link in Table 4-1 links to online features labeled "Project Status Map," "Water Management Reports," "Pertinent Data," "Hydrologic Query," "Related Sites," and "Recreation Information." The recreation information feature provides lake visitors with continually updated information regarding the current status of each recreation facility at each lake along with extensive general information regarding recreation opportunities. The "Related Sites" link leads to a collection of other relevant websites maintained by other water management agencies.

Pertinent data sheets and period-of-record plots for 25 USACE FWD multipurpose reservoirs accessed through the Pertinent Data component of the USACE FWD online information system are employed in Chapter 6 of this book. Available information includes (1) pertinent data sheets for each of the 25 reservoir projects with descriptions of the dams and appurtenant outlet structures with dimensions, elevations, discharge capacities, and reservoir storage capacities; (2) plots of historical and current storage levels; (3) water surface elevation versus duration plots; and (4) flood control operating metrics.

The USACE FWD website was selected for inclusion in Table 4-1 because it provides extensive conveniently accessible online information relevant to a significant number (25) of large multipurpose reservoirs located across Texas. The USACE Fort Worth, Galveston, and Tulsa Districts have water management responsibilities for river basins in Texas. The Corps of Engineers website in Table 4-1 covers only Fort Worth District programs and projects but provides links to the other districts. The Galveston and Tulsa Districts maintain their own websites. Nonfederal reservoir owners also maintain websites providing information about their reservoirs.

TWDB Databases and Planning Reports

Three of the many compilations of information available online from TWDB are included in Table 4-1. Each of these three TWDB compilations includes an array of types of information accessed through multiple webpage links. The rivers and reservoirs component of the TWDB website includes information regarding the 15 major river basins, eight coastal basins, and about 200 major reservoirs of the state. Descriptive information and quantitative data are provided for each river basin and each individual reservoir. Recent and historical storage volume plots and data tabulations are provided for individual reservoirs and statewide totals.

The precipitation and lake evaporation webpages access databases of monthly precipitation depths from January 1940 and monthly evaporation depths from January 1954 to near the present for 92 one-degree by one-degree quadrangles that encompass Texas. TWDB updates the databases each year to extend through the preceding year. The statewide average monthly precipitation depths plotted in Figure 1-2 were computed from the TWDB quadrangle data. The monthly quadrangle precipitation and lake evaporation data are analyzed in Chapter 5.

The SB1 regional and statewide planning process is introduced in Chapter 1 and discussed further in Chapter 2. The reports documenting the 16 regional water plans and state water plan completed in 2002 pursuant to the 1997 SB1 were supplemented by updated water plan reports in 2007, 2012, 2017, and 2022. The reports presenting the next updated water plans are scheduled for completion in 2027. All versions of the 16 regional and statewide water plans are available at the last TWDB website listed in Table 4-1. Planning procedures are explained and current membership and activities of the 16 regional planning groups are noted. Various data compilations supporting the planning reports are also provided through this website.

TCEQ Water Rights Viewer

As discussed in Chapters 1, 2, and 3, TCEQ administers a water rights permit system that regulates storage of streamflow in reservoirs and diversion of water from streams and reservoirs for beneficial use. Water rights information has been accessible through various TCEQ sources for many years. TCEQ publicly released in 2021 an online Water Rights Viewer to provide convenient access to information for each of the several thousand water rights along with other general information. The website in Table 4-1 that accesses the Texas Water Rights Viewer includes a user guide for navigating the interactive map-controlled online database system.

The interactive map accessed at the viewer website includes geographic features, such as river basins and streams, and USGS gauge sites with links to current NWIS streamflow data. A map layer includes the locations of all authorized water rights with links to information that includes a copy of the water right and the adjudication final determination document, current ownership of the water right, and recent water use data. TCEQ-adopted EFS can be accessed. Users can also view interactive map layers with administrative boundaries that include watermaster jurisdictions, regional water planning areas, and groundwater conservation districts.

Water Availability Modeling System

The last two websites in Table 4-1 provide access to the WRAP modeling system and WRAP simulation input datasets called WAMs for all the river basins of Texas. WRAP simulations with WAM input datasets for the river basins of Texas support discussions in Chapters 5, 6, and 7 of this book.

The TCEQ WAM and TAMU WRAP websites in Table 4-1 are interlinked. The WRAP modeling system is documented by six lengthy manuals published as Texas Water Resources Institute (TWRI) technical reports that are found at both the WRAP and TWRI websites (Wurbs 2008, 2019, 2022a, b, c; Wurbs and Hoffpauir 2022). Many other publications and training courses are also found at the WRAP website. The TCEQ WAM website provides WRAP simulation input datasets covering all of Texas, explanations of water right permitting procedures, and information regarding EFS and other water allocation endeavors.

The generalized WRAP with WAM datasets is applied by water right permit applicants or their civil engineering firm consultants and TCEQ staff in administration of the water rights system. TWDB staff, regional planning groups, and their consultants employ the modeling system in regional and statewide planning studies. River authorities and other organizations apply the modeling system in project feasibility and operational planning studies. The WRAP/WAM modeling system has also been employed in university and agency research studies to explore a variety of water management issues (Wurbs 2021).

Water Rights Analysis Package Modeling System

The WRAP modeling system developed at TAMU is generalized for application anywhere in the world, with input datasets being developed by model users for river systems of concern. For applications in Texas, WRAP simulation input datasets available at the TCEQ WAM website or through TWDB are altered to reflect proposed water management plans of interest, which could involve changes in water use or system operating practices, construction of new facilities, or other water management strategies. The WRAP/WAM modeling system facilitates assessments of hydrologic and institutional capabilities for satisfying requirements for water supply, hydroelectric energy generation, environmental flows, flood control, and reservoir storage. Basin-wide impacts of proposed actions are simulated.

WRAP was conceptualized and early versions developed in research at TAMU sponsored jointly by TWRI, USGS, and BRA. The modeling system has been greatly improved and is maintained at TAMU under the sponsorship of TCEQ. Various federal and state agencies have funded improvements and additions to the modeling system and applications thereof.

The WRAP modeling system consists of a (1) user interface; (2) simulation model called SIM with a monthly computational time step; (3) daily version of the simulation model called SIMD with additional capabilities for modeling reservoir flood control operations and high pulse components of instream flow standards; (4) salinity simulation model; (5) program for organizing simulation results and performing supply reliability, flow and storage frequency, and other types of analyses; and (6) program HYD for compiling and analyzing hydrology time series datasets.

Simulation models developed by USACE HEC are discussed earlier in this chapter. HEC developed a system for managing time series data called HEC-DSS (data storage system) that is incorporated in HEC simulation models and several non-HEC modeling systems including WRAP. The WRAP programs create and read DSS files. The DSS interface program HEC-DSSVue (Hydrologic Engineering Center 2009) is an integral component of the WRAP modeling system. HEC-DSSVue was used to plot WRAP simulation results presented as figures in Chapters 5, 6, and 7.

Generalized WRAP with Texas WAM Datasets

TCEQ as lead agency, TWDB, TPWD, and contractors consisting of consulting firms and university researchers adopted WRAP and developed WAM datasets pursuant to the 1997 SB1 discussed in Chapters 1 and 2 (Wurbs 2022b, 2005b; Alexander and Chenoweth 2022). The WRAP simulation input datasets from the TCEQ WAM system or modifications thereof are called water availability models.

The 15 major river basins and eight coastal basins of Texas are modeled as 20 WAMs. Operation of 3,460 dams and reservoirs and other constructed facilities in accordance with treaties between the United States and Mexico, five interstate compacts, thousands of water right permits or certifications, federal water supply storage contracts, and other institutional arrangements are simulated (Wurbs 2021, 2022b, 2024). The 15 major and eight coastal basins of Figure 1-4 are incorporated into 20 WAMs as follows. The Guadalupe-San Antonio WAM includes both the San Antonio and the Guadalupe River Basins. The San Antonio River is a tributary of the Guadalupe River though their confluence is near the coast. The San Jacinto-Brazos Coastal Basin is included in the Brazos River Basin WAM. The Lavaca-Colorado Coastal Basin is included in the Colorado River Basin WAM. The other major river basins and coastal basins are each modeled as individual WAM datasets.

Full authorization and current use scenario versions of the 20 WAM datasets are available at the TCEQ WAM website. The full authorization scenario models all water right permit holders using all water to which they are legally entitled, subject to water availability. Many water right permits include projected future water needs. The current use scenario simulates recent actual water use. WAM datasets modified by TWDB for use in planning studies reflect projections of future water needs, reservoir sedimentation, and other conditions. Model users modify WAM datasets to reflect water needs, proposed projects, and management strategies of interest.

TCEQ updates WAMs to reflect new or amended water right permits. Likewise, TWDB updates future water use projections and other future conditions in each planning cycle. TCEQ periodically updates the hydrologic periods-of-analyses for each WAM, sometimes through contracts with civil engineering firms. TWDB employs TCEQ updates of hydrologic periods of analysis but also develops additional intermediate updates as needed to support the SB1 planning process discussed in Chapters 1 and 2. The basic WAMs are maintained with state-appropriated funds. The entire water management community has access to the datasets.

Simulations and Statistical Analyses of Simulation Results

Model application includes the following tasks:

- Compiling, accessing, or modifying water management and hydrology input datasets for the simulation model component of the WRAP modeling system;
- Simulating water resources development, allocation, regulation, management, and use scenarios based on the hypothetical premise of a repetition of historical natural hydrology; and
- Developing water supply reliability and streamflow and reservoir storage frequency metrics and otherwise organizing and analyzing simulation results.

A specified water management scenario is combined with historical natural hydrology in a WRAP simulation. Historical natural hydrology is adopted to capture the hydrologic characteristics of a river basin. The water management and use scenario may reflect actual current water use, projected future conditions, the premise that all water right permit holders use their full authorized amounts, or some other scenario. Hydrologic periods-of-analyses in the WAM datasets in most cases extend from January 1940 to near the present. Hydrologic period-of-analysis sequences of naturalized streamflows and reservoir evaporation less adjusted precipitation rates are input to the simulation. The specified water management scenario is simulated during a hypothetical repetition of the hydrologic period-of-analysis sequences of naturalized stream flows and reservoir net evaporation less adjusted precipitation rates. Simulation results are organized, and frequency and reliability metrics are computed.

Various supply reliability and streamflow and reservoir storage frequency metrics representing likelihood, probability, or percent of time are adopted in the WRAP modeling. Volume reliability is computed as the percentage of the volume of water demand that is supplied during a simulation. Period reliability is the percentage of time that a specified percentage of a target is supplied in a simulation. For example, TCEQ applies the following criteria in evaluating water right permit applications. For approval of a proposed permitted increase in agricultural water use, at least 75% of the proposed new diversion target must be supplied at least 75% of the time based on the premises reflected in the model. For proposed increases in municipal water use, 100% of the water demand must be supplied 100% of the time during the simulation. The Texas water rights permit system and the WRAP simulation model protect senior permit holders from having their water supply reliabilities adversely affected by junior water rights.

Simulation Modes

Conventional planning and water right permitting applications are based on simulating a specified scenario of water development, management, and use during a hypothetical repetition of natural hydrology during a long hydrologic period of analysis such as perhaps 1940 to 2021. A single WRAP/WAM simulation includes hydrology for a single sequence comprised of all 984 months of 1940 to 2021. Supply reliability and reservoir storage and streamflow frequency metrics are computed from simulation results.

Firm yields are routinely computed for planning studies. Firm yield is the maximum annual diversion rate, which varies seasonally during the year, supplied continuously without shortage through a simulation while maintaining a specified minimum storage reserve. The SIM and SIMD simulation models include an optional iterative firm yield analysis mode. The model adjusts specified water supply or hydropower targets while automatically repeating the entire hydrologic period-of-analysis (such as 1940 to 2021) simulation in an iterative search for the firm yield.

The WRAP/WAM short-term conditional reliability modeling (CRM) mode is designed for developing reliability and frequency metrics for a future period typically ranging from a month to a year (perhaps longer) that are conditioned on present known reservoir storage levels. With a WAM 1940 to 2021 hydrology dataset, the WRAP simulation model may perform 81 or 82 annual automated simulations with each of the 12-month hydrology sequences beginning in the same selected month with the same specified beginning reservoir storage contents. Water supply reliability and reservoir storage and streamflow

frequency metrics are computed from the results of the multiple short-term simulations.

CRM allows estimation of probabilities of reservoir storage levels over the next several months, year, or longer, conditioned upon given present storage contents (Wurbs et al. 2012; Wurbs 2022b, c). Many short-term forecast simulations with different hydrology sequences begin with the same beginning reservoir storage contents. The probabilities of reservoir storage contents equaling or exceeding various levels at various future times such as one year from now, at the end of the irrigation season, or several months later in the drought are conditioned on present volumes of water in storage. Different options reflecting varying levels of computational complexity include weighting each alternative hydrologic forecast simulation sequence the same or alternatively weighting each of the hydrology sequences differently as a function of known beginning reservoir storage contents.

Routine WRAP/WAM applications are based on a monthly computational time step. An extensive experience base of monthly simulation applications focused largely on water supply has been established since 2000 by the Texas water management community. More recently developed WRAP/WAM modeling features employing a daily computational time interval add capabilities for simulating reservoir flood control operations and more complex environmental instream flow requirements including high flow pulse components (Wurbs 2021, 2024; Wurbs and Hoffpauir 2022). Simulation of flood control operations with the daily SIMD can include any configuration and number of reservoirs operated based on streamflows at any number of downstream gauges.

Development of the daily modeling features of WRAP and WAMs was motivated largely by EFS established through the process created by the 2007 SB3 discussed in Chapters 2 and 3. Monthly and daily simulations are combined as follows for use in the water right permitting process. Daily instream flow targets for EFS computed in a daily simulation are aggregated to monthly targets for inclusion in a monthly WAM to assess impacts on municipal, industrial, agricultural, and other water users. Daily versions of WAMs are employed directly, without combining with monthly WAMs, in various other planning and research applications focused on assessing capabilities for maintaining environmental instream flow requirements and/or simulating flood control reservoir system operations (Wurbs and Hoffpauir 2022, Wurbs 2024).

References

Akan, A. O., and S. S. Iyer. 2021. *Open channel hydraulics*. 2nd ed. Cambridge, MA: Elsevier.

Alexander, K., and T. Chenoweth. 2022. "Determining surface water availability." Chapter 12 in *Essentials of Texas water resources*. 7th ed., edited by M. K. Sahs and H. Heinrich. Austin, TX: State Bar of Texas.

Bedient, P. B., W. C. Huber, and B. E. Vieux. 2019. *Hydrology and floodplain analysis*. 6th ed. Upper Saddle River, NJ: Pearson Education.

Hydrologic Engineering Center. 2009. *HEC-DSSVue HEC data storage system visual utility engine user's manual*. Version 2. CPD-79. Davis, CA: USACE HEC.

Ickert A., and M. Luna. 2004. "Corps water management system (CWMS) for the Lower Colorado River Basin." In *Proceedings of Fall 2004 Conference*, Austin, TX: ASCE, Texas Section.

Martin, Q. W. 1987. "Optimal operation of surface-water systems." *J. Water Resour. Plan. Manag.* 109 (1): 58–74.

TWDB (Texas Water Development Board). 1974. *Analytical techniques for planning complex water resources systems*. Report 183. Austin, TX: Texas Water Development Board.

Wurbs, R. A. 1994. *Computer models in water resources planning and management*. Report 94-NDS-7. Alexandria, VA: USACE Institute for Water Resources.

Wurbs, R. A. 1995. *Water management models: A guide to software*. Upper Saddle River, NJ: Prentice Hall Division, Pearson Education.

Wurbs, R. A. 1996. *Modeling and analysis of reservoir system operations*. Upper Saddle River, NJ: Prentice Hall Division, Pearson Education.

Wurbs, R. A. 2005a. *Comparative evaluation of generalized reservoir/river system models*. TR-282. College Station, TX: Texas Water Resources Institute.

Wurbs, R. A. 2005b. "Texas water availability modeling system." *J. Water Resour. Plan. Manag.* 131 (4): 270–279.

Wurbs, R. A. 2008. *Salinity simulation with WRAP*. TR-317. College Station, TX: TWRI.

Wurbs, R. A. 2011. "Generalized models for river system development and management." In *Current issues in water management*, edited by U. Uhlig. InTechOpen online. https://www.intechopen.com/books/2042.

Wurbs, R. A. 2019. *Water rights analysis package river system hydrology*. TR-431. College Station, TX: TWRI.

Wurbs, R. A. 2021. "Water rights analysis package modeling system." *Tex. Water J.* 12 (1): 68–90.

Wurbs, R. A. 2022a. *Fundamentals of water availability modeling with WRAP*. TR-283, 10th ed. College Station, TX: TWRI.

Wurbs, R. A. 2022b. *Water rights analysis package modeling system reference manual*. TR-255, 14th ed. College Station, TX: TWRI.

Wurbs, R. A. 2022c. *Water rights analysis package modeling system users manual*. TR-256, 14th ed. College Station, TX: TWRI.

Wurbs, R. A. 2024. *Incorporation of environmental flow standards and other water management complexities in water availability modeling*. TR-552. College Station, TX: TWRI.

Wurbs, R. A., and R. J. Hoffpauir. 2022. *Water rights analysis package daily modeling system*. TR-430, 4th ed. College Station, TX: TWRI.

Wurbs, R. A., and W. P. James. 2002. *Water resources engineering*. Upper Saddle River, NJ: Prentice Hall Division, Pearson Education.

Wurbs, R. A., S.T. Schnier, and H. E. Olmos. 2012. "Short-term reservoir storage frequency relationships." *J. Water Resour. Plan. Manag.* 2012 (138): 597–605.

Chapter 5

River System Hydrology

Hydrologic conditions in Texas vary greatly both spatially and temporally. Spatial hydroclimatic diversity ranging from the dry western to wet eastern regions of the state is dramatic. Flows in rivers throughout the state are also highly variable over time with continuous, storm event, seasonal, and multiple-year fluctuations reflecting extremes of droughts and floods as well as more frequent but less severe fluctuations. Streamflow variability is driven by variability in precipitation and evaporation. Hydrologic variability and associated water supply reliability, flood risk, and future uncertainty are fundamental to water management. Large volumes of reservoir storage are essential for developing supplies with acceptable levels of reliability and partially controlling flood flows to reduce economic damages and protect public safety.

Stationarity, or lack thereof (nonstationarity), refers to long-term homogeneity over time as affected by permanent changes or trends. Climate change, land use change, water resources development, and other factors may affect precipitation, evaporation, river flows, and demands on water resources. Permanent changes in river flow characteristics over prolonged periods of time in Texas result primarily from changes in water use and construction of river regulation structures and water resources development projects. The impacts of human activities on low flows are typically quite different than on high flows. Regulation of rivers by dams reduces flood flows but may increase low flows at downstream locations. The effects of a dam on flows just below the dam are much less evident further downstream.

This chapter explores the characteristics of precipitation, reservoir surface evaporation, and streamflow in Texas with a particular focus on variability and stationarity. River system hydrology is fundamental to the development and operation of reservoir projects and other water management endeavors. Precipitation and watershed evapotranspiration are climatic drivers of river flows, including inflows to reservoirs. Lake surface evaporation greatly contributes to the drawdown of the volume of water stored in a reservoir. Precipitation falling on the water surface and evaporation from the water surface are major components of reservoir water budgets. Connections between surface and groundwater are important. Water supply capabilities depend on water quality as well as quantity. The last section of this chapter discusses natural pollution of surface streamflow from salts dissolved from underlying geologic formations that occur in portions of the upper watersheds of several major river basins in Texas and neighboring states.

Hydrologic Aspects of Texas Weather and Climate

Streams and lakes in Texas and elsewhere are components of the global hydrologic cycle. Water from the oceans is converted to atmospheric water vapor through evaporation. Some of the water vapor supplied to the atmosphere from the Gulf of Mexico and eastern Pacific Ocean near Mexico is transported by air masses moving inland over Texas. Evaporation from streams and lakes and evapotranspiration from land surfaces add smaller quantities to the atmospheric water supplied by evaporation from sea surfaces. The water vapor is converted to precipitation whenever suitable conditions occur. Some of the precipitation that reaches the ground replenishes soil moisture and groundwater. Rainfall flows overland or through an upper ground layer to streams. Groundwater may continue to contribute to streamflow during dry periods of no rainfall. Streamflow may seep into the underlying ground. Some of the flow from streams and rivers discharges into the Gulf of Mexico and mixes with saltwater becoming unfit for human consumption. Salts are removed as seawater is evaporated to become atmospheric water vapor.

Texas Weather and Climate

The term *weather* refers to the condition of the atmosphere at any particular time and place. Weather is always changing, perhaps more so in Texas than in most places. The climate of a particular region is the composite of weather characteristics over many years. Climate reflects weather variations, including extremes as well as more normal or average conditions. Elements of weather and climate include temperature, air pressure, humidity, clouds, precipitation, evapotranspiration, and wind.

The large land area of Texas encompasses multiple regions with distinctly different climates. Texas is viewed in this chapter as being hydrologically comprised of western, central, and eastern thirds. The western third of the state is arid to semiarid desert, and the eastern third is humid with high rainfall and associated high runoff to streams. Much of East Texas is densely forested land. West Texas tends to be sparsely vegetated due to lack of water. The middle third of Texas provides a climatic transition between the western and eastern extremes. West Texas is representative of major regions of other western states in the United States. East Texas is comparable to the humid eastern states.

Summers are hot and winters are mild in Texas. Precipitation throughout the state consists mainly of rainfall with only occasional snow and sleet. Winter snow and sleet occur most frequently in the colder northern extremes of the state. In the upper Rio Grande Basin in Colorado and New Mexico, like other regions in colder northern states, summer streamflow results from melting snowpack accumulated over the preceding winter. However, snowpack accumulation and melt are generally not a significant component of the hydrology of Texas.

North Texas and the Panhandle are part of the tornado alley that experiences tornadoes much more frequently than most other regions of the United States. About 8,200 tornadoes in Texas have been documented and classified during the 1950s to 2010s (Bomar 2017). The majority have been relatively weak, but the more severe tornadoes have caused devastating damage and loss of life.

Texas's location at the northwestern end of the Gulf of Mexico results in susceptibility to tropical storms, including hurricanes, and associated flooding (Bomar 2017). A tropical cyclone is a rapidly rotating storm system characterized by a low-pressure center, fierce winds, and a spiral arrangement of thunderstorms that produce heavy rain. A hurricane is an extraordinarily strong tropical storm. Wind and flood damage caused by these storms is most intense near the Gulf Coast but extends significant distances inland. Hurricanes since 1900 include 1900 and 1915 unnamed hurricanes that damaged Galveston, 1961 Hurricane Carla, 1967 Hurricane Beulah, 1983 Hurricane Alicia, 2005 Hurricane Rita, 2008 Hurricane Ike, and 2017 Hurricane Harvey. With rainfall totals of up to more than 40 in. in Southeast Texas, Tropical Storm Imelda in 2019 caused extreme flooding but did not reach the upgraded classification to a hurricane.

In terms of loss of human life, the greatest natural disaster in US history was the hurricane that killed more than 7,000 people in Galveston, Texas, on September 8, 1900 (Williams 1997, Bixel and Turner 2000). The death toll is uncertain, and some estimates significantly exceed 7,000. Following the hurricane, construction of a seawall and placement of fill material made Galveston Island much less susceptible to tidal and rainfall flooding. However, decades of overpumping of groundwater have caused land subsidence in the Houston and Galveston area. Highways used to evacuate Galveston Island during hurricanes are several feet lower than during the early and even late 1900s and thus more susceptible to inundation.

Hurricane Harvey in late August 2017 was the most economically costly natural disaster in Texas history and the second most costly in US history, closely behind if not tied with Hurricane Katrina in late August 2005. Much of the damage from Katrina was in and near the city of New Orleans, Louisiana. Hurricane Katrina resulted in more than 1,800 fatalities. Hurricane Harvey made landfall near Rockport, Texas, on August 25, 2017, and caused damage in more than 40 Texas counties. Rainfall exceeding 60 in. occurred over several days in some area (ASCE 2018). Much of the flood damage was in the city of Houston and throughout Harris County and surrounding areas.

Floods and Droughts

Floods and droughts are fundamental aspects of Texas weather that govern water management. On various occasions in the past, floods have occurred at a particular location in Texas concurrently with drought in other regions of the state. Conversely, all or most of the state have experienced severe drought or flood conditions at the same time. Droughts have sometimes been ended by major floods.

Bomar (2017) classifies floods as flash versus area. Flash floods are caused by intense thunderstorms that unleash massive amounts of rainfall during short periods of time over a concentrated area. Large volumes of rainfall runoff from relatively small watersheds quickly increase flows in streams with little warning or response time. Conversely, area floods are caused by widespread, prolonged rain with

slower rising stream stages that cause damages but allow warning time facilitating minimization of loss of life.

Burnett (2008) describes 118 significant floods that occurred in Texas between 1819 and 2006. The year 2015 was one of the wettest on record with multiple floods. The ASCE Texas Section explored 2017 Hurricane Harvey flooding and lessons learned therefrom (ASCE 2018). Tropical Storm Imelda flooding in 2019 and several storms during 2020 were also very costly.

A drought is an extended period during which water availability is significantly below normal. Drought is a normal, recurring feature of climate. Drought is a temporary aberration, in contrast to aridity, which is a permanent characteristic of climate in regions of low precipitation. The beginning and end of a drought and its geographical coverage are typically much more difficult to delineate than for a flood. Droughts may last from several months to several years. A drought may be limited to a local region or be statewide, national, or international in extent.

Drought is defined from various perspectives. Meteorological drought refers to a period of below-normal precipitation and above-normal temperature and evaporation rates. Hydrologic drought refers to below-normal streamflow and reservoir storage. Agricultural drought refers to conditions affecting agriculture, such as soil moisture and crop yield. The economic cost of drought is dependent on economic development as well as meteorologic and hydrologic drought severity. More recent droughts in Texas were more economically costly than the 1950 to 1957 drought due to population and economic growth that has occurred since 1957.

The hydrologically most severe drought since before 1900 for much of the state began gradually in 1950 and ended in April 1957 with one of the largest floods on record. The 2010 to 2014 drought is comparable in hydrologic severity to the 1950 to 1957 drought in some areas of the state (Winters 2013). For more than half of Texas, 2011 had the lowest annual precipitation since the beginning of official precipitation records in 1895 (Nielsen-Gammon 2012). Major droughts in the 1910s and 1930s also affected large areas of Texas. The historic Dust Bowl drought of the 1930s, with its severe impact on farmers, centered on the Texas and Oklahoma panhandles and extended into areas of New Mexico, Colorado, and Kansas.

Reservoir storage drawdowns are employed in Chapters 6 and 7 to measure drought severity. Observed storage levels are explored in Chapter 6. Hydrologic period-of-analysis reservoir storage volumes

generated in WRAP/WAM simulations are presented in Chapter 7 as indicators of drought characteristics.

Multiple-Year Cycles

El Nino-Southern Oscillation (ENSO) is the best-known of the several cyclic weather phenomena resulting from connections between oceanic and weather systems. ENSO cycles of about two to seven years have occurred during at least the last several hundred years. El Nino and La Nina are the warm and cool phases of the recurring ENSO climate pattern across the tropical Pacific Ocean. An irregular pattern occurs of shifts in ocean surface temperature and disruptions in wind and rainfall patterns across the tropics. The temporary changes in sea surface temperatures and accompanying changes in winds and atmospheric circulation cause temporary changes in weather patterns in various regions of the world. In response to ENSO conditions, particular regions of the planet may experience unusually warm or cold winters. Droughts or torrential rains may be more likely to occur within different phases of the ENSO cycle.

The ENSO cycle tends to affect weather in the various regions of Texas in somewhat predictable ways. Winters and summers tend to be cooler or warmer, hurricanes more or less likely, and rainfall may increase or decrease during the occurrence of various ENSO conditions. However, measuring ENSO conditions or cyclic phases and predicting effects on temperature, precipitation, and weather patterns are not precise and involve significant uncertainties.

Observed metrics descriptive of ENSO or other climatic cycles can potentially enhance forecasts of hydrologic conditions useful in various water management applications. For example, Wei and Watkins (2011) related naturalized streamflows from the TCEQ WAM dataset for the Colorado River Basin to various indices to improve probabilistic hydrologic forecasts that could possibly be useful to the LCRA in operating the Highland Lakes System of Texas during drought. Probabilistic regression analyses were performed to relate WAM naturalized stream flows to autoregressive hydrologic persistence, Pacific Ocean water surface temperature patterns associated with ENSO and the Pacific Decadal Oscillation, and indices for other climatic cycles that could be incorporated into seasonal streamflow forecasts.

The CRM features of the WRAP modeling system introduced in Chapter 4 are designed for forecasting water availability over the next several months conditioned upon preceding reservoir storage contents. WRAP CRM options include regressing naturalized streamflows with preceding reservoir storage contents. Bista (2015) used the LCRA reservoir system as a case study to investigate various WRAP/WAM modeling capabilities for supporting drought management. Incorporation of ENSO indices in WRAP CRM was investigated but found to contribute little, if any, additional utility to CRM forecasting capabilities.

Long-Term Climate Change Associated with Global Warming

In addition to daily, seasonal, annual, ENSO, and other cycles, global climate has undergone slow but continuous change throughout earth's history. Scientists have detected warming and cooling cycles spanning thousands of years. Effects of human activity on long-term climate change through increases in concentrations of carbon dioxide and other trace gauges, known as the greenhouse effect, have been a major issue of scientific research and political debate over the last several decades (Kadir et al. 2018, Intergovernmental Panel on Climate Change 2023). The literature exploring climate change due to global warming and the effects of humans on climate change is massive.

The year 2023 was the hottest on record for Texas and regions throughout the world. Many previous historical record highs of daily high temperature measurements were exceeded globally and in Texas during 2023.

Temperature is the variable most directly affected by global warming. Precipitation and evaporation are the climatic variables that most directly affect water management. Evaporation rates generally increase with increases in temperature. Precipitation is different. Various aspects of global warming may either increase or decrease precipitation in a particular region or have a negligible or no net effect on precipitation. Effects on precipitation may vary between intense high precipitation events, drought conditions, and periods of average or median precipitation.

Bomar (2017) notes that the continuous great fluctuations in Texas weather render any slow long-term changes in climate attributable to global warming very difficult to detect and measure. Cook et al. (2015,

2019) and others have predicted that weather will be more variable and droughts likely more severe in the American Southwest and Central Plains, including Texas, in the future due to long-term climate change. Nielsen-Gammon et al. (2020) assess future impacts and management strategies associated with droughts in Texas during the latter half of the 21st century that may be more severe than those experienced during the last hundred years or perhaps last multiple hundreds of years.

Many researchers throughout the world over the last several decades have employed the general strategy of combining complex computer models simulating global circulation and climate with other complex computer models simulating watershed hydrology and water management to investigate the effects of scenarios of future climate change on hydrology and water management. Two such investigations of rivers and reservoirs in Texas are noted as follows.

Wurbs et al. (2005) combined the WRAP/WAM modeling system described in Chapter 4 and the Soil and Water Assessment Tool (SWAT) watershed model (SWAT 2023) with precipitation and evaporation data from a global circulation model representing selected future climate scenarios to investigate possible impacts of climate change on water supply capabilities of river and reservoir systems in the Brazos and San Jacinto River Basins of Texas. This investigation was sponsored by the National Institute for Global Environmental Change of the US Department of Energy. Simulated effects of global warming on hydrology and water availability varied between regions within the two adjacent river basins. Results were inconclusive. All components of the modeling strategy were very approximate, but the global circulation model and associated scenarios of increases in greenhouse gases were considered to reflect much greater uncertainty than the river basin hydrology and water management components.

Shao et al. (2023) developed an expanded modeling framework combining the WRAP/WAM modeling system and the Distributed Soil Vegetation Model (Wigmosta et al. 1994) for simulating watershed hydrology under predicted future climate using downscaled versions of climatic data generated in the Coupled Model Intercomparison Project Phase 6 (Eyring et al. 2016) global circulation models to statistically assess effects on reservoir firm yields of future climate change resulting from global warming. The National Science Foundation, Pacific Northwest National Laboratory, TWRI, and TWDB sponsored this research. The effects on firm yields of six reservoirs in the upper Trinity

River Basin of Texas were modeled. Firm yields were found to generally decrease with global warming due primarily to increases in evaporation. Shao et al. (2023) also provide a literature review with a lengthy list of references dealing with modeling techniques for predicting future impacts of global warming on hydrology and water management.

Precipitation and Reservoir Evaporation Databases

The Texas Water Development Board maintains annually updated datasets of monthly precipitation rates and monthly reservoir surface evaporation rates for 92 one-degree latitude by one-degree longitude quadrangles composing a grid that encompasses the state. The precipitation data date back to January 1940. The reservoir evaporation database began in January 1954. The monthly quantities are in units of inches each month. The TWDB database is accessed through the website referenced in Table 4-1 of the preceding chapter. Methods employed in compiling the database are explained at the website.

TWDB obtains the precipitation data from the NWS gauge network. The number of gauges for which NWS data has been included in the TWDB compilation has increased over time, with 3,959 gauges used in 2016. TWDB hydrologists employ a Thiessen polygon network to develop area-weighted means of monthly precipitation depths for each of the 92 quadrangles for each month from January 1940 through the most recent preceding year.

The TWDB dataset of monthly evaporation rates includes reservoir surface net evaporation less precipitation depths as well as the gross evaporation depths discussed in this chapter. The gross and net monthly evaporation data extends from January 1954 and is updated annually. The monthly evaporation and precipitation datasets employ the same grid of 92 one-degree latitude by one-degree longitude quadrangles. TWDB obtains lake evaporation data from standard pan measurements at monitoring stations maintained by NWS and other cooperators that include reservoir operators, government agencies, and research institutes. The number of gauges used in the TWDB compilation has varied, ranging between 78 and 102 evaporation stations during each year since 1999.

The reservoir evaporation depths are estimated based on measurements from standard evaporation pans and lake/pan multiplier coefficients accessible at the TWDB website that vary over the 12 months of the year and with location. TWDB staff use a Thiessen polygon network to develop area-weighted means of monthly evaporation depths for each of the 92 quadrangles for each month from January 1954 through the most recent preceding year.

TWDB is creating during 2022 to 2024 an additional database of daily evaporation and precipitation depths covering January 1981 to the present for 188 major reservoirs. This new TWDB database will also be conveniently accessible online to the water management community.

Analyses of the Monthly and Annual Precipitation and Evaporation Data

The WRAP modeling system introduced in Chapter 4 includes features for organizing, managing, and performing statistical analyses, and displaying data from the TWDB monthly precipitation and reservoir evaporation database (Wurbs 2019b, 2021b). Statistics are computed and time series plots are prepared for each of the 92 quadrangles and area-weighted statewide average precipitation and reservoir evaporation rates. Monthly quantities, annual totals, and annual series of the minimum and maximum monthly value each year or moving averages for any specified number of months may be computed and plotted. Methods employed to compute the precipitation and evaporation statistics presented in this chapter are explained in detail by Wurbs (2019b). Statistical tabulations were compiled with the WRAP hydrologic time series analysis program HYD. All time series plots throughout this book, including those in this chapter, were prepared by the author using HEC-DSSVue.

The 92 quadrangles that encompass Texas are delineated in Figures 5-1 and 5-2, with each grid cell representing a quadrangle. Land areas encompassed by the 1-degree quadrangles vary slightly due to the curvature of the earth but are each about 4,000 sq. mi. The quantity in the top of each grid cell in Figure 5-1 is the 1940 to 2022 mean annual precipitation rate in inches/year for that quadrangle. The bottom number is the linear trend slope of annual precipitation expressed as a percentage of the mean annual precipitation (top number). The quantity in the top of each grid cell in Figure 5-2 is the 1954 to 2022 mean annual reservoir evaporation rate in inches/year for that quadrangle. The bottom number is the linear trend slope of annual evaporation expressed as a percentage of the mean annual

			16.7	17.7	18.6	20.9	24.0						
			-0.098	-0.270	-0.169	0.039	0.030						
			16.5	18.2	20.2	23.8	27.5						
			-0.357	-0.104	-0.096	-0.025	-0.142						
			16.5	18.1	20.5	22.0	25.9	30.8					
			0.024	0.037	-0.025	0.094	0.057	0.040					
			15.9	17.9	22.5	22.8	25.4	29.8	34.8	41.0	46.3	49.2	51.3
			-0.066	0.022	-0.243	0.133	0.087	0.126	0.171	0.124	-0.053	0.068	0.047
			15.3	17.0	20.5	22.6	26.3	30.2	33.9	38.9	43.8	47.9	50.9
			-0.199	-0.193	-0.185	0.015	0.056	0.131	0.122	0.122	0.023	0.081	0.085
10.8	14.3	14.6	11.6	13.4	18.0	21.3	24.7	28.9	33.1	38.6	44.3	49.4	53.7
-0.149	-0.185	-0.166	0.058	-0.069	-0.174	0.077	0.177	0.091	0.227	0.109	0.074	0.090	0.106
9.28	15.4	13.7	14.9	13.6	18.7	22.4	25.2	30.2	33.1	39.8	46.8	54.6	57.2
0.036	-0.312	-0.019	-0.088	0.069	-0.267	0.034	0.006	0.005	0.165	0.012	0.017	0.136	0.212
		19.7	14.4	11.7	16.8	24.3	26.5	31.1	34.3	41.6	47.6	49.3	57.4
		-2.441	-1.055	-0.116	-0.004	0.002	0.133	0.042	0.060	0.044	0.223	0.356	0.130
						20.6	21.9	25.3	34.9	39.8	44.6		
						-0.148	-0.088	0.004	-0.094	0.028	0.189		
							20.3	23.9	29.4	34.8			
							-0.010	0.040	0.111	-0.020			
							17.9	21.8	25.9				
							0.273	0.071	0.013				
									26.4				
									0.152				

Figure 5-1. Quadrangle 1940 to 2022 mean annual precipitation in inches (top number) and annual precipitation linear trend slope (bottom number) as a percentage of annual mean.

evaporation (top number). These metrics are discussed in the next two sections of this chapter.

Statewide averages of the monthly precipitation and evaporation are plotted in Figures 5-3 and 5-4. Statewide average annual precipitation and evaporation are compared in Figure 5-5. The statewide average 1940 to 2022 mean annual precipitation is 28.03 in. The statewide average 1954 to 2022 mean annual reservoir evaporation depth is 59.43 in. Comparisons of precipitation and reservoir evaporation are relevant to discussions of reservoir management issues found later in this book. Reservoir surface evaporation significantly exceeds precipitation throughout most of Texas. The differences are dramatic in West Texas.

Precipitation is somewhat seasonal though the timing of rainfall also reflects great annual variability. Reservoir evaporation depths have a very distinct seasonal pattern. Seasonality characteristics are illustrated by Figures 5-3 and 5-4 and Tables 5-1 and 5-2.

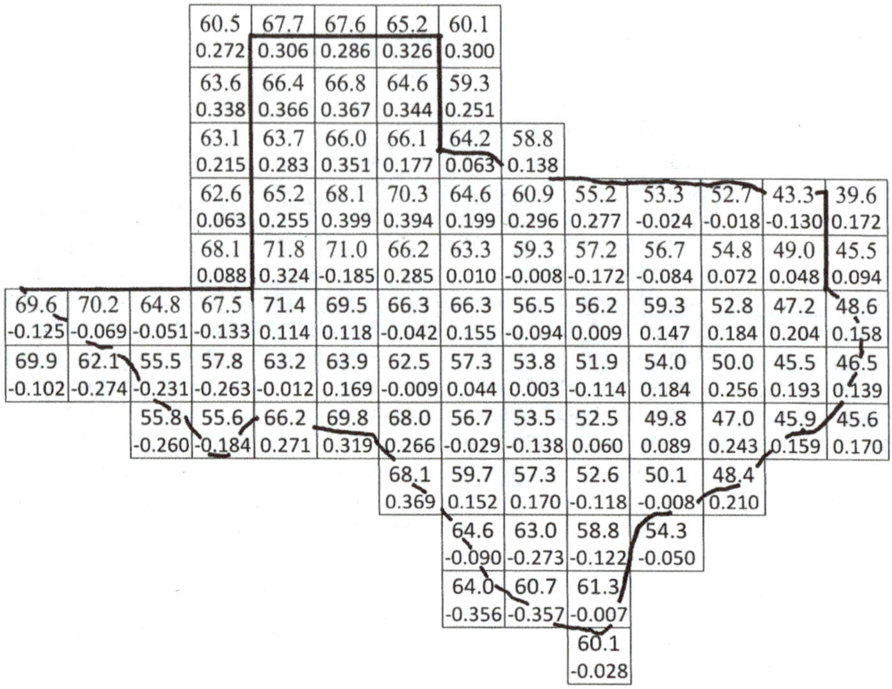

Figure 5-2. Quadrangle 1954 to 2022 mean annual reservoir evaporation in inches (top number) and annual evaporation trend slope (bottom number) as a percentage of annual mean.

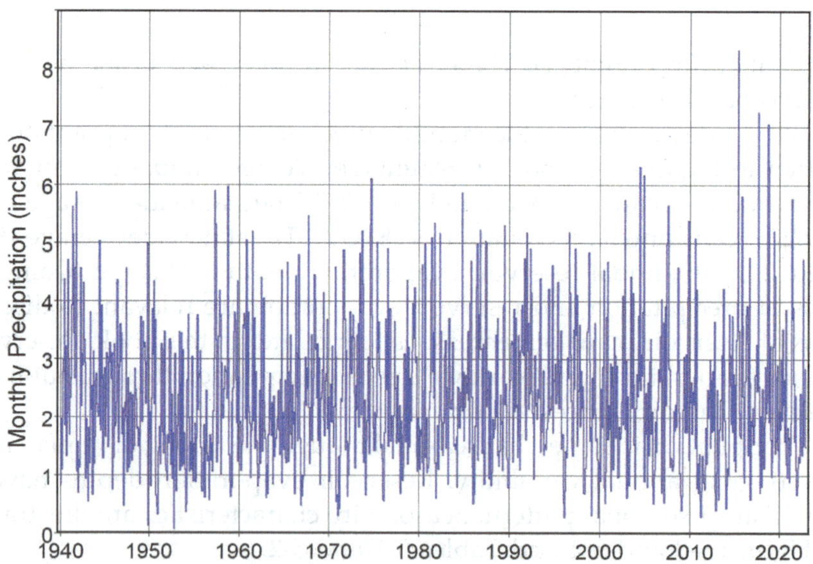

Figure 5-3. Statewide average of 1940 to 2022 monthly precipitation depths.

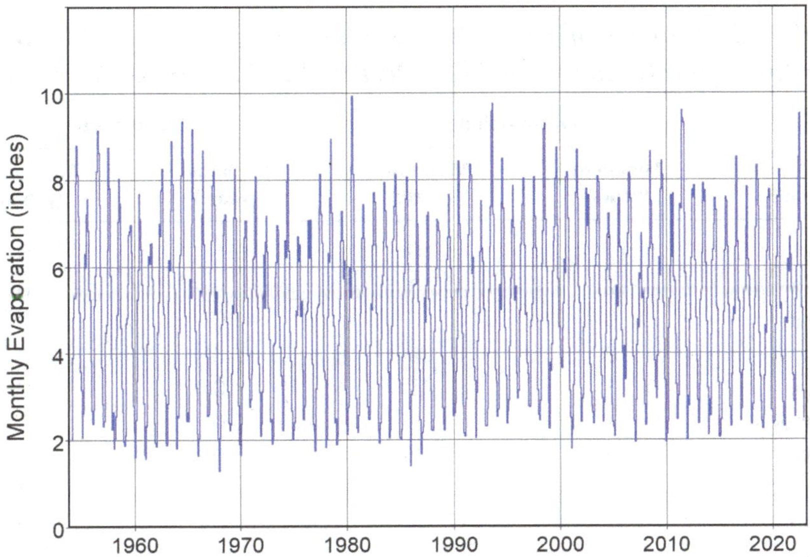

Figure 5-4. Statewide average of 1954 to 2022 monthly reservoir evaporation depths.

Table 5-1. Statewide Average Monthly and Annual Precipitation and Evaporation.

	Precipitation		Evaporation	
Month	**1940–2022 (inches)**	**1954–2022 (inches)**	**1954–2022 (inches)**	**1954–2022 (% Precip.)**
Jan	1.63	1.61	2.36	147
Feb	1.67	1.66	2.80	169
Mar	1.80	1.80	4.44	247
Apr	2.22	2.15	5.44	253
May	3.32	3.27	5.79	177
June	3.00	3.04	7.20	237
Jul	2.44	2.43	7.86	323
Aug	2.58	2.59	7.30	282
Sep	3.10	3.19	5.71	179
Oct	2.66	2.72	4.73	174
Nov	1.87	1.88	3.30	176
Dec	1.75	1.74	2.51	144
Annual	28.03	28.08	59.43	212

Table 5-2. Number of Years during which Minimum and Maximum Monthly Quantities for the Year Occurred in Each of the 12 Months.

	Precipitation		Evaporation	
Month	**Minimum (years)**	**Maximum (years)**	**Minimum (years)**	**Maximum (years)**
Jan	17	0	35	0
Feb	12	0	6	0
Mar	6	0	0	0
Apr	6	5	0	0
May	1	23	0	0
June	1	8	0	9
Jul	2	7	0	43
Aug	1	9	0	17
Sep	1	14	0	0
Oct	6	13	0	0
Nov	13	2	1	0
Dec	17	2	27	0
Total	83	83	69	69

Statewide average 1940 to 2022 and 1954 to 2022 mean monthly precipitation for each of the 12 months of the year are tabulated in Table 5-1 along with annual means. Statewide average 1954 to 2022 mean monthly and annual reservoir evaporation in inches and as a percentage of the precipitation for each of the 12 months of the year are also included in Table 5-1. The mean annual precipitation was 28.03 in. during 1940 to 2022. The 1954 to 2022 mean annual evaporation of 59.43 in. is 212% of the 1954 to 2022 mean annual precipitation of 28.08 in.

Table 5-2 shows the number of years during 1940 to 2022 that the minimum and maximum monthly rainfall depth occurred in each of the 12 months of the year. The likelihood of very high precipitation is highest in May, and the driest season of the year tends to be November through February. The maximum monthly precipitation occurred in May during 23 of the 83 years.

Table 5-2 also shows the number of years during 1954 to 2022 that the minimum and maximum monthly reservoir evaporation depth occurred in each of the 12 months of the year. The maximum evaporation rates occurred during June, July, or August during all 69 years of the 1954 to 2022 analysis period. The minimum evaporation was in January or December in 35 and 27, respectively, of the 69 years.

Variability and Stationarity

Variability and stationarity of precipitation, reservoir evaporation, and river flows are explored in later sections of this chapter. Several relevant general considerations regarding variability and stationarity are noted in the following paragraphs.

Temporal and spatial variability is decreased by averaging over larger time intervals or spatial areas. This chapter focuses on precipitation and evaporation depths in inches and streamflow rates in cubic feet per second (ft^3/s). These variables may fluctuate from instant to instant. This continuous variability is dampened by averaging or summing over a fixed period. Monthly precipitation and evaporation depths fluctuate much more than annual totals. Likewise, average precipitation and evaporation depths over a 4,000 sq. mi quadrangle fluctuate more than these same quantities averaged over the entire state of Texas. The daily streamflow rates discussed later are quantities averaged across a stream cross-section area at a gauge site and averaged over a day. Instantaneous flow rates vary more than daily averages, which vary more than monthly or annual averages.

Stationarity refers to long-term homogeneity over time with no permanent changes or trends. Insights regarding stationarity as well as variability are provided by time series plots of precipitation, evaporation, observed river flows, and naturalized river flows comprised of observed flows adjusted to remove the effects of water development. Least-squares linear regression provides a simple analysis tool to complement graphical visual analysis of time series datasets in assessing stationarity or lack thereof.

Linear regression is described in numerous numerical methods and statistics textbooks. A horizontal trend line with slope at or near zero indicates no trend or long-term change. A positive linear trend slope suggests an increase in precipitation, evaporation, streamflow, or other time series variable. A negative regression slope indicates a long-term decrease in the time series variable over the analysis period. Linearity of the trend line is a major simplifying approximation that reduces the validity of the results. However, linear regression trend analysis does provide useful general insight regarding stationarity. Standard least-squares regression analyses of precipitation and evaporation data applied as described by Wurbs (2019b, 2021b) yield results discussed in the next two sections.

The length of the period of analysis is a key consideration in interpreting either graphical time series plots or regression trend analysis metrics. The regression slope may vary greatly with different periods-of-record or analysis periods. For example, means and linear regression slopes for statewide average annual precipitation and reservoir evaporation are compared in Table 5-3 for alternative analysis periods. Widespread extreme drought conditions were experienced throughout Texas during 1950 to 1957, which followed a wetter than normal 1940s decade and ended with massive widespread flooding in April–May 1957.

No permanent changes or multiple-decade long trends in precipitation are evident in the 1940 to 2022 time series plots of monthly and annual depths. Any changes in monthly or annual precipitation resulting from global warming or other phenomena are hidden by the great variability to the extent of being undetectable by the limited analyses reflected in this book. The two 1940 to 2022 time series datasets consisting of the maximum and minimum monthly precipitation in any month of each year may have changed as discussed in the next section of this chapter.

Increases in evaporation rates are expected with global warming (Cook et al. 2019, Nielsen-Gammon et al. 2020, Intergovernmental Panel on Climate Change 2023, Shao et al. 2023). Increases in evaporation in Texas would be consistent with the published literature on global

Table 5-3. Annual Means and Regression Slopes for Different Analysis Periods.

	Annual Precipitation		Annual Evaporation	
Analysis Period	**Mean**	**Regression Slope**	**Mean**	**Regression Slope**
	(in./yr)	(in./yr/yr)	(in./yr)	(in./yr/yr)
1940–2022	28.034	0.00555	–	–
1954–2022	28.014	0.01071	59.435	0.05322
1957–2022	28.463	−0.02396	59.194	0.08351
1940–1999	28.199	0.03715	–	–
1954–1999	28.321	0.08211	58.905	0.07780
2000–2022	27.603	0.00728	60.540	−0.38101

warming. Based on the data analysis discussed later in this chapter, evaporation rates appear to have increased. However, any long-term increases in evaporation in Texas during 1954 to 2022 are obscured by the variability of the evaporation data and thus are difficult to accurately measure.

Significant nonstationarities are evident in observed river flows at some of the gauge sites discussed later in this chapter. Observed stream flows from watersheds reflecting different levels of development and naturalized flows computed by adjusting observed flows have been investigated in application of the statewide WAM system. Precipitation and reservoir evaporation rates represent climatic conditions. Permanent changes in precipitation and evaporation rates are reflective of climate change. The nonstationarities in daily and annual river flow characteristics explored later in the chapter result primarily, if not essentially entirely, from factors other than climate change. Water resources development, control, and use have resulted in significant nonstationarities in river flows.

Precipitation

Precipitation in Texas is mainly rainfall. Snow and sleet occur infrequently. Bomar (2017) cites a twentieth-century statewide Texas mean annual precipitation of 27.92 in. determined by the NWS using NWS data and spatial averaging methods. The small difference between the NWS 1900 to 1999 mean of 27.92 in. and the 1940 to 2022 mean of 28.03 in. estimated based on the database described in the preceding section can be attributed to different periods of analysis and spatial averaging methods and variations in the number of rain gauges employed.

Precipitation in Texas is extremely variable both spatially and temporally. The extreme spatial variability is illustrated by the quantities in Figure 5-1. The lowest annual precipitation for a quadrangle enclosed completely in Texas is 11.6 in./year, which is 41.4% of the statewide mean of 28.03 in./year. This dry quadrangle is in West Texas. On the other extreme, a quadrangle in East Texas has an annual precipitation of 54.6 in./year, which is 195% of the statewide mean of 28.03 in./year. Isohyets of mean annual precipitation extend generally north and south. Traveling in a straight line directly from the city of El Paso in extreme West Texas eastward to the Sabine River which serves as the border between Texas and Louisiana, the mean annual precipitation increases about 1 in. per 15 mi.

Statewide mean annual January through December precipitation during each year of 1940 through 2022 is plotted in Figures 1-3 and 5-5. The 10 largest values in inches for statewide mean annual precipitation during 1940 to 2022 are 40.57 (1941), 39.95 (2015), 39.14 (2004), 37.06 (1991), 36.29 (1973), 35.80 (1957), 35.39 (2007), 34.45 (1997), 33.95 (2018), and 33.76 (1949). The 10 smallest values for statewide mean annual precipitation in inches during 1940 to 2022 are 13.60 (2011), 16.68 (1956), 19.06 (1954), 19.90 (1963), 21.85 (1999), 21.91 (1948), 22.28 (2005), 22.40 (1943), and 22.64 (1988 and 2022).

Statewide averages of 1940 to 2022 monthly precipitation depths are plotted in Figure 5-3. Mean monthly precipitation in each of the 12 months of the year is shown in Table 5-1. Metrics in Table 5-2 demonstrate the likelihood of the minimum and maximum precipitation depths in any year occurring in a specific month of the year.

The statewide maximum and minimum monthly precipitation depths for each year of 1940 to 2022 are plotted in Figure 5-6. The two plots in Figure 5-6 show the minimum (blue circles) and maximum (red squares) precipitation depth occurring in any month during each of

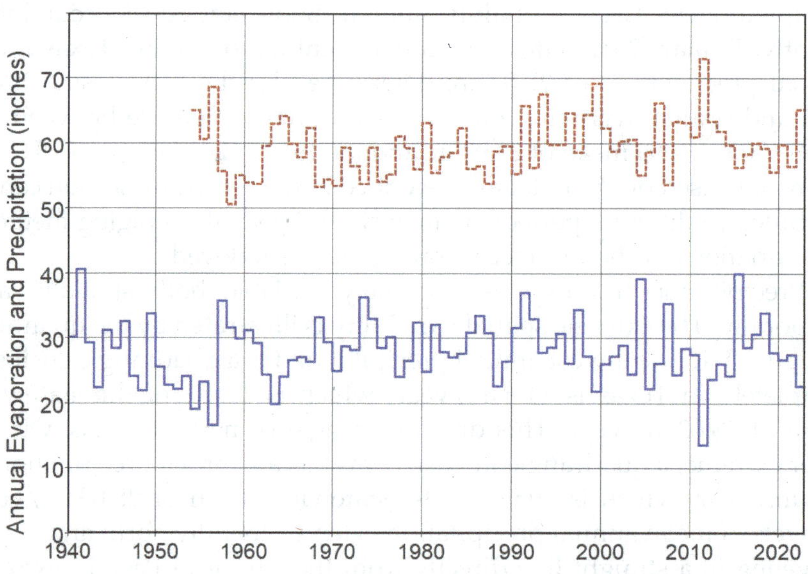

Figure 5-5. Statewide average of 1954 to 2022 mean annual evaporation (red dashed line) and 1940 to 2022 precipitation (blue solid line).

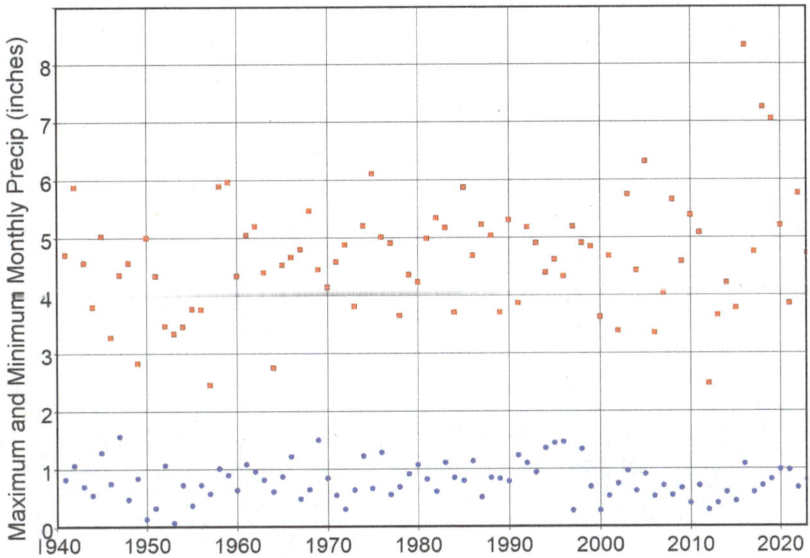

Figure 5-6. Statewide maximum (red squares) and minimum (blue circles) monthly precipitation for each year of 1940 to 2022.

the 83 years. The five smallest monthly precipitation depths in inches in Figure 5-6 are 0.071 (1952), 0.14 (1949), 0.28 (1996), and 0.29 (1999 and 2011). The seven largest monthly precipitation depths in inches in Figure 5-6 are 8.3 (2015), 7.2 (2017), 7.1 (2018), 6.3 (2004), 6.1 (1974), 6.0 (1958), and 5.9 (1957). The minimum monthly precipitation in each of the 83 years appears to be stationary or close thereto. The maximum monthly quantities during 2015, 2017, and 2018 are extremely high, suggesting a possible recent increase in extreme rainfall events.

The mean and regression slope for the maximum monthly precipitation depth in each year plotted in Figure 5-6 are 4.64 in./month and 0.01057 in./month per year. The slopes for the annual monthly maxima for 57 of the quadrangles are positive. Regression slopes are negative for the other 26 quadrangles. The mean and regression slope for the annual monthly minimum precipitation depths are 0.790 in./month and 0.000600 in./month per year. The slopes for the minimum monthly precipitation each year for 24 quadrangles are positive and for 59 quadrangles are negative.

The driest and wettest quadrangles located entirely within Texas are in West and East Texas, respectively, and have mean annual precipitation depths of 11.6 and 54.6 in. Monthly precipitation depths for these driest and wettest quadrangles are plotted in Figures 5-7 and 5-8. Annual

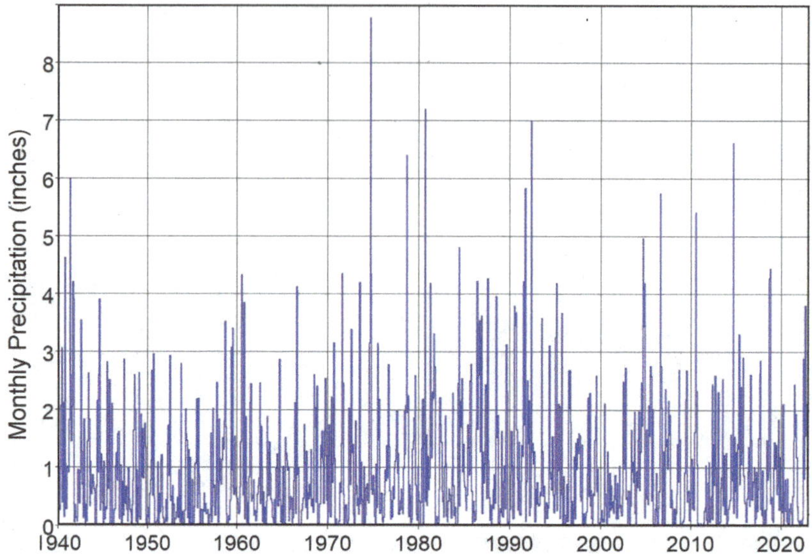

Figure 5-7. Monthly precipitation for West Texas quadrangle in Figure 5-1 with annual mean of 11.6 in.

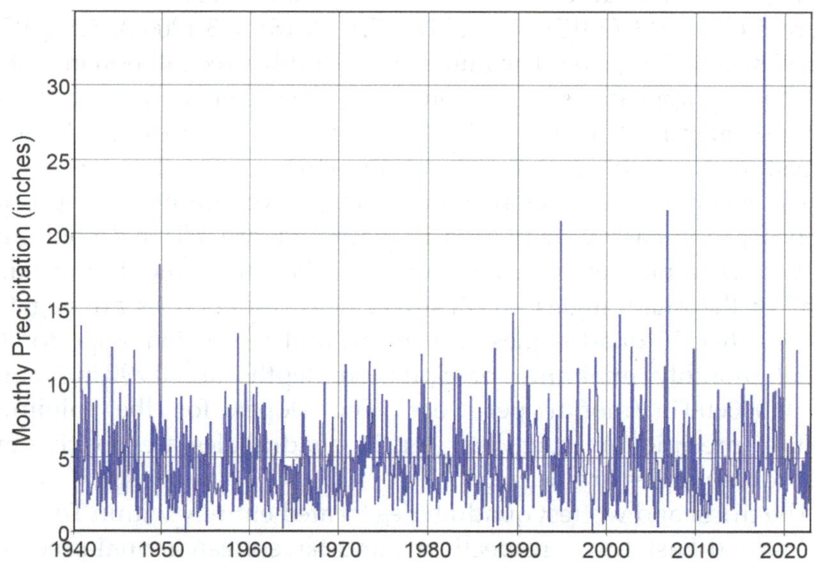

Figure 5-8. Monthly precipitation for East Texas quadrangle in Figure 5-1 with annual mean of 54.6 in.

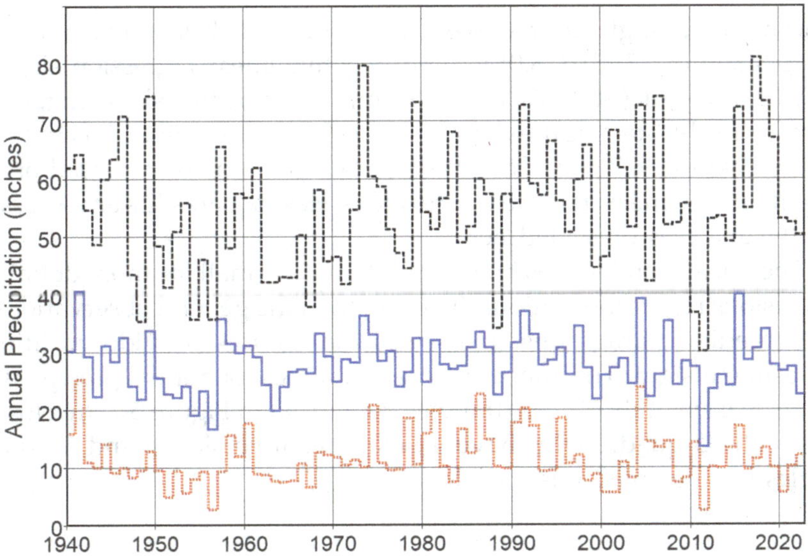

Figure 5-9. Annual precipitation for wettest quadrangle in East Texas (black dashes), statewide average (blue solid line), and driest quadrangle in West Texas (red dots).

precipitation depths for these driest (red dots) and wettest (black dashes) quadrangles are compared to the statewide average (blue solid line) in Figure 5-9.

Monthly and annual precipitation quantities appear stationary in the 1940 to 2022 time series plots. No long-term permanent changes in precipitation characteristics reflected in the monthly and annual totals are evident. Any changes in monthly or annual precipitation resulting from global warming or other phenomena are hidden by the great continuous variability. As previously noted, the annual time series of the maximum monthly precipitation depth in each year exhibits a possible recent trend of increasing precipitation extremes. A period that included major storm events producing floods in 2015, 2017, and 2018 followed drought conditions during the period 2011 to 2014, which included the 2011 minimum rainfall on record for much of Texas.

The bottom number in each grid cell of Figure 5-1 is the linear trend slope of annual precipitation for that quadrangle expressed as a percentage of the mean annual precipitation (top number). For example, a quadrangle in Central Texas has a mean annual precipitation of 21.322 in. (21.3 in.) and positive linear regression slope of 0.01648 in./year, which is 0.077% of 21.322 in., implying an increase in mean (statistical expectation value) annual precipitation from 20.63 in. in 1940 to 22.00 in. in 2022. The adjacent quadrangle to the west in Figure 5-1 has a mean annual precipitation of 17.981 (18.0) in. and negative linear regression slope of −0.03120 in./year (−0.174% of 17.981 in.). The negative slope suggests a decrease in the statistical expected value (average) annual precipitation during 1940 to 1922 over this 4,000 sq. mi portion of Texas based on the simplifying premises reflected in textbook least-squares linear regression and the analyses presented here.

The linear trend slopes in Figure 5-1 are too small for more definitive conclusions regarding stationarity other than the general observations of the preceding paragraphs. However, the metrics provide additional interesting insight. The annual regression slopes for annual precipitation depths in each of the 92 quadrangles delineated in Figure 5-1 are close to zero with 33 quadrangles having small negative slopes and 59 quadrangles having small positive slopes.

Reservoir Surface Evaporation

Well over half of the rainfall that reaches the land surfaces of the earth is returned to the atmosphere through evapotranspiration. Raindrops often completely evaporate before reaching the ground. Evaporation is the transformation of water from its liquid to its vapor phase. Evaporation occurs from water bodies, soils, and various wet surfaces. Transpiration is evaporation from plants. Plants remove moisture from the soil and release it to the air as water vapor. Evapotranspiration is the combined total of evaporation and transpiration.

Key meteorological factors that affect evaporation rates include temperature, vapor-pressure gradient, and wind. Solar radiation provides the required energy. Evaporation increases with increased water temperature. As the air becomes saturated, evaporation will slow and stop. Thus, wind plays a key role in removing water vapor from the zone near the liquid water surface.

Shallow water bodies are more rapidly heated than deeper reservoirs and thus incur greater evaporation rates. The rate of evaporation from a saturated soil surface is about the same as from an adjacent lake surface at the same temperature. Evaporation decreases as the soil dries. This chapter focuses on evaporation from the surfaces of major reservoirs. However, reservoir surface evaporation provides an index of maximum potential evaporation rates for a saturated land area.

The 1954 to 2022 reservoir evaporation depths in inches/year of each individual quadrangle are tabulated in the upper half of each of the grid cells of Figure 5-2. Statewide averages of monthly evaporation rates are plotted in Figure 5-4. Statewide average annual evaporation (red dashed line) and precipitation depths (blue solid line) are compared in Figure 5-5. The plots in Figure 5-10 for two quadrangles reflecting high and low extremes illustrate differences in annual evaporation rates between West and East Texas.

The statewide 1954 to 2022 average reservoir evaporation rate is 59.43 in./year. Evaporation rates exceed precipitation at all locations except far East Texas. Evaporation rates greatly exceed precipitation in West Texas. The West Texas quadrangle (solid line) in Figure 5-10 has a mean annual evaporation of 71.4 in. and annual precipitation of 13.4 in., as contrasted with the East Texas quadrangle (dashed line) with an annual evaporation of 45.5 in. and annual precipitation of 54.6 in. Evaporation rates increase from east to west but not nearly as much as precipitation rates decrease. Reservoir evaporation rates exhibit a distinct annual seasonal cycle of variability as illustrated by Tables 5-1 and 5-2. Evaporation is highest in the hot summer months and much lower in cold winter months.

Trend analysis methods applied to precipitation, discussed earlier in this chapter, are also applied to evaporation rates. Linear regression

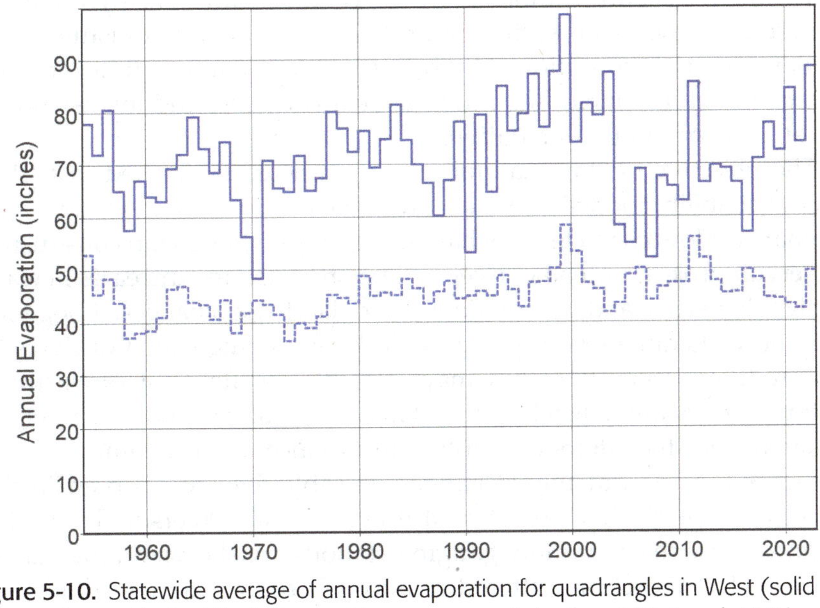

Figure 5-10. Statewide average of annual evaporation for quadrangles in West (solid line) and East (dashed line) Texas with mean annual depths of 71.4 and 45.5 in.

analysis results in a trend line slope of 0.0532 in./year for statewide annual evaporation rates, which is 0.0895% of the statewide mean of 59.43 in. The 1954 to 2022 mean in inches and linear trend line slope in percent of the mean for each of the 924,000 sq. mi individual quadrangles are shown Figure 5-2. Thirty-four quadrangles have small negative slopes. The other 58 have small positive slopes. Slopes vary significantly for different short subperiods of the overall 1954 to 2022 analysis period.

Statewide average evaporation depths and depths for many of the individual quadrangles appear to have increased, which is consistent with the global warming literature. However, continuous variability makes permanent changes difficult to definitively detect and measure. TWDB has occasionally refined data compilation procedures, such as the number and location of evaporation pans and factors for converting evaporation pan measurements to actual lake evaporation estimates, which could possibly also contribute to variations over time.

Observed Historical River Flows

Actual flows over the last 80 years or longer at stream gauge sites throughout Texas are explored in this section. Naturalized river flows discussed in the next section reflect computational adjustments to simulate natural conditions. Both actual and natural flows exhibit extreme variability including multiple-year droughts and intense floods as well as continuous fluctuations.

The objective of flow naturalized adjustments performed in WRAP water availability modeling, which are discussed later, is to remove non-stationarity. Flow characteristics have changed over time with construction of reservoir projects and other river regulation structures, increases in water supply diversions and return flows, and land use changes. Long-term streamflow alterations vary greatly with location. Regulation of rivers by dams reduces flood flows but may increase low flows at downstream locations. Flows immediately below dams are greatly affected by reservoir operations, but the effects diminish with distance downstream.

Flows recorded at the 22 gauges in Table 5-4 are adopted in this chapter to explore characteristics of flows of major rivers in Texas. The locations of these streamflow gauging stations are shown on the map of Figure 5-11. This set of scattered locations is representative of flows of

Table 5-4. Selected Streamflow Gauging Stations.

Map ID	Gauge ID	Location River and Nearest City	Record Begins	Watershed Area Total	Watershed Area Contributing	Mean Flow
			–	(square miles)		(cubic feet/second)
GP	08-3640.00	Rio Grande at El Paso	5/1899	–	29,270	681
GF	08-4613.00	Rio Grande below Falcon Dam	1/1958	–	165,000	3,107
GB	08-4750.00	Rio Grande at Brownsville	1/1934	356,000	176,000	1,541
CA	07227500	Canadian River near Amarillo	4/1938	19,445	15,376	228
CC	07228000	Canadian River near Canadian	4/1938	22,866	18,178	229
RD	07331600	Red River below Denison Dam	1/1924	39,720	31,792	4,918
RA	07335500	Red River at Arthur City, Texas	10/1936	44,445	36,517	9,011
SR	08030500	Sabine River near Ruliff	10/1924	9,329	same	8,237
NR	08033500	Neches River near Rockland	7/1904	3,636	same	2,394
NE	08041000	Neches River near Evadale	4/1921	7,951	same	6,228
TD	08057000	Trinity River at Dallas	10/1903	6,106	same	1,832
TO	08065000	Trinity River near Oakwood	10/1923	12,833	same	5,460
TR	08066500	Trinity River at Romayor	5/1924	17,186	same	8,064
WC	08068000	West Fork San Jacinto, Conroe	8/1939	828	same	504
BR	08114000	Brazos River at Richmond	10/1922	45,107	35,541	7,529
NE	08110500	Navasota River at Easterly	3/1924	968	same	428
BW	08096500	Brazos River at Waco	1/1900	29,559	19,993	2,357
CS	08147000	Colorado River near San Saba	11/1915	31,217	19,819	939
CA	08158000	Colorado River at Austin	3/1898	39,009	27,606	2,039
CC	08161000	Colorado River at Columbus	5/1916	41,640	30,237	2,920
SF	08183500	San Antonio River, Falls City	5/1925	2,113	same	489
NT	08210000	Nueces River at Three Rivers	7/1915	15,427	same	709

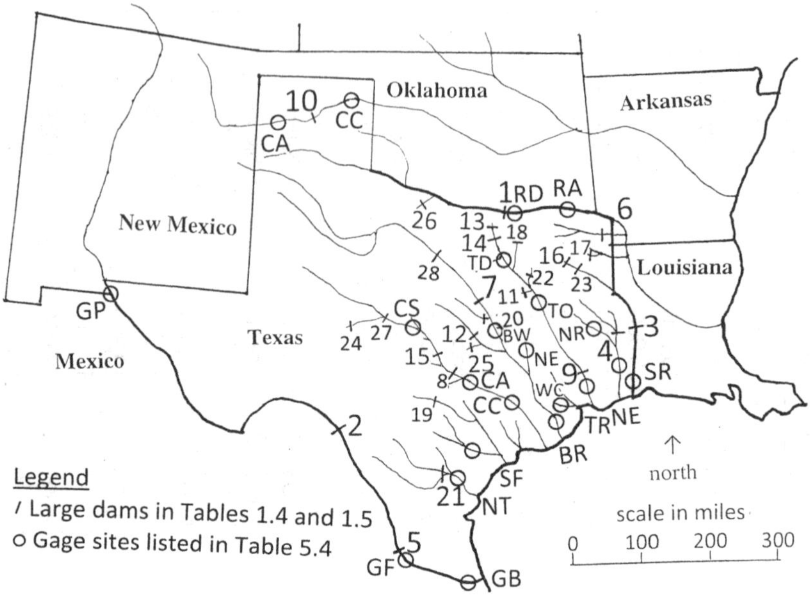

Figure 5-11. Sites of streamflow gauges and large dams.

major rivers throughout Texas. These selected gauges have long periods of record.

Figure 5-11 is identical to Figure 1-9 except for the addition of the 22 streamflow gauge sites listed in Table 5-4. The gauge site letter identifiers on the map of Figure 5-11 reference the first column of Table 5-4. The numbers labeling dam sites on the map reference the first column of Tables 1-4 and 1-5.

The NWIS maintained by the USGS includes 1,145 gauges in Texas with historical daily data. These data are accessible at the NWIS website included in Table 4-1. Streamflow gauges on the Rio Grande are maintained by the IBWC rather than the USGS. Daily flows of the Rio Grande at the first three gauge sites listed in Table 5-4 were downloaded from an IBWC website. Daily flows at all the other gauge sites were downloaded from the USGS NWIS website. The gauge number identifiers assigned by IBWC and USGS are tabulated in the second column of Table 5-4. The third column defines the gauge location in the standard USGS and IBWC format of stream name and nearest city.

USGS NWIS is updated daily. IBWC websites include a variety of information including near-current gauged stream flows that are updated often. However, historical daily flow observations at the IBWC website have not been updated in recent years.

The flow data recorded for the 19 USGS gauges downloaded from the NWIS website extend through December 2023. The daily flows at the IBWC gauges on the Rio Grande were also downloaded from the IBWC website in 2023. However, the flows extend through December 2018 for the El Paso gauge and December 2011 for the Falcon and Brownsville gauges. The beginning of the period of record for the downloaded data for each of the 22 gauges is tabulated in the fourth column of Table 5-4. The observed data is complete for the period of record indicated in Table 5-4 for all gauges, with one exception of an October 1989 through November 1996 gap of missing data for the Red River below Denison Dam. Period-of-record mean flow rates are tabulated in the last column.

The total and contributing watershed areas are tabulated in the fifth and sixth columns of Table 5-4. NWIS includes both the total watershed area above a gauge site and the portion of this watershed area that actually contributes to streamflow. Portions of the river basins in dry flat West Texas and New Mexico contribute essentially no runoff to streamflow. Therefore, the contributing area may be significantly less than the total area of the river basin.

Numerous small playa lakes found in the Texas High Plains are located within the noncontributing areas of the watersheds. Playa lakes are shallow hollows in the ground in the Southern High Plains of the United States that may contain water following rainfall events and may serve as wetlands. Water collected in the playa lakes during rainfall events evaporates and seeps into the ground without contributing runoff to streams.

HEC-DSSVue (Hydrologic Engineering Center 2009) is introduced in Chapter 4. All precipitation, evaporation, and flow data investigated in Chapter 5 are stored in DSS files managed using HEC-DSSVue. The streamflow data discussed here were downloaded by the author from NWIS using HEC-DSSVue. Likewise, the time series were plotted and arithmetic operations were performed with HEC-DSSVue. Daily mean flows in cubic feet per second are downloaded from NWIS and also expressed in units of ac-ft/day in WRAP modeling and later in this book. One ft^3/s (cfs) is equivalent to 0.50417 ac-ft/day.

With the exception of a dry riverbed with no flow, flow rates tend to fluctuate continually. Daily flow rates published by USGS in NWIS are averages over the day. Variability is dissipated with averaging over a larger time interval. Daily means are less variable than instantaneous rates. Computing monthly means as the average of daily means tends to average out flow variations. Averaging over a year further reduces variability.

Flow Characteristics of Rivers in Arid West Texas

Dramatic decreases in the flow of the Rio Grande illustrate the impacts of irrigated agriculture and large reservoirs in a dry climate. The Lower Rio Grande Valley is the dominant region of Texas for irrigated agriculture supplied by surface water. The productive agricultural economy in this dry region relies on water pumped from the Rio Grande. The river is controlled by International Amistad and Falcon, the second and fifth largest reservoirs located partially or wholly in Texas (Tables 1-4 and 1-5).

The Rio Grande Basin (Figure 2-1) encompasses 356,000 sq. mi, but much of this area is flat desert that contributes no runoff to the river. The watershed drainage area that contributes flow to the Rio Grande is about 182,000 sq. mi. The first three gauge sites listed in Table 5-4 are on the Rio Grande at El Paso, Falcon Dam, and Brownsville. Daily flows at these sites are plotted in Figures 5-12 through 5-14. Annual flows are compared in Figure 5-15 [El Paso (blue solid), Falcon Dam (black dashes), and Brownsville (red dots)].

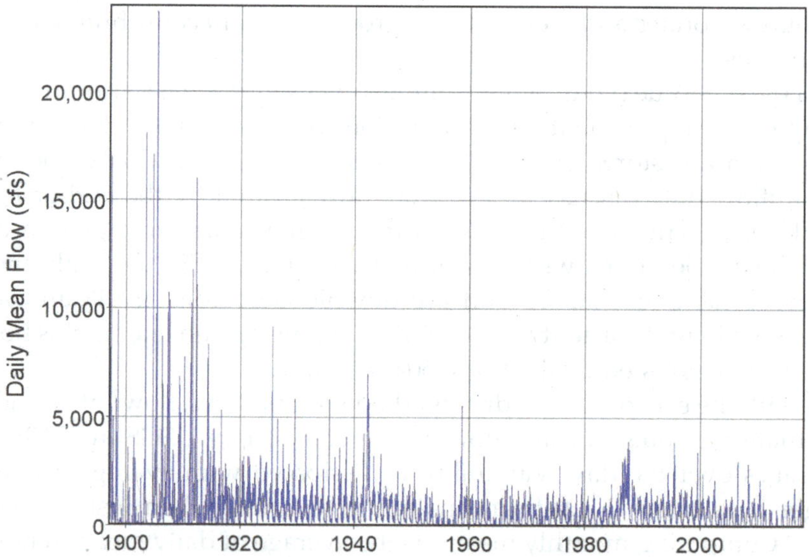

Figure 5-12. Daily flow of Rio Grande at El Paso.

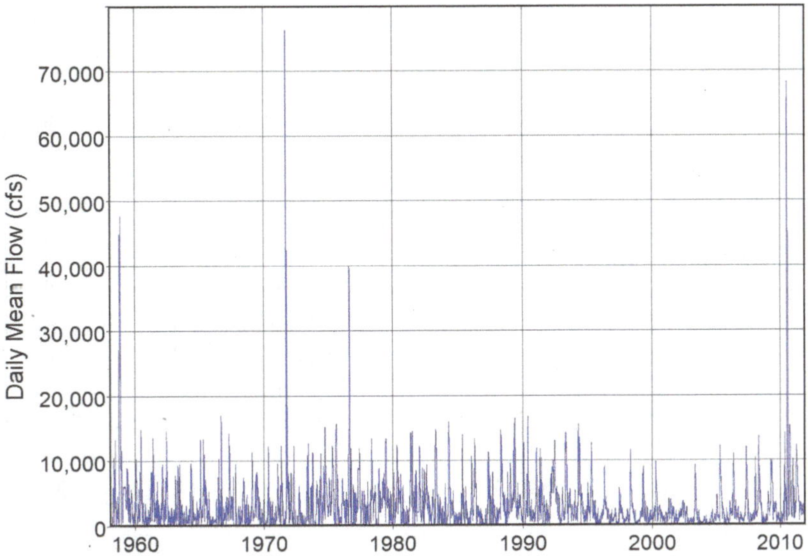

Figure 5-13. Daily flow of Rio Grande below Falcon Dam.

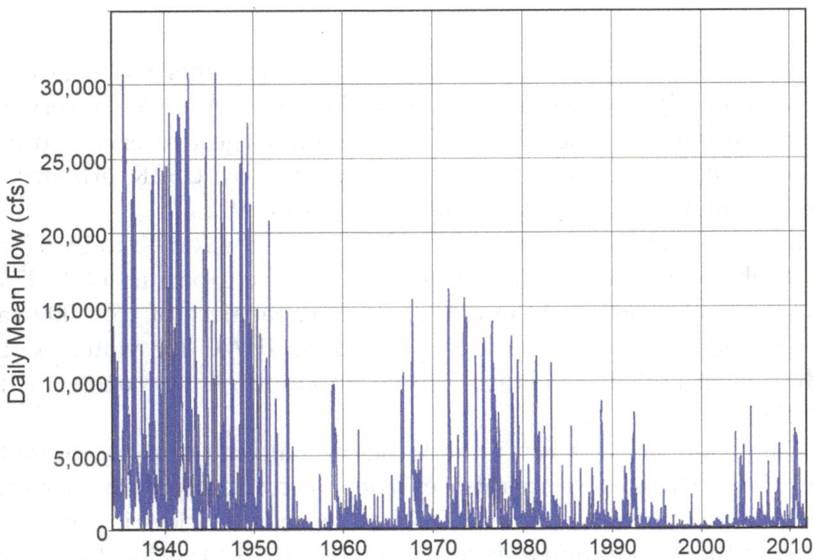

Figure 5-14. Daily flow of Rio Grande at Brownsville.

The IBWC gauging station on the Rio Grande at the city of El Paso is located 1,256 river miles above the river outlet at the Gulf of Mexico and 1.7 mi above the American Dam at El Paso. Elephant Butte Reservoir on

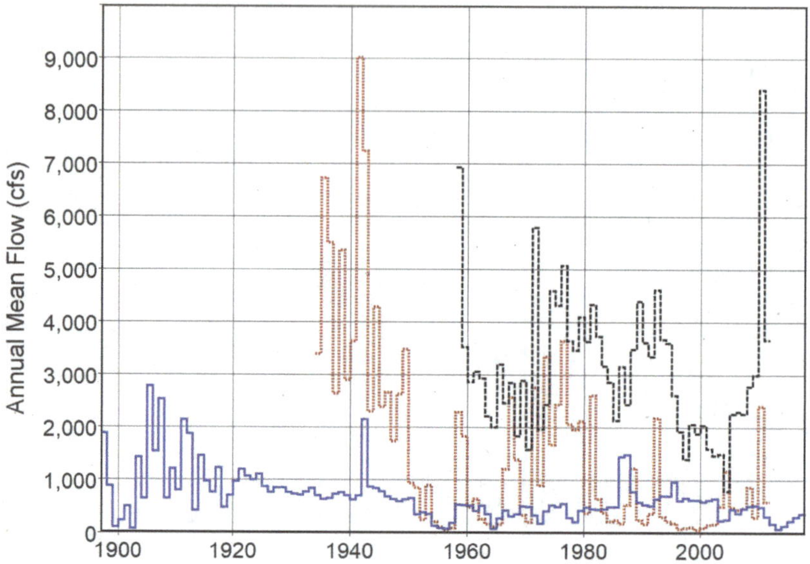

Figure 5-15. Annual flows of Rio Grande at El Paso (blue solid), Falcon Dam (black dashes), and Brownsville (red dots).

the Rio Grande 125 mi upstream of El Paso accounts for most of the reservoir storage controlling flows at El Paso. With a conservation storage capacity of 2,065,000 ac-ft, this is the largest reservoir in New Mexico. Elephant Butte Reservoir is operated by USBR primarily to supply irrigation. Initial impoundment was in 1915.

International Falcon and Amistad Dams on the Rio Grande 275 and 574 mi above its outlet have conservation storage capacities of 2,647,000 and 3,276,000 ac-ft and flood control capacities of 501,000 and 1,734,000 ac-ft. The projects are operated by IBWC for water supply, hydropower, and flood control. Initial impoundment occurred in 1953 and 1969.

Daily flows of the Rio Grande below Falcon Dam are plotted in Figure 5-13. These flows are releases from Falcon Reservoir to the river to supply downstream water needs and generate hydroelectric power. The gauge on the Rio Grande at Brownsville (Figure 5-14) is located 49 river miles above the river outlet at the Gulf of Mexico, 7 mi downstream of the international bridge between Brownsville, Texas, and Matamoros, Tamaulipas, and 226 mi below Falcon Dam. Under current conditions, unlike a century ago, the flow of the Rio Grande into the Gulf of Mexico is minimal except during flood events.

The Canadian River is another extreme case of flows decreasing dramatically over the last several decades due largely to development of irrigated agriculture. However, whereas agricultural production from surface water irrigation is concentrated in the Lower Rio Grande Valley, agriculture in the High Plains relies almost completely on groundwater.

During the 1950s to 1990s, irrigated agriculture in the High Plains from water supplied from the Ogallala Aquifer was a dominant component of total water use in Texas. During recent decades, municipal and industrial water use in larger population centers has grown rapidly concurrently with a leveling off of agricultural use. Depleting groundwater reserves has driven a shift to surface water supplies statewide. However, surface water is not a viable alternative in the Canadian River Basin of Texas.

Flows of the Canadian River at the Amarillo and Canadian gauges are plotted in Figures 5-16 and 5-17. The gauge at Amarillo is 30 mi upstream of Sanford Dam at Lake Meredith and 80 mi downstream of the New Mexico border. The gauge near the city of Canadian is 70 mi downstream of Sanford Dam and 20 mi upstream of the Oklahoma border.

The Canadian River has practically stopped flowing, and the storage content of Lake Meredith remains continuously at extremely low levels. Lake Meredith and Sanford Dam are listed in Table 1-4 as having the tenth largest storage capacity in Texas. The reservoir project is owned and operated by the Canadian River Municipal Water District

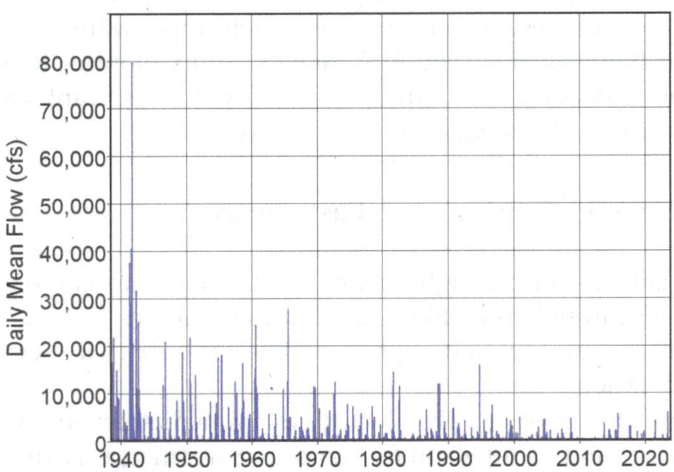

Figure 5-16. Daily flows of Canadian River at Amarillo.

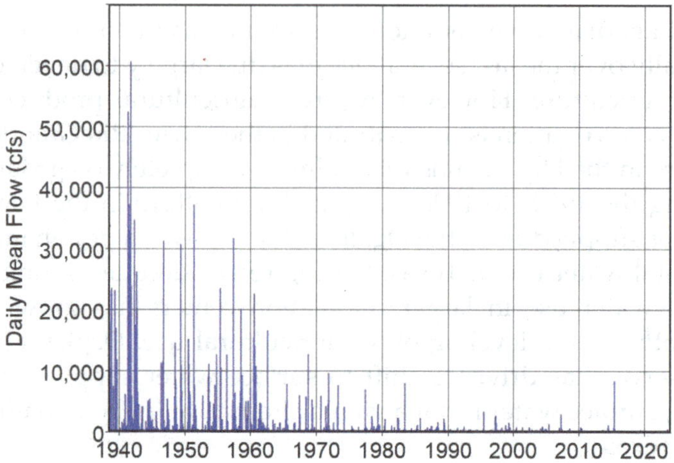

Figure 5-17. Daily flows of Canadian River at Canadian.

(CRMWD) to supply water for its 11 member cities through a 358 mi aqueduct system. Lake Meredith has been in operation since 1941. With declining water supply from Lake Meredith, CRMWD has developed significant groundwater supplies since 2001. Although surface and groundwater supplies are managed conjunctively, in recent years, CRMWD has supplied its member cities primarily from groundwater since little water has been available from Lake Meredith.

The upper watersheds of the Red, Brazos, and Colorado River Basins also extend into dry West Texas. Table 5-4 indicates that large areas of these three river basins do not contribute to streamflow. The noncontributing areas of these river basins are in their upper watersheds in West Texas and New Mexico. The Red, Brazos, and Colorado River Basins have large watersheds extending across Texas. Significant volumes of flow enter the rivers in Central and East Texas.

River Flows in Central and East Texas

Flows at gauge sites throughout Central and East Texas are plotted in Figures 5-18 through 5-36. Flows at all these locations are characterized by extreme variability. Stationarity of flow rates or lack thereof varies between locations.

Mean daily flows of the Red River immediately downstream of Denison Dam and about 90 mi further downstream near Arthur City are plotted in Figures 5-18 and 5-19. The gauge at Arthur City, Oklahoma, is

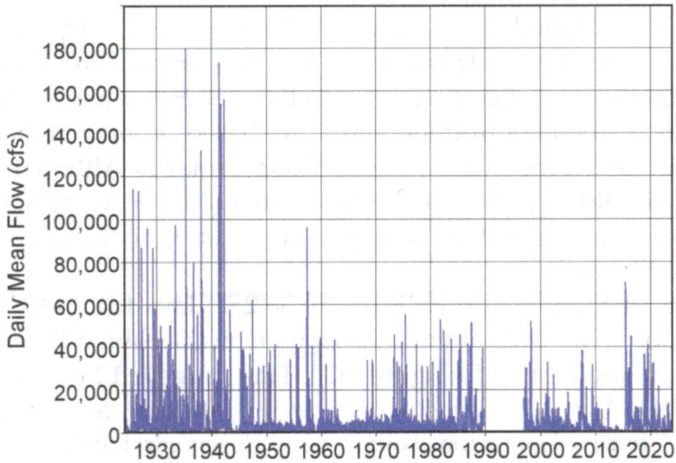

Figure 5-18. Daily flows of Red River below Denison Dam.

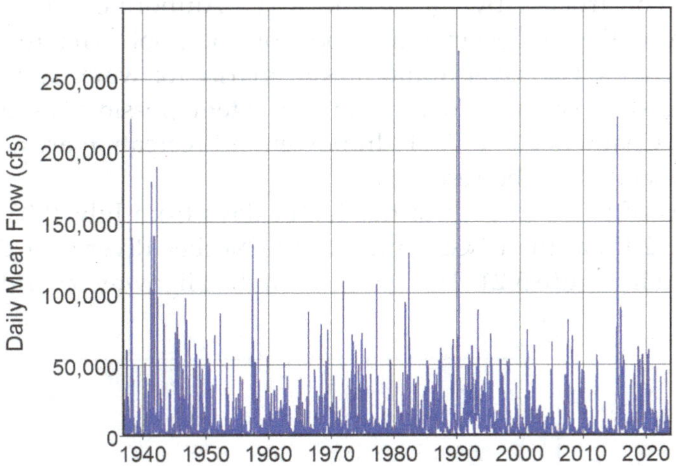

Figure 5-19. Daily flows of Red River at Arthur City.

about 60 mi upstream of the border between Oklahoma and Arkansas and 15 mi north of the city of Paris, Texas. Lake Texoma impounded by Denison Dam is the largest reservoir located wholly or partially in either Texas or Oklahoma. Impoundment began in 1943. As indicated in Table 1-4, Lake Texoma has by far the largest flood control pool in Texas and the fifth largest conservation pool. The USACE Tulsa District operates the project for flood control, hydroelectric power generation, water supply, and recreation. Water supply use is severely constrained by salinity problems discussed later in this chapter.

The flows immediately below the dam plotted in Figure 5-18 are comprised primarily of hydropower releases from Lake Texoma and spills when the conservation pool is full and overflowing. As discussed in Chapter 3, flood control operations are based on making no releases that would contribute to downstream flood damages. Although no flood damages should occur from the flows in Figure 5-18, rainfall runoff entering the Red River downstream may result in flooding in the flows plotted in Figure 5-19.

Means of the flows in each of the 36,251 days from October 1924 through December 2023 at the USGS gauge on the Sabine River near the city of Ruliff are plotted in Figure 5-20. This gauge is 12 mi upstream of Interstate Highway 10, which connects the cities of Beaumont, Texas, and Lake Charles, Louisiana. Toledo Bend Reservoir, located a few miles upstream, has the largest conservation storage capacity of any reservoir in Texas or the southern United States. Toledo Bend is a nonfederal reservoir constructed by the Sabine River Authorities of Texas and Louisiana with no designated flood control pool. The reservoir is operated by the two river authorities primarily for water supply while mitigating downstream flooding to the extent possible, as discussed further in Chapters 6 and 7. Hydropower and recreation are also major purposes served by the reservoir.

Means of flows in each of the 44,014 days from July 1903 through December 2023 at the USGS gauge on the Neches River near Rockland are plotted in Figure 5-21. This gauge is on the Highway 69 bridge 20 mi

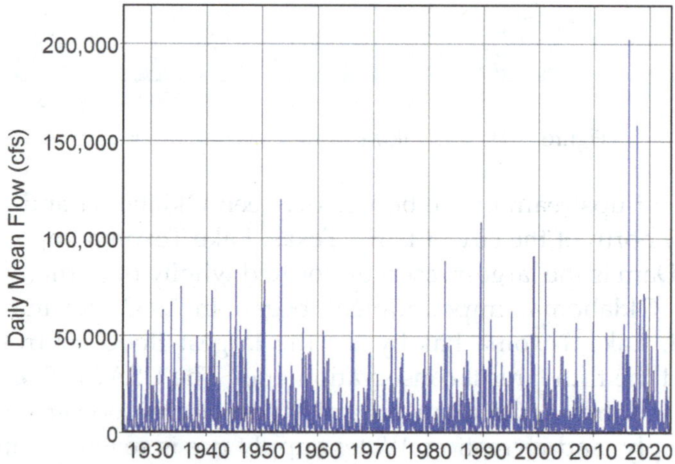

Figure 5-20. Daily flows of Sabine River at Ruliff.

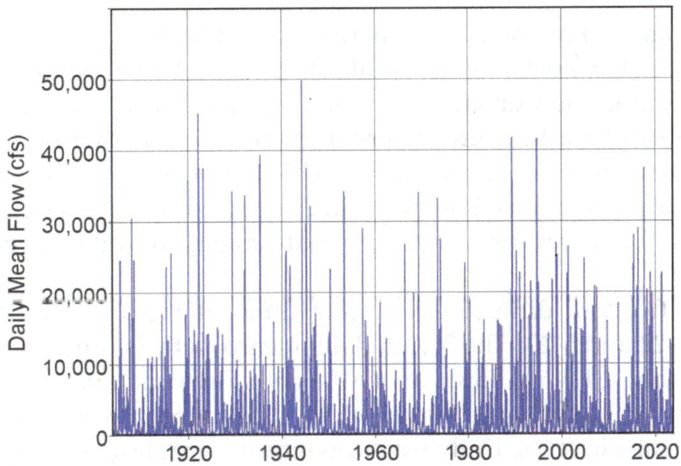

Figure 5-21. Daily flows of Neches River at Rockland.

upstream of the confluence of the Angelina River with the Neches River. The only reservoir with a capacity exceeding 32,000 ac-ft located upstream of this gauge site is Lake Palestine in the far upper basin with a storage capacity of 411,300 ac-ft.

The Neches River daily flows in Figure 5-22 were observed at a USGS gauge at the Highway 96 bridge 25 mi upstream of IH 10 in Beaumont. A maximum allowable non-damaging discharge of 20,000 ft^3/s at this gauge

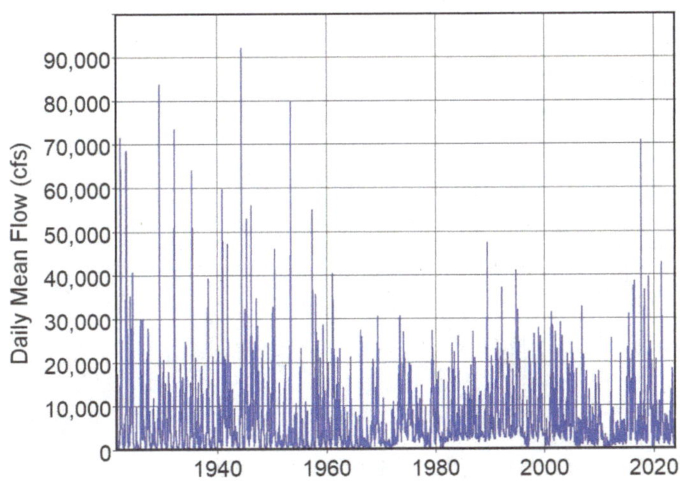

Figure 5-22. Daily flows of Neches River at Evadale.

site is designated by the USACE Fort Worth District for purposes of flood control operations of the federal multiple-purpose Sam Rayburn Reservoir (Figure 7-12) located upstream on the Angelina River. The effects of the 3,998,000 ac-ft Sam Rayburn Reservoir with initial impoundment in 1965 are evident in the daily flows of Figure 5-22. Flood flows are reduced by flood control operations. Low flows are increased by hydroelectric power releases and releases for water supply diversions from the lower Neches River.

The effects of different averaging time intervals are illustrated in Figures 5-22 and 5-23. The 38,145 daily means in ft^3/s during April 1921 to December 2023 of flows of the Neches River at Evadale from the USGS gauge records are averaged to 1,224 monthly means and 102 annual means in ft^3/s. Daily (red dots) means are shown in Figure 5-22. Monthly (blue solid line) and annual (red dots) means are compared in Figure 5-23.

Daily flows of the Trinity River at the cities of Dallas, Oakwood, and Romayor are presented in Figures 5-24 through 5-26. Annual flows at these three gauge sites are plotted in Figure 5-27 [Dallas (blue solid), Oakwood (red dots), and Romayor (black dashes)]. Figure 7-4 is a map of the Trinity River Basin.

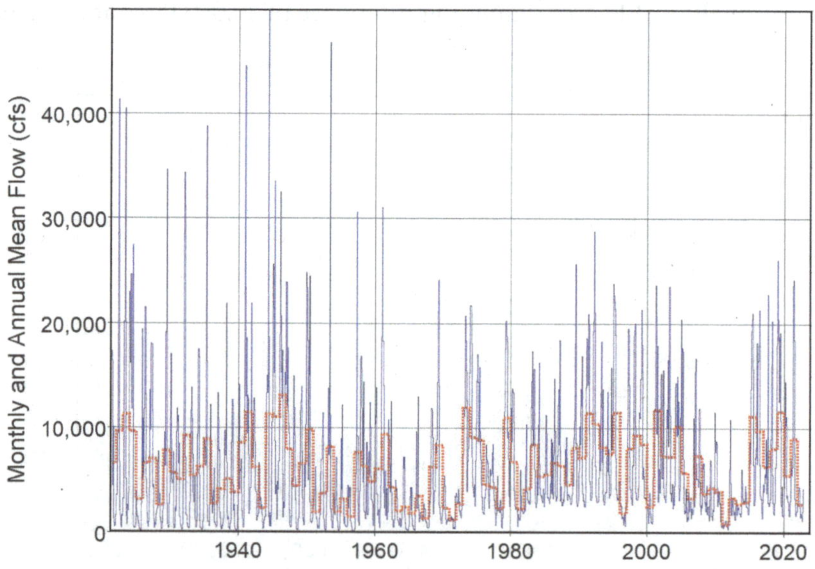

Figure 5-23. Monthly (blue solid line) and annual (red dots) flows of Neches River at Evadale.

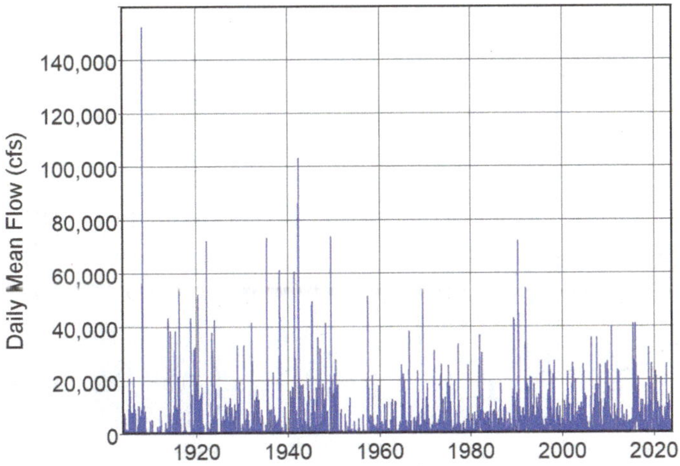

Figure 5-24. Daily flows of Trinity River at Dallas.

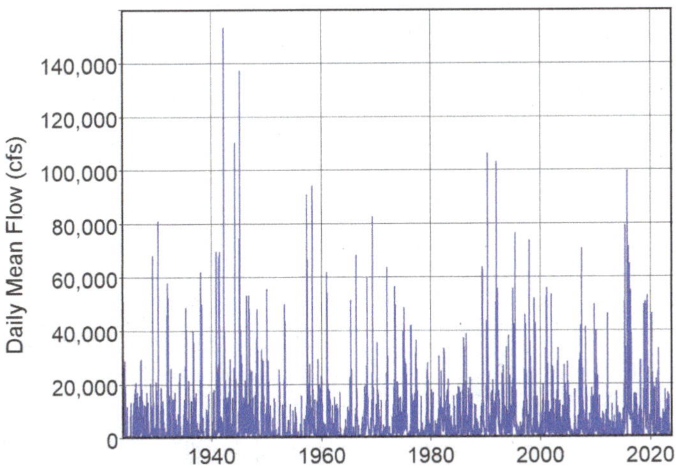

Figure 5-25. Daily flows of Trinity River at Oakwood.

The Figure 5-24 gauge is at West Commerce Street just west of downtown Dallas. A maximum allowable nonflooding discharge of 13,000 ft^3/s at this gauge site is designated by the Corps of Engineers for purposes of reservoir flood control operations. The USACE FWD uses this gauge along with other gauge sites in operating the flood control pools of the federal multiple-purpose Lakes Benbrook, Joe Pool, Ray Roberts, Lewisville, and Grapevine. A number of nonfederal water supply reservoirs are also located upstream of this site.

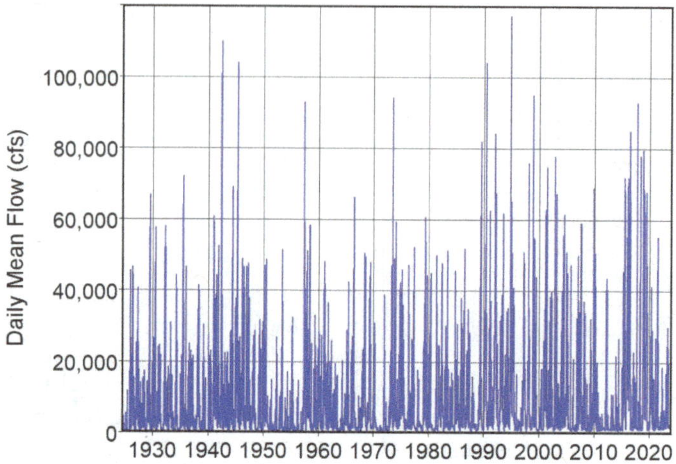

Figure 5-26. Daily flows of Trinity River at Romayor.

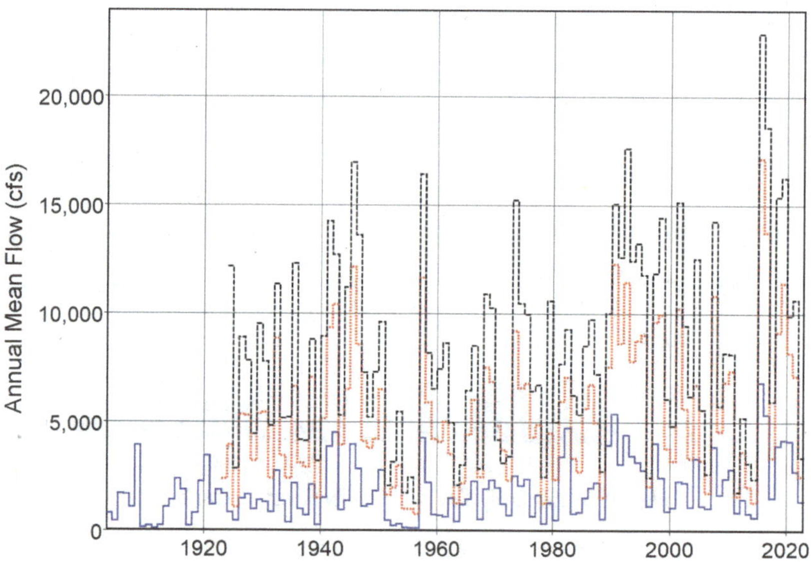

Figure 5-27. Annual flows of Trinity River at Dallas (blue solid), Oakwood (red dots), and Romayor (black dashes).

The gauge on the Trinity River at Oakwood is at Highway 79 about 40 mi below Richland-Chambers Reservoir. The Romayor gauge at Farm-to-Market Road 787 is 20 mi below Livingston Dam and 50 mi above the river outlet at Galveston Bay. Lake Livingston operated by the

Trinity River Authority is the largest reservoir in the Trinity Basin and the ninth largest in Texas.

The Dallas and Fort Worth metropolitan area in the upper Trinity River Basin had a 2020 population of 6.8 million people and has been one of the fastest growing areas in the nation during the last several decades. Many reservoir projects were constructed on the Trinity River and its tributaries during the 1950s to 1980s. Houston, another large continually growing city in the adjoining San Jacinto River Basin, transports water by pipeline from Lake Livingston on the lower Trinity River. Low flows in the Trinity River have increased with increases in wastewater treatment plant discharges. High flows have decreased with flood control operations of eight Corps of Engineers reservoirs. Long-term mean flows have decreased with increased water use.

Daily flows of the West Fork of the San Jacinto at Conroe are shown in Figure 5-28. This gauge is at IH 10 ten miles below the dam at Lake Conroe. The 429,890 ac-ft Lake Conroe with impoundment beginning in January 1973 is operated by the San Jacinto River Authority to supply water for the city of Houston and other smaller cities.

Daily flows of the Colorado River at San Saba, Austin, and Columbus are plotted in Figures 5-29 through 5-31. Annual flows at the three gauges [San Saba (blue solid), Austin (red dots), and Columbus (black dashes)] are presented in Figure 5-32. The San Saba gauge is at Highway 190 about 60 mi upstream of Buchanan Dam. The Columbus gauge is at

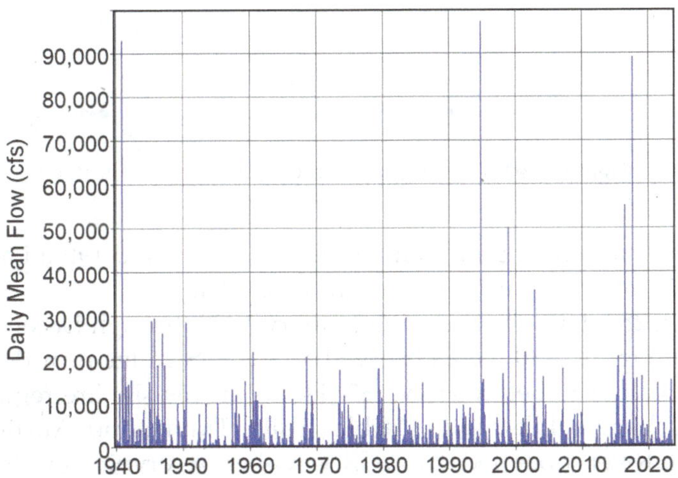

Figure 5-28. Daily flows of the West Fork of the San Jacinto River at Conroe.

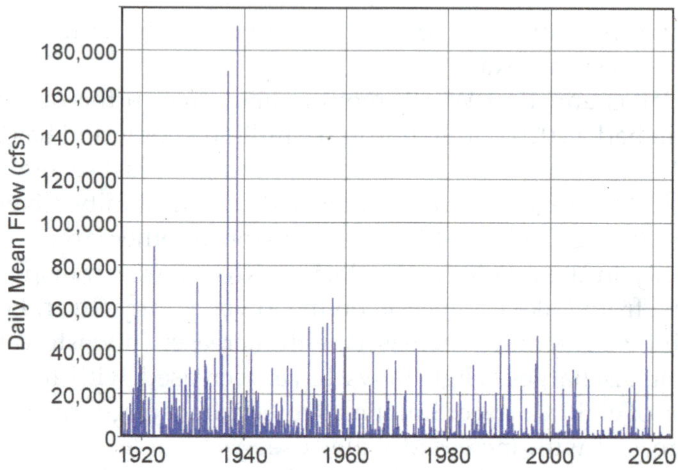

Figure 5-29. Daily flows of the Colorado River at San Saba.

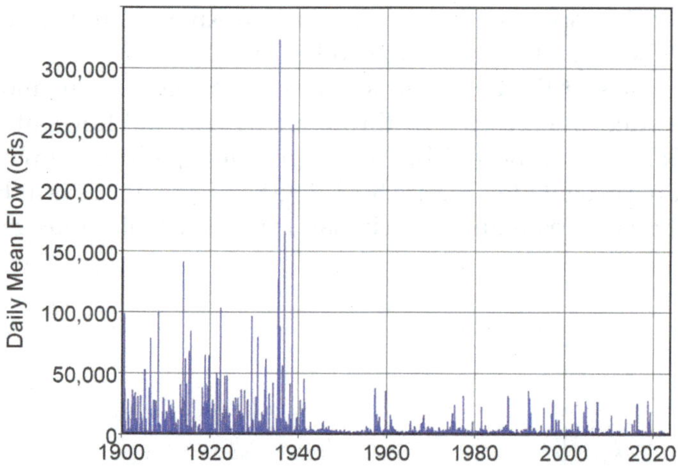

Figure 5-30. Daily flows of the Colorado River at Austin.

Highway 90 about a hundred miles below Austin and 190 mi above the river outlet at Matagorda Bay south of Bay City.

The 45,289 daily flows during January 1900 through December 2023 plotted in Figure 5-30 were observed at a USGS gauge near downtown Austin a half mile below Hwy 183. Flows at this site are regulated by Lakes Buchanan, Inks, LBJ, Marbles Falls, Travis, and Austin on the Colorado River operated by LCRA. Other reservoirs on tributaries entering the Colorado River upstream of Austin are operated by other

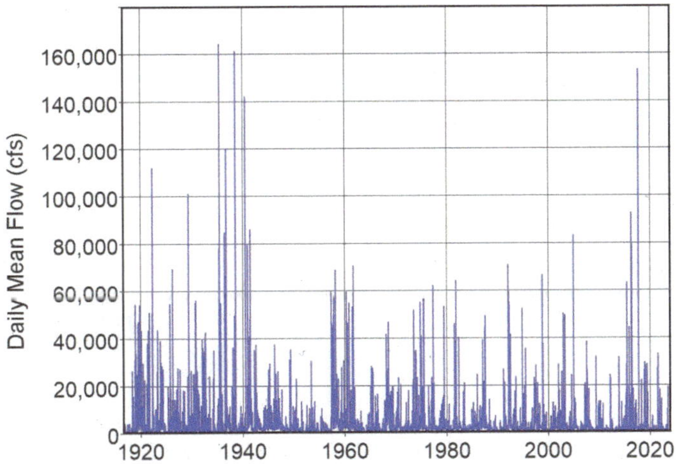

Figure 5-31. Daily flows of the Colorado River at Columbus.

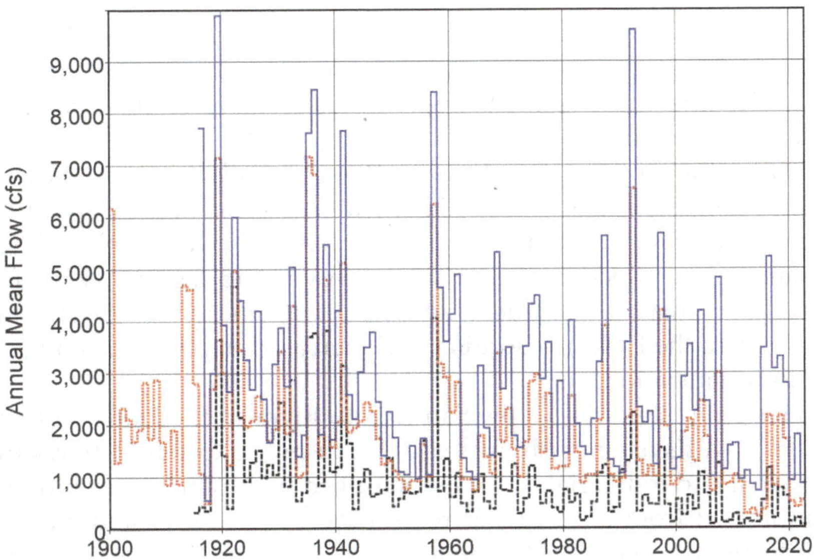

Figure 5-32. Annual flows of the Colorado River at San Saba (blue solid), Austin (red dots), and Columbus (black dashes).

entities. The maximum daily flow during 1900 to 2023 was 323,000 ft^3/s on June 15, 1935. Impoundment of flows in Lakes Travis and Buchanan (Figures 7-7 and 7-8) began in 1940 and 1937, respectively. The effects of

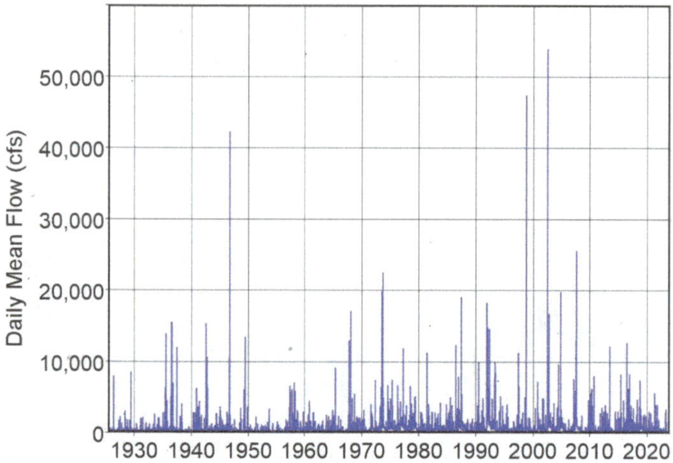

Figure 5-33. Daily flows of the San Antonio River at Falls City.

these two large reservoirs on the daily flows of the Colorado River in Austin are shown by Figure 5-30 to be dramatic. The effects of the reservoirs on the annual mean flows at this gauge site plotted in Figure 5-32 are significant but not dramatic. The dams are storing high flows and maintaining a much more uniform river flow rate than provided by nature.

Daily flows of the San Antonio River at Falls City are shown in Figure 5-33. This gauge is at FM Highway 791 about 50 mi downstream of downtown San Antonio. Flow of the San Antonio River below the city of San Antonio increased significantly over the last 80 years from wastewater treatment effluent accompanying increased groundwater supply from the Edwards Aquifer and increased impervious land cover due to urbanization.

Flows of tributaries of the San Jacinto River in the Houston metropolitan area have similarly increased in response to return flows from municipal and industrial water use supplied by groundwater and interbasin import from Lake Livingston on the Trinity River and increased runoff due to urban development.

Flows of the Nueces River in Figure 5-34 were observed at a gauge just below the Frio River confluence downstream of the city of Three Rivers. Choke Canyon Reservoir located upstream of Three Rivers has a storage capacity of 663,000 ac-ft with impoundment beginning in 1982. Choke Canyon Reservoir is operated by the city of Corpus Christi for water supply as discussed in Chapter 7.

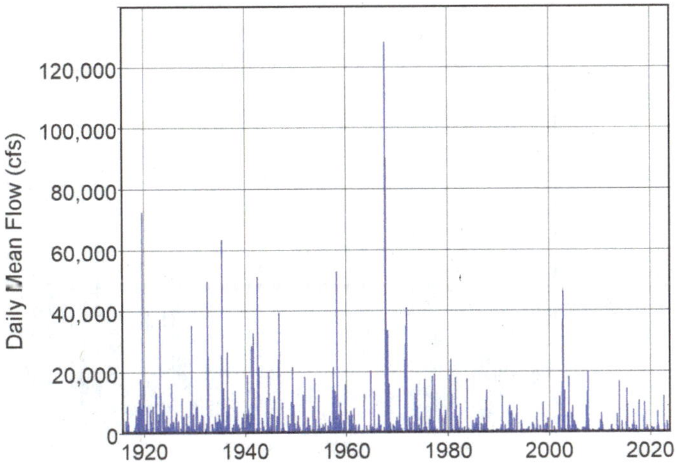

Figure 5-34. Daily flows of the Nueces River at Three Rivers.

As discussed later in this chapter and in Chapters 6 and 7, much of the flow of the Nueces River and its tributaries flows into the ground recharging the Edwards Aquifer. Water from the Nueces River Basin entering the Edwards Aquifer recharge outcrop may eventually find its way to the San Antonio River through discharges from wastewater treatment plants in the city of San Antonio.

Daily flows of the Navasota River at Easterly in Figure 5-35 reflect a relatively undeveloped watershed. Limestone Reservoir and Dam located 11 mi upstream of this gauge site is owned and operated by the BRA

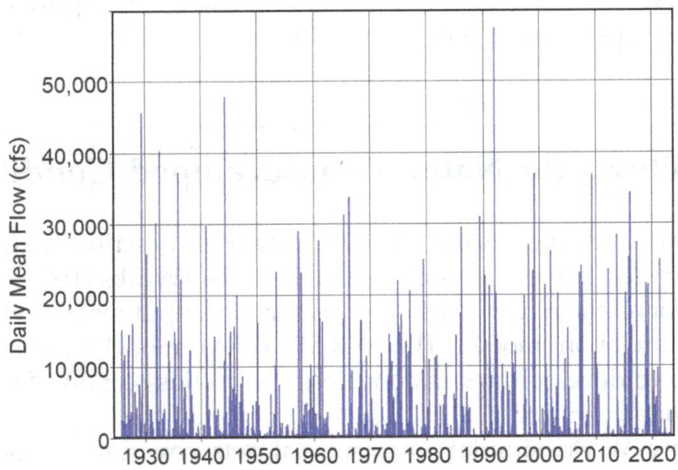

Figure 5-35. Daily flows of the Navasota River at Easterly.

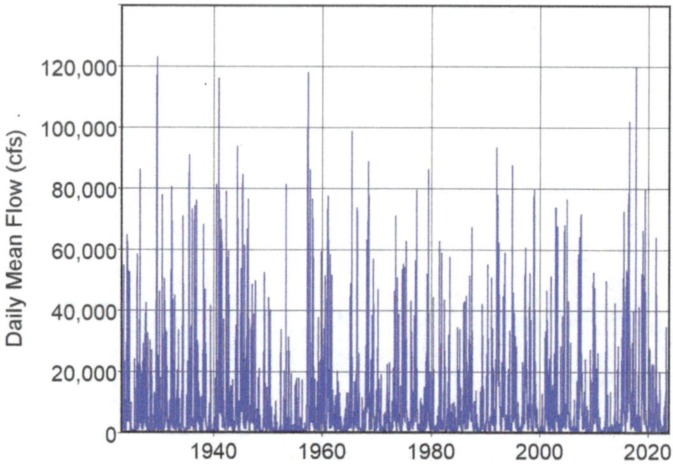

Figure 5-36. Daily flows of the Brazos River at Richmond.

primarily to release water for downstream water supply diversions from the lower Brazos River.

Daily flows at a gauge on the Brazos River at Richmond are plotted in Figure 5-36. This gauge is located near Highway 90 about 60 mi above the Brazos River outlet near Freeport. A maximum allowable non-damaging discharge of 60,000 ft^3/s at the Brazos River gauge at Richmond is designated by USACE FWD for reservoir flood control operations. USACE FWD uses this gauge along with other gauge sites shown in Figure 7-2 of Chapter 7 in operating the flood control pools of the system of nine federal multipurpose reservoirs located on the Brazos River and its tributaries. Many other nonfederal water supply reservoirs are located upstream of this gauge site.

River Flows for Natural Undeveloped Conditions

The WRAP modeling system and WAM datasets described in Chapter 4 can be accessed at websites included in Table 4-1 (Wurbs 2021b, c). WAM datasets for all the river basins of Texas serve as input for the generalized WRAP simulation model. The WRAP/WAM modeling system was employed to develop information presented in this section of Chapter 5 and in Chapters 6 and 7.

Twenty WAM simulation input datasets covering Texas include sequences of monthly naturalized stream flows at about 470 gauge sites that are combined in the WRAP simulation model with watershed parameters to synthesize naturalized flows at about 14,000 ungauged locations. Naturalized flows represent flows that would have occurred if people had not developed and used the water resources of the river basins. These flows approximating natural undeveloped river basin conditions are created by computationally adjusting observed flows to remove the effects of reservoirs, water supply diversions, return flows, and other relevant factors (Wurbs 2006, 2019b, 2021a). Trend analyses of naturalized flows in WAM datasets have shown that the naturalized flows are essentially stationary at most locations (Wurbs and Zhang 2014; Wurbs 2019b, 2021b).

Actual historical streamflows are significantly different than natural flows under hypothetical undeveloped conditions for many river reaches in Texas. Conversely, the differences between natural and actual flows are negligible in many other river reaches. Storage and water use associated with major reservoirs account for most of the differences between natural and actual flows on major rivers of Texas. Major rivers with large watersheds are different in this regard than streams in smaller urban watersheds where urban land use changes dominate changes in streamflow characteristics.

Flows are expressed in units of acre-feet/month and acre-feet/day in the monthly and daily versions of the WRAP/WAM modeling system. An acre-foot is a volume equivalent to covering an area of one acre to a depth of one foot. An acre-feet/day is equivalent to 0.5042 ft^3/s.

Illustrative Site in Central Texas

Observed daily mean and annual mean flows at a gauge on the Brazos River at Waco in Central Texas are plotted in units of ft^3/s in Figures 1-5 and 1-6 of Chapter 1. This gauge is at Highway 340 just downstream of Waco and about five miles downstream of the Bosque River confluence shown in Figure 7-2. The observed daily flows are converted to units of acre-feet/day in Figure 5-37 and averaged to monthly flows in acre-feet/month in Figure 5-38.

The Corps of Engineers makes no releases from upstream flood control reservoirs that would contribute to flows at the Brazos River gauge at Waco exceeding a maximum allowable non-damaging flow of 25,000 ft^3/s (49,590 ac-ft/day). USACE FWD uses this gauge along with other gauges on the Brazos River in operating the flood control pools of

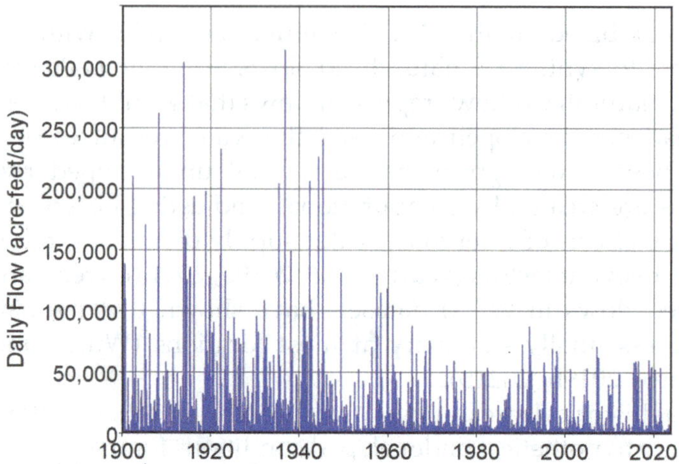

Figure 5-37. Daily flows of the Brazos River at Waco in acre-feet/day.

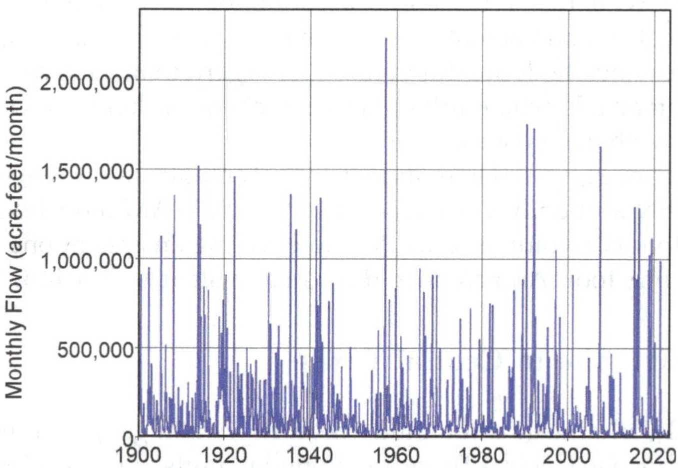

Figure 5-38. Monthly flows of the Brazos River at Waco in acre-feet/month.

the multipurpose Lakes Waco, Aquilla, and Whitney located upstream of this site. The USACE system includes six other reservoirs on tributaries that enter the Brazos River downstream of the Waco gauge. Many other water supply reservoirs are also located upstream of the Waco gauge as well as throughout the river basin. The effects of the upstream reservoirs are evident in the daily gauged flows of Figure 5-37. However, the effects of reservoir flood control operations are dissipated in the

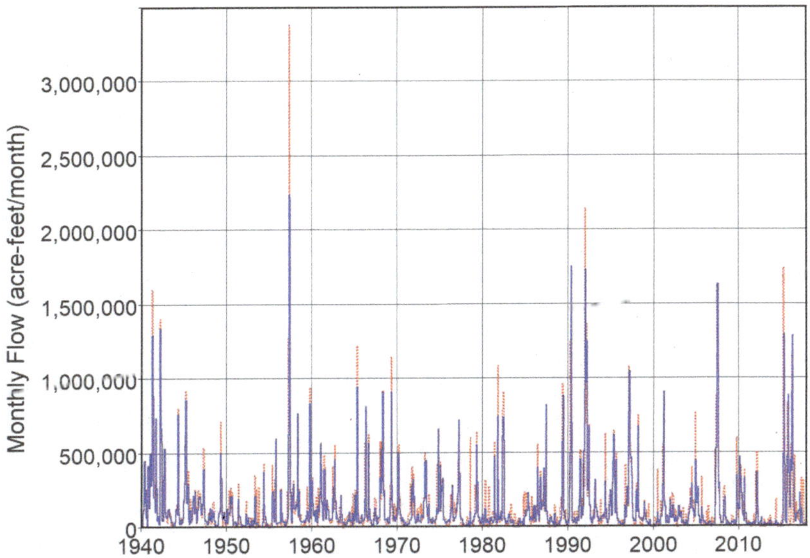

Figure 5-39. Comparison of monthly observed (blue solid line) and naturalized (red dots) flows of the Brazos River at Waco.

process of averaging the daily mean flows to obtain the monthly mean flows of Figure 5-38.

The daily WAM dataset for the Brazos River Basin includes 1940 to 2017 monthly naturalized flows for the gauge on the Brazos River at Waco that have also been disaggregated to daily flows and aggregated to annual flows (Wurbs 2019a). The monthly and annual means of the naturalized and observed flows are plotted for comparison in Figures 5-39 and 5-40. The blue solid lines represent actual observed stream flows. The red dotted lines represent flows after adjustments to remove the impacts of water development, regulation, and use. Daily naturalized flows representing undeveloped conditions tend to be higher than the observed flows for higher flows but may be either lower or higher for lower flows. Annual totals of naturalized flows tend to be higher than the corresponding observed flows but not dramatically higher.

Natural Flow at River Basin Outlets

WAM-based estimates of mean naturalized flow at the outlets of all Texas river basins except the Rio Grande, Canadian, and Red Basins and coastal basins are presented in Table 5-5. The naturalized flows in

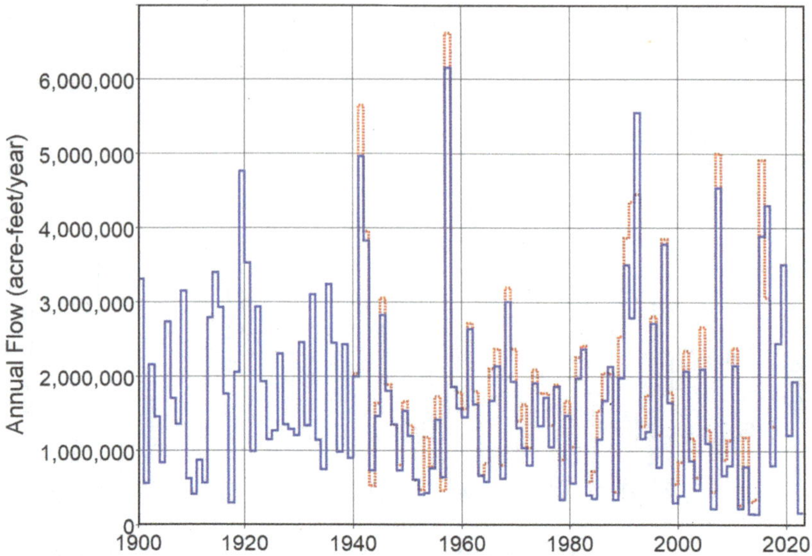

Figure 5-40. Comparison of annual observed (blue solid line) and naturalized (red dots) flows of the Brazos River at Waco.

Table 5-5 are compared later with simulated regulated flows in Table 6-7 of Chapter 6. River basin drainage areas and 1940 to 2021 mean annual precipitation are tabulated in the second and third columns of Table 5-5. The WAM periods of analysis for the naturalized flows are listed in the fourth column.

The last four columns of Table 5-5 contain four alternative expressions of long-term average naturalized flow at the outlets of the rivers. As previously explained, naturalized flows represent natural conditions without water resources development and water use. The average naturalized flows from WAMs are shown in the fifth and sixth columns in units of ft^3/s and ac-ft/year. The mean annual flow volume in acre-feet divided by the total river basin drainage area in acres provides a depth in feet which is converted to inches. The resulting inches of precipitation runoff are tabulated in the seventh column. The annual natural flow at the outlet is a volume equivalent to covering the river basin to a depth shown in the seventh column of Table 5-5. The last column is the natural flow volume in the seventh column expressed as a percentage of the mean annual volume of precipitation falling on the river basin which is tabulated in the third column as the average annual precipitation depth.

Table 5-5. Mean Flow at River Outlets under Natural Undeveloped Conditions.

River Basin	Basin Area (square miles)	Annual Precip (inches)	WAM Analysis Period	Mean of Naturalized Streamflow at the River Outlet			
				(cubic feet/second)	(acre-feet/ year)	(inches/ year)	(% Precipitation)
Nueces	16,700	25.0	1934–2021	695.5	503,845	0.57	2.26%
Guadalupe	10,133	32.5	1934–2021	3,086	2,235,750	4.14	12.7%
Lavaca	2,309	40.0	1940–2021	1,366	989,297	8.03	20.1%
Colorado	41,480	24.7	1940–2016	3,369	2,440,780	1.10	4.47%
Brazos	45,573	29.6	1940–2017	8,445	6,117,871	2.52	8.50%
San Jacinto	3,936	46.9	1940–2021	3,547	2,569,336	12.2	26.1%
Trinity	17,913	39.7	1940–2018	9,114	6,602,863	6.91	17.4%
Neches	9,937	49.0	1940–2019	8,503	6,160,296	11.6	23.7%
Sabine	9,756	48.1	1940–2021	9,465	6,750,392	13.0	27.0%
Cypress	2,929	47.5	1948–2021	2,442	1,769,341	11.3	23.8%
Sulphur	3,767	46.9	1940–2021	3,370	2,441,490	12.2	25.9%
Total	164,433	32.7		53,402	38,581,261	4.40	13.4%

The river flows in Table 5-5 represent water that would be leaving Texas under natural undeveloped conditions. The outlets for the flows of the Sulphur and Cypress Rivers are the border of Texas with Arkansas and Louisiana. The Sulphur River and Cypress Creek in Northeast Texas are tributaries of the Red River which flows to the Mississippi River. The outlets for all the other rivers in Table 5-5 are at the Gulf of Mexico in Texas.

Precipitation and streamflow vary greatly from year to year and within each year. The discussion here focuses on long-term averages. An estimated average precipitation depth of 32.72 in. falls on the 164,433 sq. mi area of the 11 river basins, which is equivalent to a precipitation volume of 286,961,460 ac-ft. The long-term average of the estimated natural river flow from the 11 river basins is 38,581,260 ac-ft, which is 13.4% of the volume of the precipitation falling on the watershed drainage areas.

The mean annual volume of natural streamflow leaving the river basin at its outlet is expressed as a percentage of the mean annual volume of precipitation falling on the watershed in the last column of Table 5-5. Outlet streamflow as a percentage of rainfall ranges from 2.26% in the Nueces River Basin to 27.0% in the Sabine River Basin. Loss of streamflow to groundwater recharge is a large component of the water budget of the Nueces River Basin, as discussed later, and may be significant in other river basins. Evapotranspiration accounts for the majority if not essentially all the remainder of the precipitation falling on the river basins.

Under natural undeveloped conditions, the precipitation falling on a river basin leaves the basin in primarily two ways: streamflow at the outlet and evapotranspiration. Precipitation also infiltrates into the ground, and conversely groundwater contributes to streamflow. In the long term as reflected by annual means, ground and surface water interactions may tend toward a generally balanced equilibrium in many though not all regions. Groundwater discharge into the Gulf of Mexico may be a significant component of long-term natural water budgets in some areas.

Although the data in Table 5-5 are necessarily approximate, general insight is provided regarding river system hydrology of Texas under natural undeveloped conditions. Precipitation increases going from west to east across the state. Streamflow resulting from precipitation expressed as a percentage of precipitation also increases from west to east across the state. The Neches, Sabine, Cypress, and Sulphur are the wettest river basins in terms of both precipitation falling on the watershed and natural flow at the outlet expressed as a percentage of precipitation. The three river basins with the lowest precipitation and the lowest natural flow as a percentage of basin-wide mean precipitation are the Nueces, Colorado, and Brazos.

Table 5-5 indicates that the estimated long-term average volume of natural streamflow at the outlets of the Colorado and Brazos are 4.47% and 8.50%, respectively, of the mean volume of precipitation falling in these river basins. Large upstream regions of the Brazos and Colorado River Basins in West Texas contribute little rainfall runoff to the upper river systems of which essentially none reaches the outlets.

The natural flow at the outlet of the Nueces River is much lower than at the outlets of the other rivers in Table 5-5 in terms of both total flow and flow volume as a percent of precipitation volume. Loss of streamflow as recharge to groundwater aquifers is the major factor in reducing natural flows of the Nueces River as discussed in the next section. The Nueces River and its tributaries flow into the highly fractured outcrop of the Edwards Aquifer which is comprised of cavernous limestone as well as across recharge zones of the Carrizo–Wilcox, Gulf Coast, and other aquifers. Streamflow is lost to aquifer recharge before reaching the basin outlet at Nueces Bay.

The Rio Grande and Canadian River Basins are not included in Table 5-5 due to difficulties in compiling accurate data. Under undeveloped conditions with no reservoirs and no agricultural irrigation or other water use, the natural flow of the Rio Grande and Canadian River would be a significant component of the total river flow leaving Texas each year. However, the total flows at their outlets from Texas have been negligible in recent decades. Interactions between groundwater and surface water and potential climate change effects complicate understanding the relationships between natural and actual streamflows in these large international and interstate river basins. Various other factors also affect the accuracy of estimates of natural flows in these river basins.

The Red River has high natural and actual flows but is omitted from Table 5-5 due to interstate complexities in available information. Natural flows are high relative to other Texas rivers. Natural flows are close to the same as observed flows for much of the length of the Red River that serves as the border between Texas and Oklahoma and between Texas and Arkansas.

The eight coastal basins between the major river basins are also omitted from Table 5-5. The coastal basins drain into the Gulf of Mexico through multiple small streams. The means of naturalized flows are about the same or higher than the corresponding means of observed flows at most locations throughout the major river systems. However, in the coastal basins, the actual observed flows measured at gauge sites may be higher than naturalized flows due to return flows from water use supplied by inter-basin transport from the major river basins.

Connections between Surface and Groundwater

Streams throughout Texas gain water from and lose water to groundwater. The quantities of water involved in these components of the hydrologic cycle vary greatly between river basins. Groundwater occurs under water table and artesian conditions. Under water table conditions, the water is unconfined and does not rise in wells above the water table. The term *water table* refers to the upper surface of the zone of saturation. Under artesian conditions, the water is confined within a water-bearing stratum by an overlying relatively impermeable stratum. Due to water being under pressure, it will rise in wells above the level at which it is encountered in drilling the wells. Rainfall infiltrating the ground surface replenishes soil moisture deficits before percolating by gravity to the water table. Much or most of the recharge of some groundwater aquifers is through rainfall and streamflow entering outcropping recharge zones as discussed later in this section.

Lower reaches of the larger rivers in Central and East Texas are perennial, meaning streamflow occurs throughout the year in most years. Streams that flow regularly during the wet seasons of the year but not year-round during dry seasons are called intermittent. Ephemeral streams flow only during and for a short duration after heavy rainstorms. Ephemeral flows occur in headwaters of perennial streams, in smaller streams, and in arid regions such as West Texas.

In perennial streams, the groundwater table is consistently above the bottom of the streambed, groundwater contributes to streamflow, and streamflow increases in a downstream direction. Perennial streams are commonly called effluent or gaining. Base flows are provided from the saturated ground through the bed and banks of the stream even during long periods without rainfall.

In ephemeral streams, the underlying water table of groundwater is consistently lower than the bottom of the streambed. Streamflow seeps into the bed and banks, maintaining moisture in the underlying strata. Ephemeral streams are called influent or losing.

Water stored in a reservoir likewise interacts with water stored in the underlying and surrounding ground. Groundwater with a high water table may result in spring flow or seepage into a lake. Conversely, seepage from a lake may contribute to groundwater. Lake levels fluctuate. Lake seepage may contribute to raising water tables of underlying aquifers. As reservoir storage is drawn down by water use, water may flow from adjacent saturated ground into the reservoir. The adjacent ground is saturated again later when the reservoir refills.

Investigations of Interactions between Surface and Groundwater

Creation of the statewide WAM system introduced in Chapter 4 and employed in the preceding section of Chapter 5 was authorized by the Texas legislature in its 1997 SB1. Early development of the WAM system included an assessment sponsored by the TNRCC, later renamed TCEQ, of the interactions between surface and groundwater in each of the major river basins and coastal basins of Texas excluding the Rio Grande (Parsons Engineering Science 1999). A later TCEQ-administered study by the Bureau of Economic Geology (2005) provides a detailed review of available information on surface and groundwater interactions throughout the state. A recent TWDB-sponsored study focuses on modeling interactions between surface and groundwater in Central and West Texas (WSP Engineering 2022). These and other technical reports exploring surface and groundwater interactions for specific aquifers, river basins, or regions are available from TCEQ and TWDB.

TCEQ-sponsored research by the Bureau of Economic Geology at the University of Texas at Austin in 2023 to 2024 is exploring groundwater aquifer storage and recovery strategies. WRAP with WAM datasets are being used in this research to assess available unappropriated stormwater streamflow that could be used for groundwater recharge.

WAM datasets include channel loss parameters used in a WRAP simulation to model losses of water in stream channel reaches due to seepage and evapotranspiration. Channel loss computations are not activated in several WAMs because the actual losses are considered negligible. In other WAMs, channel losses are computed for some but not all stream reaches. The WAM for the Nueces River Basin illustrates the case of extremely high losses of streamflow to recharge of groundwater (Wurbs 2023). The upper reaches of the Nueces River and its tributaries cross the recharge zone of the Edwards Aquifer where the entire flow of the stream may be lost to groundwater recharge. For middle and lower reaches of the Nueces River and its tributaries located downstream of the Edwards recharge zone, channel seepage losses per mile of stream length are estimated to generally range between 0.30% and 0.70% of the streamflow discharge per mile (HDR Engineering 1998).

Unlike the Edwards Aquifer, most groundwater aquifers in Texas and elsewhere supplying large quantities of water consist largely of sand and gravel. The Edwards Aquifer is comprised of caverns through limestone that are essentially underground streams. The principal recharge zone of the Edwards Aquifer is a 1,500 sq. mi area of fractured and cavernous limestone exposed on the surface allowing large quantities of water to flow into the aquifer. This recharge zone extends across the upper portions of the Nueces, San Antonio, and Guadalupe River Basins in the Texas Hill Country just north of the cities of Uvalde, Hondo, San Antonio, and New Braunfels.

Interconnections between surface and groundwater are significant throughout Texas. Base flow from saturated banks of perennial streams maintains low flows during dry periods. Streamflow is lost through seepage in ephemeral streams. Spring flows contribute to streamflow. Losses of streamflow to aquifer recharge as streams flow over recharge zones can involve particularly large quantities of water. The effects of aquifer recharge zones and spring flows on streamflow are most significant in the Guadalupe, San Antonio, and Nueces River Basins and the western portions of the Colorado and Brazos River Basins. The Edwards Aquifer has particularly notable effects on streamflow in the Guadalupe, San Antonio, and Nueces River Basins from the perspectives of both aquifer recharge and spring flow to streams.

Conjunctive surface and groundwater management issues are relevant throughout Texas (Young et al. 2018). Management of the Edwards Aquifer has unique dimensions. Surface streams flow through the recharge zone into underground streams in the cavernous Edwards limestone. Protection of the water quality of the Edwards Aquifer is dependent on protecting the water quality of the rivers that recharge the aquifer. Surface water development projects can affect the quantities of aquifer recharge. For example, flood flows of the Nueces, Frio, and other rivers crossing the recharge zone exceed the recharge capacity. Dams upstream of the recharge zone can be employed to store high flows for later release at optimal rates for aquifer recharge.

TWDB Groundwater Availability Models

TWDB maintains groundwater availability models (GAMs) for the nine major and 21 minor aquifers of Texas pursuant to the 1997 SB1 (Kelley et al. 2008). GAM software, datasets, and documentation can be downloaded through the TWDB groundwater webpage. The GAMs were developed and are maintained and updated by TWDB staff and consulting firm contractors.

The generalized MODFLOW groundwater model developed by USGS and extensively applied throughout the United States is employed for the Texas GAMs. MODFLOW is widely used and tested, well documented, and in the public domain.

The calibrated GAMs simulate aquifer geology, hydraulics, recharge, and pumping. The models provide water availability information for SB1 regional and statewide planning studies, activities of the groundwater conservation districts, and various other TWDB programs.

Natural Salinity in the Permian Basin Region

Dissolved solids or salts are the inorganic solutes that occur in all natural waters because of weathering of rocks and soils. The terms *total dissolved solids* (TDS) and *salinity* are used interchangeably. Evapotranspiration produces essentially pure water vapor and increases salinity concentrations of the remaining liquid water. Human activities such as irrigated agriculture and construction of storage reservoirs increase evaporation and consequently increase salinity of water resources. Elevated salinity levels are detrimental to agricultural, municipal, and industrial water use. Seawater is a major source of salt in coastal areas. Saltwater encroachment from the ocean into the lower coastal reaches of rivers is a concern in Texas and elsewhere. The river systems of Texas and neighboring states shown in Figure 5-41 share the common problem of extremely high salt concentrations in upper stream reaches resulting from geologic formations occurring in small isolated sub-watersheds (Rought 1984, Wurbs 2002).

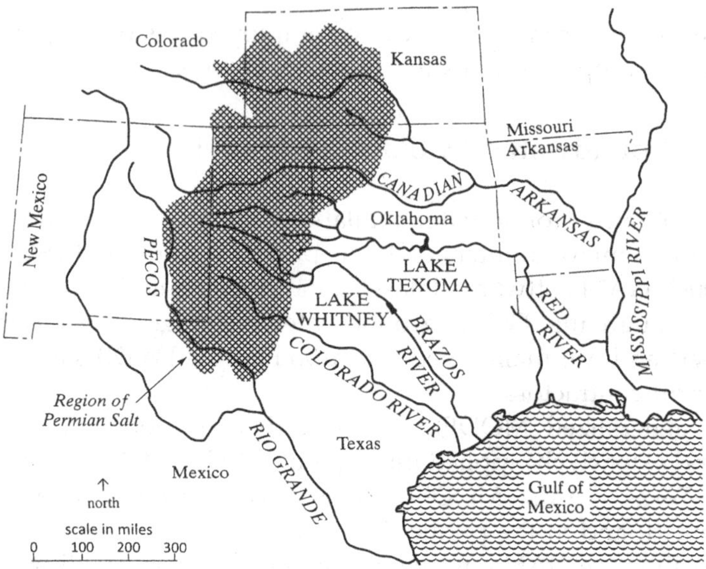

Figure 5-41. Rivers subject to Permian Basin natural salt pollution.

Capabilities for supplying water for municipal, industrial, and agricultural use from the Arkansas, Brazos, Canadian, Pecos, and Red Rivers are significantly constrained by natural salt pollution in the geologic region of Permian salt shown in Figure 5-41 (Wurbs 2002, Wurbs and Lee 2009). This region was covered by an inland sea during the Permian age 230 million years ago (Rought 1984). Deposits of halite were formed as evaporating seawater precipitated salts. This semiarid region now consists of gypsum and salt-encrusted rolling plains containing numerous salt springs and seeps. Small tributary streams in primary salt areas have dissolved solids concentrations that sometimes exceed that of seawater.

Water percolates to shallow salt-bearing strata creating salt brine. The brine moves laterally or vertically until it is discharged at a saline spring or along a streambed. Evaporation of water at the land surface forms a crust of salt over salt flats or salt plains. Rainfall runoff transports the salts to streams. The mineral pollutants consist largely of sodium chloride with moderate amounts of calcium sulfate and other dissolved minerals. Salt concentrations in the downstream reaches of the rivers decrease with dilution from low-salinity tributary inflows.

McCrory (1984) and Wurbs and Lee (2009, 2011) document studies of the effects of salinity on water supply capabilities in the Brazos River Basin. The mean concentrations in Table 5-6 are based on USGS field

Table 5-6. Salinity Concentrations in the Brazos River Basin.

USGS Gauge Location	Drainage Area (square miles)	Mean Flow (cubic feet/second)	Mean Salt Concentration TDS (milligram/liter)	Chloride (milligram/liter)	Sulfate (milligram/liter)
Salt Croton Creek at Aspermont	64	4	56,920	32,860	2,270
Salt Fork Brazos River at Aspermont	5,130	60	12,400	6,070	1,230
Brazos River at Seymour	15,538	269	3,590	1,480	696
Brazos River below Possum Kingdom Dam	23,596	712	1,530	601	309
Brazos River below Whitney Dam	27,189	1,230	928	342	178
Brazos River at College Station	39,599	4,530	438	112	77
Brazos River at Richmond	45,007	6,868	339	79	56
Yequa Creek below Somerville Dam	1,009	252	167	30	48
Navasota River below Limestone Dam	239	161	131	22	13
Navasota River near Bryan	1,454	600	144	38	33

measurements from 1964 through 1986 funded by USACE FWD to support natural salt pollution control studies. Site drainage area, mean flow rate, and salinity concentrations in milligrams per liter are tabulated. Salinity is expressed in terms of TDS and TDS components chloride and sulfate. Figure 7-2 is a map of the Brazos River Basin showing the locations of reservoirs and USGS gauges.

Long-term means are tabulated in Table 5-6. Salinity concentrations, like streamflow rates, are extremely variable over time as well as spatially. Concentration–duration statistics for TDS are tabulated in Table 5-7. TDS concentrations at selected gauging stations equaled or exceeded specified percentages of time are presented in Table 5-7. USGS compiled salt concentration datasets for USACE as monthly mean concentrations for each of the 264 months of a 1964 to 1986 period of record (Wurbs and Lee 2009). The means of Table 5-6 and exceedance frequency metrics of Table 5-7 are computed from this USGS/USACE monthly data.

The USGS salinity sampling station on Salt Croton Creek near the town of Aspermont is at the outlet of a small sub-watershed that contains a primary salt source area. This site on Salt Croton Creek has a small mean flow rate and large mean TDS concentration of 56,920 mg/L which is much higher than seawater. Salt Croton Creek is a tributary of the Salt Fork of the Brazos River. The water quality measurement site on the Salt Fork of the Brazos River is downstream of the Salt Croton Creek confluence and has a much greater watershed area and mean flow rate and a mean TDS concentration of 12,400, which is still extremely high. Salt loads from Salt Croton Creek and other primary source areas in the watershed of the Salt Fork of the Brazos are diluted by flows from large areas of the watershed that have much lower salinity concentrations.

The USGS gauging station on the Brazos River at Seymour is downstream of the confluence of the Double Mountain Fork and Salt Fork of the Brazos River and upstream of Possum Kingdom (PK) Reservoir. Five gauge sites on the Brazos River are listed in the tables in upstream-to-downstream sequence. Two of these sites are immediately downstream of Possum Kingdom and Whitney Reservoirs. Salinity concentrations are diluted with low-salinity inflows as the salt loads travel downstream. The gauge sites on Yequa Creek and the Navasota River are included in Table 5-6 to provide a comparison of salinity concentrations on streams that are not downstream of the Permian Basin salt source region. The Navasota River and Yequa Creek are tributaries of the lower Brazos River. Mean TDS concentrations of 167, 131, and

Table 5-7. Duration–Concentration Relationships for Monthly Means of Total Dissolved Solids Concentrations.

Exceedance Frequency	TDS Concentration of Flows of Brazos River at USGS Gauging Stations at				
	Seymour	PK Dam	Whitney Dam	College Station	Richmond
(%)	(milligram/liter)	(milligram/liter)	(milligram/liter)	(milligram/liter)	(milligram/liter)
1	14,500	2,710	1,560	1,040	902
10	11,900	2,290	1,250	763	635
30	8,530	1,890	1,000	557	426
50	6,220	1,620	945	505	382
70	4,320	1,420	750	412	317
90	2,420	1,130	666	313	235
100	618	475	456	212	153

144 mg/L at these three downstream tributary locations are typical of river reaches that are not connected to the Permian Basin salt source region.

The TDS concentrations in Tables 5-6 and 5-7 can be referenced to the following quantities for comparison. Seawater typically has a TDS concentration between 31,000 and 38,000 mg/L, averaging about 35,000 mg/L. US Environmental Protection Agency secondary drinking water standards recommend treatment of municipal water supply as necessary to prevent TDS, chloride, and sulfate concentrations from exceeding 500, 250, and 250 m/L, respectively. The most suitable TDS levels for drinking water are between about 50 and 100 mg/L. Desalination is technologically practical but very expensive.

The frequency metrics of Table 5-7 are relevant as well as the long-term mean concentrations of Table 5-6 when referencing the 500 mg/L TDS standard for drinking water. All locations on the Brazos River have TDS concentrations exceeding 500 mg/L much of the time even though the flows at College Station and Richmond have long-term mean TDS concentrations of less than 500 mg/L.

The Brazos River Authority has contracted for water supply storage capacity in nine multiple-purpose USACE reservoirs and owns three other water supply reservoirs. As discussed in Chapter 7, operations of these USACE/BRA reservoirs in the Brazos River Basin are significantly affected by the natural salt pollution.

Lake Texoma on the Red River is another example of the major reservoirs for which water supplies are severely constrained by salinity. The USACE Tulsa District's multiple-purpose Lake Texoma is the largest reservoir in Texas and Oklahoma. The project has been used primarily for flood control, hydroelectric power generation, and recreation, but municipal water supply use with desalination has increased with intensifying water demands. Recreation is enhanced by salinity from the perspective of striped bass anglers. The lake is popular with anglers who enjoy catching striped bass, which are found only in water bodies with high salinity.

The USACE Fort Worth District, USACE Tulsa District, USBR, USGS, state water agencies, river authorities, and university researchers have conducted extensive natural salt pollution control studies focused on salt containment strategies for these affected river basins over the last several decades. Many salinity control projects have been proposed, and some have been implemented. These endeavors are discussed further in Chapter 7.

References

ASCE. 2018. *Addressing flood risk: A path forward for Texas after Hurricane Harvey.* Austin, TX: ASCE Texas Section Task Committee on Post-Hurricane Harvey Recommendations.

Bista, A. 2015. "River/reservoir system water availability modeling support for drought management." Master of Science thesis. Texas A&M University, Dept. of Civil and Environmental Engineering.

Bixel, P. A., and Turner, E. H. 2000. *Galveston and the 1900 storm: Catastrophe and catalyst*. Austin, TX: University of Texas Press.

Bomar, G. W. 2017. *Weather in Texas: The essential handbook*. Austin, TX: University of Texas Press.

Bureau of Economic Geology. 2005. *Groundwater-surface water interactions in Texas.* Austin, TX: University of Texas, Texas Commission on Environmental Quality.

Burnett, J. 2008. *Flash floods in Texas.* College Station, TX: Texas A&M University Press.

Cook, B. I., T. R. Ault, and J. E. Smerdon. 2015. "Unprecedented 21st century drought risk in the American Southwest and Central Plains." *Sci. Adv.* 1 (1): e1400082.

Cook, B. I., R. Seager, A. P. Williams, M.J. Puma, et al. 2019. "Climate change amplification of natural drought variability: The historic mid-twentieth-century North American drought in a warmer world." *J. Clim.* 32 (17): 5417–5436.

Eyring, V., S. Bonny, G. A. Meehl, C. A. Senior, et al. 2016. "Overview of the coupled model intercomparison project phase 6 (CMIP6) experimental design and organization." *Geosci. Model Dev.* 9 (5): 1937–1958.

HDR Engineering. 1998. *Trans-Texas Water Program, West Central Study Area, Phase II: Edwards aquifer recharge analyses.* Austin, TX: San Antonio River Authority, San Antonio Water System, Guadalupe-Blanco River Authority, Lower Colorado River Authority, Nueces River Authority, Texas Natural Resource Conservation Commission, Texas Water Development Board.

Hydrologic Engineering Center. 2009. *HEC-DSSVue HEC data storage system visual utility engine user's manual.* Version 2. CPD-79. Davis, CA: USACE HEC.

Intergovernmental Panel on Climate Change. 2023. *Synthesis Report of the IPCC Sixth Assessment Report.* New York: United Nations.

Kadir, T., E. Fernandes-McDade, B. Lock, C. Milanes, et al. 2018. *Recent research on climate change: A bibliography with an emphasis on California.* Sacramento, CA: California Environmental Protection Agency.

Kelley, V., R. Mace, and N. Deeds. 2008. "Groundwater availability modeling: The Texas experience." *Water Rep.* (54): 1–9.

McCrory, J. A. 1984. "Natural salt pollution control, Brazos River Basin, Texas." In *Salinity in watercourses and reservoirs,* edited by R. H. French. Boston: Butterworth.

Nielsen-Gammon, J. W. 2012. "The 2011 Texas drought." *Tex. Water J.* 3 (1): 59–95.

Nielsen-Gammon, J. W., J. L. Banner, B. I. Cook, D. M. Tremaine, et al. 2020. "Unprecedented drought challenges for Texas water resources in a changing climate: What do researchers and stakeholders need to know?" *Earth's Future* 8 (8): e2020EF001552.

Parsons Engineering Science. 1999. *Surface water/groundwater interaction evaluation for 22 Texas river basins*. Austin, TX: Texas Natural Resource Conservation Commission.

Rought, B. G. 1984. "The southwestern salinity situation: The Rockies to the Mississippi River." In *Salinity in watercourses and reservoirs*, edited by R. H. French. Boston, MA: Butterworth.

Shao, M., N. Fernando, J. Zhu, G. Zhao, et al. 2023. "Estimating future water availability through an integrated climate-hydrology-management modeling framework at a basin scale under CMIP6 scenarios." *Water Resour. Res.* 59 (7): e2022WR034099.

SWAT (Soil and Water Assessment Tool). 2023 "Soil and water assessment tool." Accessed November 28, 2023. https://swat.tamu.edu/.

Wei, W., and D. W. Watkins. 2011. "Probabilistic streamflow forecasts based on hydrologic persistence and large-scale climate signals in central Texas." *J. Hydroinf.* 13 (4): 760–774.

Wigmosta, M. S., L. W. Vail, and D. P. Lettenmaiser. 1994. "A distributed hydrology-vegetation model for complex terrain." *Water Resour. Res.* 30 (6): 1665–1679.

Williams, J. 1997. *The weather book: An easy-to-understand guide to the USA's weather*. New York: Random House.

Winters, K. E. 2013. *A historical perspective on precipitation, drought severity, and streamflow in Texas during 1951–56 and 2011*. Scientific Investigations Report 2013-5113. Reston, VA: US Geological Survey.

WSP Engineering. 2022. *Estimates of recharge and surface water—groundwater interactions for aquifers in central and west Texas*. Contract 2048302455. Austin, TX: Texas Water Development Board.

Wurbs, R. A. 2002. "Natural salt pollution control in the Southwest." *J. Am. Water Work. Assoc.* 94 (12): 58–57.

Wurbs, R. A. 2006. "Methods for developing naturalized monthly flows at gaged and ungaged sites." *J. Hydrol. Eng.* 11 (1): 55–64.

Wurbs, R. A. 2019a. *Daily water availability model for the Brazos River Basin and San Jacinto-Brazos Coastal Basin*. TR-513. College Station, TX: TWRI.

Wurbs, R. A. 2019b. *Water rights analysis package river system hydrology*. TR-431. College Station, TX: TWRI.

Wurbs, R. A. 2021a. "Monthly river flows in Texas for natural and developed conditions." *Water Cycle J.* 2 (2021): 1–14.

Wurbs, R. A. 2021b. Storage and regulation of river flows by dams and reservoirs. *Tex. Water J.* 12 (1): 10–39.

Wurbs, R. A. 2021c. "Water rights analysis package modeling system." *Tex. Water J.* 12 (1): 68–90.

Wurbs, R. A. 2023. *Daily water availability model for the Nueces River Basin*. Contract No. 582-18-80410. Austin, TX: TCEQ.

Wurbs, R. A., and C. H. Lee. 2009. *Salinity budget and WRAP salinity simulation studies of the Brazos river/reservoir system*. TR-352. College Station, TX: TWRI.

Wurbs, R. A., and C. H. Lee. 2011. "Salinity in water availability modeling." *J. Hydrol.* 407 (2): 451–459.

Wurbs, R. A., R. S. Muttiah, and F. Felden. 2005. "Incorporation of climate change in water availability modeling." *J. Hydrol. Eng.* 10 (3): 375–384.

Wurbs, R. A., and Y. Zhang. 2014. *River system hydrology in Texas*. TR-461. College Station, TX: TWRI.

Young, S. C., R. F. Mace, and C. Rubinstein. 2018. "Surface water-groundwater interaction issues in Texas." *Texas Water J.* 9 (1): 129–149.

Chapter 6

River/Reservoir System Development

The preceding chapter highlights the extreme spatial and temporal variability of precipitation, evaporation, and streamflow in Texas. Dams and reservoirs are essential for managing hydrologic fluctuations encompassing the extremes of floods and droughts along with seasonal and continuous variability. Numerous projects are operated throughout the state to store and control river flows for beneficial purposes. Most reservoir projects in Texas were constructed during the 1940s to 1980s era of large-scale water resources development nationwide. This chapter explores river basin development statewide. The next chapter focuses on specific key aspects of river and reservoir system management in each of the individual river basins of the state.

Dams and Reservoirs

Major reservoirs located partially or completely in Texas are listed in Appendix A with information regarding project location, ownership, purpose, age, and size. Additional information for each major reservoir is provided in Appendix B. The 195 projects in the two appendixes have conservation storage capacities and/or flood control storage capacities of 5,000 ac-ft or larger. Referencing Table 1-3 of Chapter 1, the 195 major reservoirs contain estimated conservation, flood control, and total storage capacities of 40,129,620 ac-ft, 18,743,090 ac-ft, and 58,872,710 ac-ft.

The 28 reservoirs with storage capacities exceeding 500,000 ac-ft are listed with descriptive information in Tables 1-4 and 1-5. Their locations are shown on the map in Figure 1-9. The 28 largest reservoirs contain 74.1%, 85.1%, and 77.6%, respectively, of the conservation, flood control, and total storage capacity of the 195 major reservoirs.

Thousands of small ponds, water supply storage tanks, stormwater detention basins, flood-retarding structures, and other storage facilities are located throughout Texas. Most of the total storage capacity is contained in the larger major reservoirs. Counts of the number of reservoirs included in various databases vary depending on the criteria adopted for inclusion in the database.

Inventories of Dams and Reservoirs

As of January 2024, Texas had 7,384 of the 91,843 dams included in the NID maintained by USACE in support of federal and state dam safety programs. The average age of the 7,384 dams located in Texas is 59 years compared to a national average of 62 years. Inclusion in NID is based on dam height and safety hazard criteria described in Chapter 2. Numerous smaller storage facilities are not included in NID. NID is discussed in Chapters 2 and 4 and is accessed through a website listed in Table 4-1.

As discussed later in this chapter, 3,464 reservoirs or allocated portions of reservoirs with conservation storage capacities equaling or exceeding 200 ac-ft are authorized by water rights administered by TCEQ. About 98% of the licensed conservation storage capacity is contained in about 210 reservoirs with authorized storage capacities of at least 5,000 ac-ft. Although most of the authorized reservoirs are existing, having been constructed over the last 100 years, several authorized reservoir projects have not yet actually been constructed.

TWDB provides information about river basins and major reservoirs through websites listed in Table 4-1 and discussed in Chapter 4. TWDB online compilations of reservoir information referenced throughout this book include 188 reservoirs with at least 5,000 ac-ft of water supply storage capacity and 20 other major reservoirs that serve other purposes without water supply. A set of historical and current condition databases provide observations for a variety of quantities. The TWDB reservoir storage dataset explored in this chapter comprises historical daily storage quantities for 122 large reservoirs.

The USACE FWD maintains an array of information at its website listed in Table 4-1 regarding the 25 multipurpose reservoirs that FWD owns and operates. Data from this website is also employed later in this chapter. The 25 reservoirs in the USACE FWD online information system are also included in the 188 reservoirs in the TWDB online information. The TWDB and USACE websites each include certain information regarding individual reservoirs not accessible at the other.

Reservoir Storage Pools

The total storage capacity of a multipurpose reservoir is commonly divided into active and inactive conservation storage, a designated flood control pool, and surcharge storage as illustrated by Figure 3-4 and discussed in Chapter 3. The top of conservation (bottom of flood control) and top of flood control pools are defined by specified water surface elevations.

Water levels encroach into surcharge storage when conservation and flood control storage capacities are exceeded. Surcharge storage volumes are not included in the storage capacity data cited in this book. A few vertical feet of surcharge above the gate-controlled storage capacity may represent a large volume since the horizontal water surface area grows larger as the water surface elevation rises toward the top of the dam.

Storage capacities are defined as volumes below a specified elevation or between two elevations. The common assumption of a horizontal, flat water surface reflected in estimates of storage volumes is reasonable. However, the water surface along the length of a reservoir actually slopes in the downstream direction as inflows flow through the reservoir and outlet structures.

The top of conservation pool elevation or normal target operating level may be precisely fixed or alternatively some degree of flexibility may be allowed to enhance operations. Storage capacity has been reallocated in several USACE reservoirs by seasonally, temporarily, or permanently varying the designated top of conservation elevation, which also defines the bottom of the flood control pool. Several permanent reallocations of flood control storage capacity to water supply by raising the top of conservation pool are noted in Chapters 3 and 6.

Storage capacities of reservoirs in Texas vary over time significantly though typically not dramatically due to accumulation of sediment deposits. Erosion and sediment deposition are highly dependent on flow rates and consequently highly dependent on the unpredictable timing and magnitude of infrequent flood or high flow events. Relationships between reservoir water surface elevation versus water surface area and storage volume are occasionally updated by site surveys to reflect loss of storage volume due to sedimentation. Original preconstruction determinations and later reevaluations of elevation/area/volume relationships and storage capacities are necessarily approximate. Stated storage capacities of federal reservoirs may be adjusted to reflect volumes reserved for predicted future sediment deposits as well as past sediment accumulation.

Volumes of Water Stored in Large Reservoirs

Water availability and supply reliability throughout Texas are governed by reservoir storage capabilities and characteristics examined in this section of this chapter. Likewise, flood flows of rivers throughout the state are greatly reduced by reservoirs. Reservoir storage variability and stationarity like most aspects of hydrology and water management in Texas vary greatly with location, particularly in traversing between western and eastern regions of the state.

TWDB Database

Variability, stationarity, and other characteristics of reservoir storage contents are explored here employing a database of observed daily storage volumes and associated quantities created and maintained by the staff of TWDB. Water conditions data compilations publicly accessible through a TWDB website listed in Table 4-1 include daily storage volumes for 122 reservoirs as well as an array of other water-related information. The 122 reservoirs included in the TWDB daily storage volume database contain 96% of the conservation storage capacity of the presently existing 188 major water supply reservoirs with storage capacities of at least 5,000 ac-ft located partially or completely in Texas. The data includes historical observations of the volume of water in acre-feet stored in each of the 122 reservoirs in each day of the period from the initial impoundment of water in each individual reservoir to the present. The reservoir storage database also includes storage capacities and other information.

TWDB obtains reservoir storage level observations from partner organizations that include USGS, USACE, IBWC, USBR, and river authorities. Measurements of water surface elevations by these agencies are automatically transmitted to TWDB in near real-time via the internet. Multiple observations during each day are combined to obtain averages for the day. The automated TWDB database system uses the latest updated storage volume versus water surface elevation relationships to convert observed water surface elevations to storage volumes. These relationships are occasionally updated by field surveys of sediment accumulation.

The following discussion incorporates three of the several variables included in the TWDB reservoir storage database: (1) Daily total storage contents, (2) daily active conservation storage contents belonging to Texas, and (3) active conservation storage capacity belonging to Texas. All three variables are volumes in units of acre-feet.

The daily total reservoir storage content is the actual observed total volume of water stored in a reservoir on a particular day, regardless of ownership. Associated volumes that include only water in active conservation storage committed to water users in Texas are computed and also recorded in the database. For reservoirs shared by Texas with Mexico or neighboring states, these active conservation storage values include only the Texas share of the conservation capacity and contents of that capacity. Inactive conservation storage for hydropower head or below the lowest outlet inverts is also excluded from the conservation storage amounts.

The daily storage time series plots and other metrics presented in this chapter were prepared by the author by downloading data from the TWDB website, converting the data from common separated values (CSV) text format to a DSS file, and plotting with the HEC-DSSVue software introduced in Chapter 4 and employed in Chapters 5 through 7. The following alternative compilations of daily time series of storage volume observations are investigated: (1) Individual total storage contents of each of 10 selected large reservoirs located across the state, (2) summation of total storage contents and also the active conservation portion of storage contents and associated active conservation capacities of multiple reservoirs in each of three river basins (Colorado, Brazos, Trinity), and (3) summations of these same three daily time series quantities for 122 reservoirs located throughout the state.

Historical Daily Storage in Selected Reservoirs

The 10 representative reservoirs listed in Tables 6-1 and 6-2 are selected to illustrate storage characteristics of large reservoirs located throughout Texas. The 10 reservoirs listed in Tables 6-1 and 6-2 are included in the 28 reservoirs located entirely or partially in Texas that have storage capacities of at least 500,000 ac-ft listed in Tables 1-4 and 1-5. The 28 dam site locations are shown in Figures 1-9 and 5-11. The map numbers in the first column of Table 6-1 refer to the dam locations shown on the maps of Figures 1-9 and 5-11. The reservoirs are numbered in order of total storage capacity, with the largest reservoir listed first.

The last two columns of Table 6-1 are the date of initial impoundment and the date at which the conservation pool was completely filled for the first time. Several or sometimes many months were required to fill each reservoir to the top of conservation pool after first closing the gates after completion of construction. The daily data series in the TWDB database begins with initial impoundment upon completion of construction. Datasets extending through December 2023 were downloaded for this chapter.

The total storage capacity below the top of conservation and top of flood control pools of the 10 selected representative reservoirs are shown in the second and third columns of Table 6-2. These are the same storage

Table 6-1. Ten Reservoirs Illustrative of Diverse Storage Characteristics across Texas.

Map No.	Plot Figure	Reservoir	River of Dam	Owner	Began Storage	First Filled
1	–	Texoma	Red River	USACE	10/1943	10/1944
2	6.1	Amistad	Rio Grande	IBWC	5/1968	8/1972
3	6.2	Toledo Bend	Sabine River	SRA	10/1966	6/1967
4	6.3	Sam Rayburn	Angelina River	USACE	1/1965	5/1965
5	6.1	Falcon	Rio Grande	IBWC	8/1953	9/1958
8	–	Travis	Colorado River	LCRA	9/1940	9/1942
9	6.4	Livingston	Trinity River	TRA	10/1968	11/1971
11	6.5	Richland-Chambers	Richland Creek	TRWA	11/1988	7/1989
13	6.6	Ray Roberts	Elm Fork Trinity	USACE	6/1987	3/1990
28	6.7	Possum Kingdom	Brazos River	BRA	3/1941	5/1941

Table 6-2. Storage Volumes for the 10 Large Reservoirs.

Reservoir	Top of Conservation	Top of Flood Control	Mean Storage	Minimum Storage (Date of Minimum)	Maximum Storage (Date of Maximum)
	(acre-feet)	(acre-feet)	(acre-feet)	(% of conservation)	(acre-feet)
Texoma	2,516,000	5,393,000	2,533,306	54.4% (9/1964)	6,011,000 (5/1990)
Amistad	3,276,000	5,010,000	2,418,515	22.0% (7/2022)	4,856,720 (9/1974)
Toledo Bend	4,477,000	None	4,031,861	57.6% (11/2011)	4,897,331 (3/2016)
Sam Rayburn	2,876,000	3,998,000	2,643,673	55.1% (8/1996)	4,284,515 (3/1992)
Falcon	2,647,000	3,148,000	1,422,457	7.51% (6/2002)	3,422,842 (7/2010)
Travis	1,135,000	1,922,000	915,103	27.3% (8/1951)	1,816,653 (1/1990)
Livingston	1,742,000	None	1,686,219	77.2% (10/1988)	2,080,000 (2/1989)
Richland-Chambers	1,113,000	None	1,056,079	62.9% (12/2014)	1,262,000 (5/1990)
Ray Roberts	788,490	1,064,600	737,915	53.8% (10/2000)	1,174,684 (5/1990)
Possum Kingdom	538,000	None	507,863	50.5% (2/1953)	746,529 (4/1957)

capacities tabulated in Table 1-5 and include all active and inactive conservation capacity regardless of ownership. These volumes are generally the latest published estimates of storage capacities, which vary from the volumes upon completion of construction due to sediment deposition in the reservoirs and in some cases later storage reallocations. The mean storage volumes in the fourth column of Table 6-2 are the average storage contents over the entire period from initial filling to full conservation storage capacity following completion of construction through December 2023.

Amistad, Sam Rayburn, Falcon, Travis, and Ray Roberts have flood control pools. The flood control pool capacity is the difference between the volumes below designated top of flood control and top of conservation pool elevations. Water levels also significantly exceed the conservation capacities of the five other reservoirs, thus encroaching into surcharge storage.

The fifth column of Table 6-2 shows minimum storage contents and the month and year of occurrence of the daily minimum during the period from initial full filling of the conservation capacity through December 2023. The minimum storage volume is expressed as a percent of the conservation storage capacity shown in the second column. The month and year in which the minimum daily storage volume occurred is shown in parenthesis in the fifth column of Table 6-2.

The minimum storage volume of International Falcon Reservoir to occur during the period from initial filling in September 1958 through December 2023 was 7.51% of the conservation storage capacity of 2,647,000 ac-ft. This minimum level occurred in June 2002. On the other extreme, the storage in Lake Livingston has never dropped below 77.2% of its capacity of 1,742,000 ac-ft. The minimum storage level of 77.2% of capacity at Lake Livingston occurred during October 1988. The minimum daily storage content of Lake Travis of 309,960 ac-ft (27.3% of 1,135,000 ac-ft) occurred during August 1951.

The maximum mean daily storage volume in acre-feet to occur on any day since the construction of the dam and reservoir project is shown in the last column of Table 6-2. The month and year in which the maximum daily storage level occurred is also shown in parenthesis. The peak storage levels in the five reservoirs with flood control storage are discussed later in the flood control storage section of this chapter.

Total observed daily storage contents since initial impoundment for eight of the 10 reservoirs in Tables 6-1 and 6-2 are plotted in Figures 6-1 through 6-7. Plots are not provided for Lakes Texoma and Travis

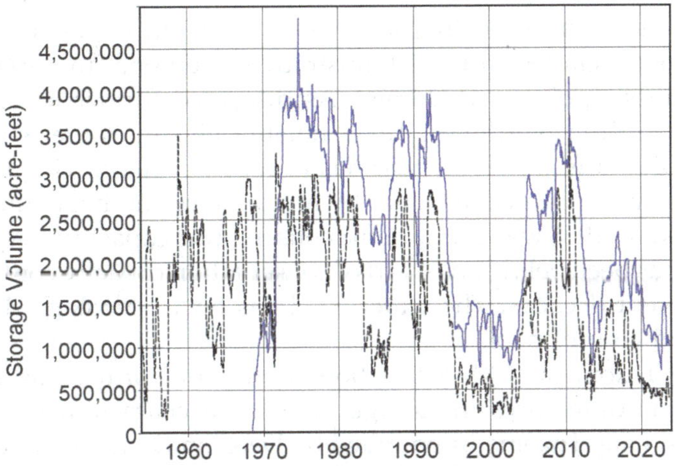

Figure 6-1. Storage in Amistad (blue solid line) and Falcon (black dashes) Reservoirs.

due to gaps of missing observed data, though the database does include estimated values for missing observations. The plots reflect complete continuous records. The storage plots of Figures 6-1 through 6-7 along with Table 6-2 illustrate the differences in reservoir storage drawdown and refilling characteristics in different regions of the state including Southwest Texas and East Texas and the span between crossing the state.

International Falcon (black dashes) and Amistad (blue solid line) Reservoirs are both included in Figure 6-1. The two-reservoir system on the Rio Grande is operated by IBWC to supply water for farmers and communities in the Lower Rio Grande Valley. The conservation pools of Falcon and Amistad were often full during the first several decades after construction but have been severely drawn down during much of the last 30 years. Two periods of dramatic storage depletions each exceed a decade in duration. As of January 2024, the two-reservoir system is experiencing an extreme drawdown that has already exceeded 12 years in duration with the future uncertain.

Other large reservoirs in West Texas that have been severely depleted continuously during the last several decades include Lakes Twin Buttes and O. H. Ivie with initial impoundment in 1962 and 1990, respectively. Water supply storage in both Twin Buttes and Ivie was at or near capacity for several years within the decade following construction but drawn down dramatically below conservation capacity continuously since 1978 and 1998, respectively.

Lake Meredith listed in Table 1-4 as the tenth largest reservoir in Texas is not included in the 10 illustrative reservoirs due primarily to gaps of missing observed storage. As discussed further in Chapter 7, Lake Meredith in Northwest Texas notably illustrates the unfortunate extreme of a large reservoir that has never been filled. The maximum volume of water stored in Lake Meredith since the beginning of impoundment in 1965 was 69.5% of the conservation capacity in April 1973. The lake has been below one-third of conservation capacity continuously during recent decades. The water level has never encroached into the large flood control pool.

Lakes Texoma and Meredith, both located in North Texas, illustrate the extreme differences in storage fluctuation characterizing different reservoirs. Lake Texoma impounded by Denison Dam on the Red River border between Texas and Oklahoma is the largest reservoir in either state. After significant drawdowns during the 1950s to 1960s, which included its most severe depletion to 54.4% capacity in September 1964, the conservation pool of Lake Texoma has since been full or almost full continuously. Storage levels often rise into the flood control pool. As discussed in Chapter 7, water supply use of Lake Texoma is constrained by natural salt pollution. Hydroelectric energy generation represents a major use of the conservation pool.

Water levels in Sam Rayburn and Toledo Bend Reservoirs in the Neches and Sabine River Basins of East Texas are at or above top of conservation much of the time as illustrated by Figures 6-2 and 6-3. Their

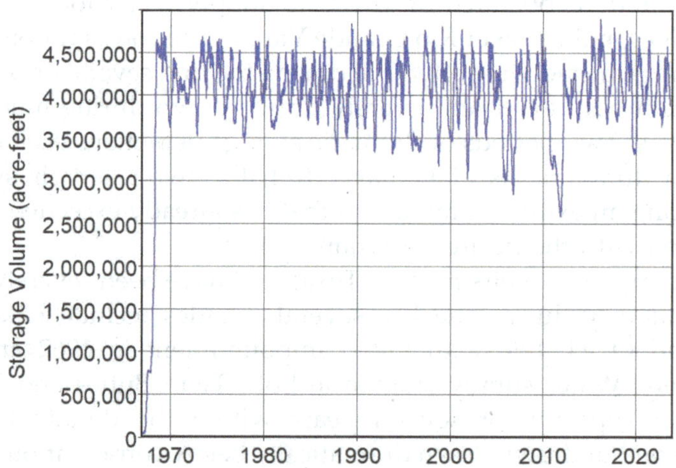

Figure 6-2. Storage in Toledo Bend Reservoir.

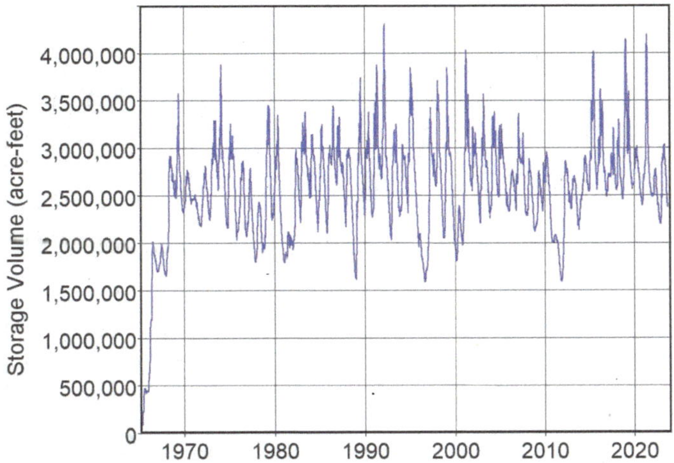

Figure 6-3. Storage in Sam Rayburn Reservoir.

seasonal and drought-driven drawdowns have never lowered storage severely below normal operating levels. The reasonably moderate drawdowns in the other reservoirs listed in Tables 6-1 and 6-2 and plotted in Figures 6-4 through 6-7 fall between extremes experienced in West and East Texas and are representative of many other Central Texas river and reservoir systems.

The volume of water stored in all reservoirs of a multiple-reservoir system is a more complete measure of water availability than the

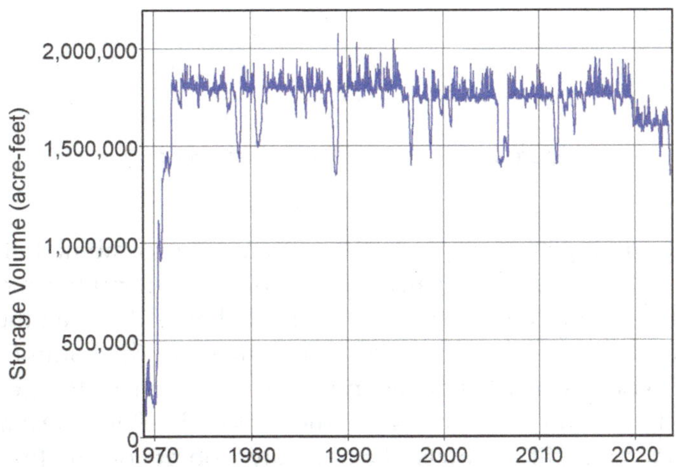

Figure 6-4. Storage in Livingston Reservoir.

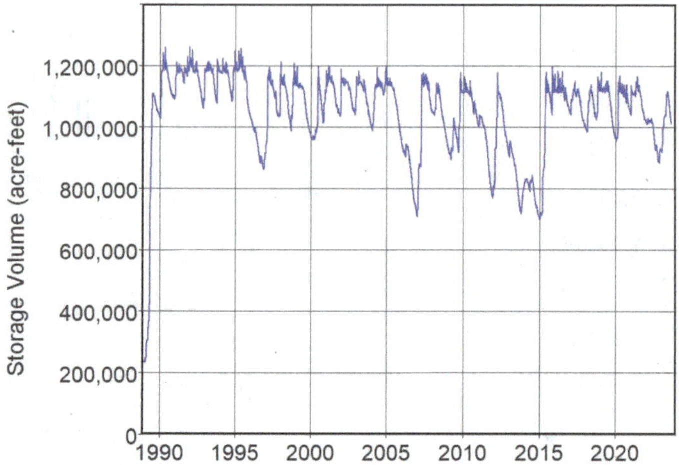

Figure 6-5. Storage in Richland-Chambers Reservoir.

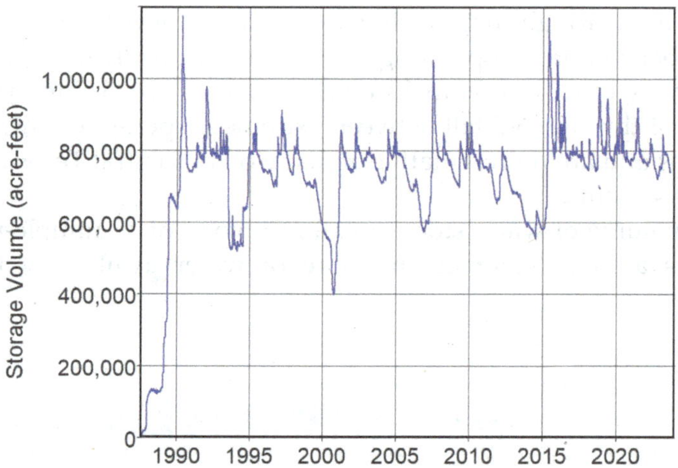

Figure 6-6. Storage in Ray Roberts Reservoir.

contents of a single reservoir of the system. The combined storage in International Amistad and Falcon Reservoirs is generally a more accurate measure of the volume of water controlled by IBWC than the content of either of the two individual reservoirs. Likewise, the volume of water available to LCRA customers depends on the combined storage contents of Lakes Buchanan and Travis on the Colorado River. Dallas Water Utilities operates Lakes Lewisville and Ray Roberts on the Elm Fork of the Trinity River as a system to supply its customers. Some systems

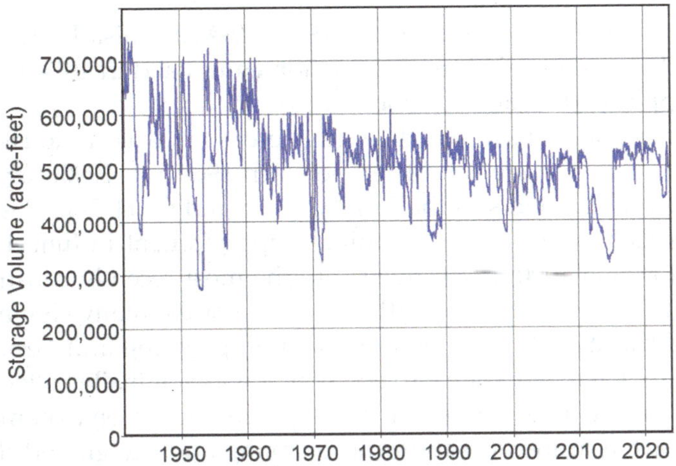

Figure 6-7. Storage in Possum Kingdom Reservoir.

include three or more reservoirs, with the storage content of each individual reservoir dependent on operating decisions that affect the allocation of storage and releases between the system reservoirs.

Estimates of storage capacities are occasionally adjusted based on site surveys of sediment accumulation. The effects of adjustments for sedimentation on the storage capacity can be seen in several of the storage plots. The estimated storage volume of Lake Livingston (Figure 6-4) below the normal operating elevation decreased a little as a result of a January 1991 sediment survey and much more after another survey in November 2019. The storage plot for Richland-Chambers Reservoir in Figure 6-5 reflects storage capacity adjustments for sediment deposition based on site surveys in December 1994, October 2007, and December 2012.

Possum Kingdom Reservoir (Figure 6-7) has lost a particularly large volume of storage capacity due to sedimentation since initial impoundment in March 1941. The most recent surveys of Possum Kingdom Reservoir were in June 1994, January 2005, and December 2016. The initial conservation storage capacity of Possum Kingdom of 724,700 ac-ft at completion of construction in 1941 had decreased to 538,100 ac-ft in the 2016 sediment survey.

The oldest of the 28 largest reservoirs (Tables 1-4 and 1-5) are Lakes Kemp, Buchanan, Travis, and Possum Kingdom with initial impoundment in 1922, 1937, 1940, and 1941, respectively. Lake Kemp on the Wichita River in the Red River Basin is the oldest of the 28 largest reservoirs. Initial impoundment of a smaller Lake Kemp constructed solely for water supply occurred in 1922. USACE Tulsa District greatly enlarged Lake Kemp with reconstruction completed in 1974 to increase water supply and to add a flood control pool.

Construction of 11 of the 28 largest reservoirs was completed before 1960. The other 17 were completed between 1960 and 1990. Ray Roberts and Richland-Chambers in the Trinity Basin and O. H. Ivie Reservoir in the upper Colorado Basin, with initial impoundment in June 1987, July 1987, and March 1990, respectively, are the most recently constructed of the 28 largest reservoirs. The author was one of the many civil engineers employed by USACE FWD who worked on planning and design of the Ray Roberts Lake and Dam project, which was originally called Aubrey Dam and Reservoir. A common naming practice has been to name federal projects for the nearest town during planning and design and then later near or after construction to rename the project for a political official.

Summations of Historical Daily Storage in Selected River Basins

The time series plots of the summation of daily storage volumes in 19, 28, and 24 reservoirs presented in Figures 6-8 through 6-10 illustrate the history of reservoir development in the Colorado, Brazos, and Trinity River Basins. Most of the larger major reservoirs, but not all major reservoirs, in each river basin are included in the summations.

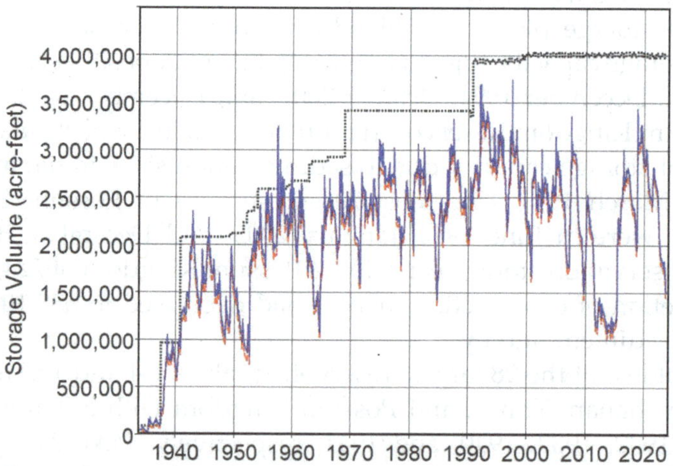

Figure 6-8. Total storage contents (solid blue line), active conservation storage contents (red dashes), and active conservation storage capacity (black dots) of 19 reservoirs on the Colorado River and tributaries.

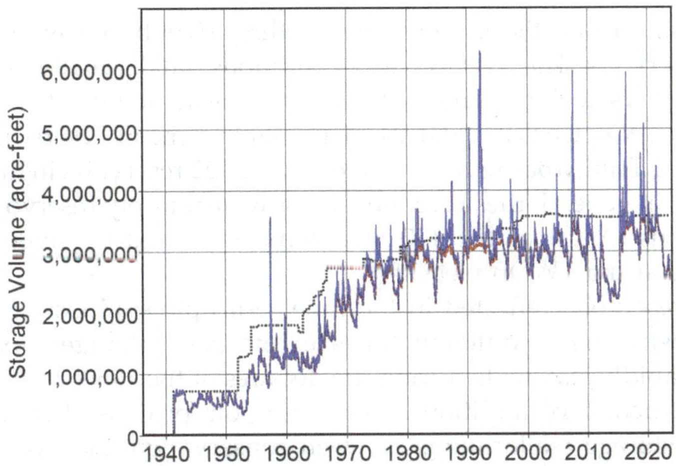

Figure 6-9. Total storage contents (solid blue line), active conservation storage contents (red dashes), and active conservation storage capacity (black dots) of 28 reservoirs on the Brazos River and tributaries.

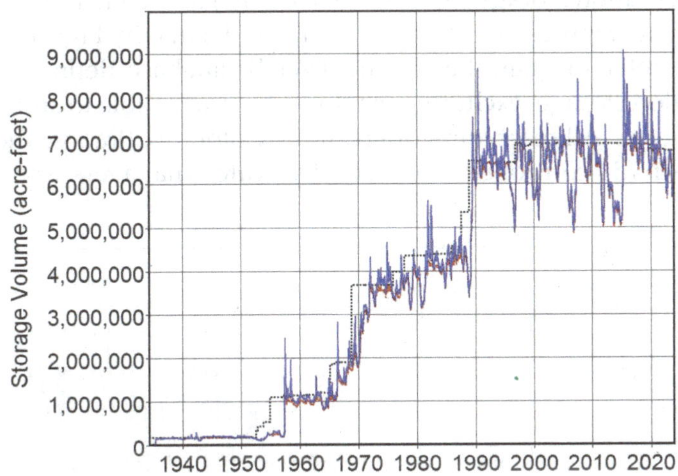

Figure 6-10. Total storage contents (solid blue line), active conservation storage contents (red dashes), and active conservation storage capacity (black dots) of 24 reservoirs on the Trinity River and tributaries.

A total of 195 major reservoirs located in 15 major river basins and three coastal basins are described in Appendixes A and B. Five other coastal basins have no major reservoirs. Twenty-eight, 42, and 31 of these 195 major reservoirs are located in the Colorado, Brazos, and Trinity

River Basins, respectively. These are the three river basins with the most major reservoirs. The TWDB water conditions database allows data for 122 reservoirs to be aggregated by river basin or by other types of groupings. Nineteen, 28, and 24 of the largest major reservoirs in the Brazos and Trinity Basins are included in the 122 reservoirs in the TWDB database. In general, these are the largest water supply reservoirs in the river basins. A list of the 122 reservoirs organized by river basin is accessible at the TWDB website.

The reservoirs reflected in the time series plots all exist today but were constructed at different times in the past. Storage capacity increased rapidly statewide during the decades of the 1960s to 1970s with the construction of additional new reservoir projects. Permanent or seasonal storage reallocations between conservation and flood control pools have been implemented at several reservoirs as discussed later in this chapter. The conservation storage capacity quantities in the TWDB database are also adjusted for occasional site surveys to estimate sediment accumulation.

Conservation storage capacity (black dots) and content (red dashes) presented in Figures 6-8, 6-9, and 6-10 and later in Figure 6-11 and Table 6-3 refer to quantities in the TWDB database defined as active conservation storage, excluding inactive pools, available to supply water users in Texas. All reservoir storage in the Colorado, Brazos, and Trinity River Basins is controlled and used by water managers and users in

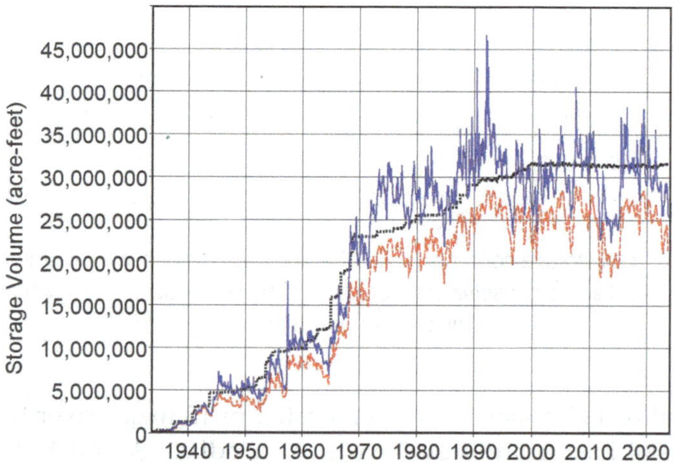

Figure 6-11. Total storage contents (solid blue line), active conservation storage contents (red dashes), and active conservation storage capacity (black dots) of 122 reservoirs.

Table 6-3. Statewide Averages of Annual Precipitation and Evaporation Depths and Summations of the Active Conservation Storage Contents of 122 Reservoirs.

Year	Annual Precip. (inches)	Annual Evap. (inches)	End-of-Month Conservation Storage Contents (% of Active Conservation Capacity)			
			March	June	September	December
Min	13.6	53.6	64.6	65.6	60.0	60.8
Mean	27.4	61.0	81.2	81.6	76.2	77.4
Max	39.9	73.0	91.2	90.2	90.3	89.7
1998	27.3	64.4	86.2	78.8	73.0	82.0
1999	21.9	69.2	82.5	84.2	74.7	70.1
2000	26.0	62.3	70.6	77.1	65.5	78.1
2001	27.2	58.7	84.8	82.4	75.9	80.0
2002	28.9	60.4	81.2	78.9	76.4	83.2
2003	24.6	60.6	83.8	83.1	76.5	76.1
2004	39.1	55.2	82.5	87.7	83.5	88.9
2005	22.3	58.7	90.1	85.9	80.0	76.0
2006	26.1	66.2	80.0	76.5	67.7	72.0
2007	35.4	53.6	80.8	90.1	90.3	88.0
2008	24.4	63.3	91.2	87.4	82.4	80.6
2009	28.4	63.2	81.6	82.6	77.3	83.1
2010	27.5	61.0	87.2	86.1	83.7	80.2
2011	13.6	73.0	79.6	73.4	60.7	60.8
2012	23.6	63.4	78.4	75.4	68.4	64.8
2013	26.1	61.9	66.1	65.6	60.0	64.3
2014	24.2	59.7	64.6	67.6	63.7	62.6
2015	39.9	56.4	69.7	83.9	79.1	85.5
2016	28.6	58.4	86.5	87.6	84.4	83.6
2017	30.6	60.0	85.4	86.4	84.3	82.2
2018	34.0	59.2	85.5	80.0	79.3	89.7
2019	27.8	55.6	89.3	90.2	82.0	80.9
2020	26.7	59.8	87.3	85.5	81.7	80.1
2021	27.5	56.5	81.9	85.2	80.8	78.1
2022	22.6	65.1	78.4	75.2	69.4	71.2
2023	–	–	74.6	76.5	65.9	68.8

Texas. Several reservoirs have inactive storage pools defined by hydroelectric power operations or outlet invert elevations. The blue solid line in these figures is the total storage contents including all flood control and conservation storage.

Summations of values for specified groups of reservoirs are plotted as three time series variables in Figures 6-8 through 6-10, and later likewise in Figure 6-11. For each of these three time series variables, Figures 6-8 through 6-10 show summations for 19, 28, and 24 reservoirs in the Colorado, Brazos, and Trinity River Basin, respectively. Summations for 122 reservoirs located throughout Texas are plotted in Figure 6-11. The difference between total storage and active conservation storage is comprised of the combined volumes of inactive conservation, flood control, and surcharge storage. Conservation storage content never exceeds conservation capacity in the daily volume quantities recorded in the TWDB database by definition.

Summations of Daily Storage in 122 Major Reservoirs

The storage volume plots in Figure 6-11 are summations for 122 reservoirs located totally or partially in Texas representing 96% of the Texas active conservation storage capacity of all major water supply reservoirs. Several of the 122 reservoirs in the TWDB database are operated to supply water users in neighboring states and Mexico as well as Texas. As discussed earlier, the summations of conservation volumes for the 122 reservoirs are adjusted to reflect only active conservation storage capacity and contents belonging to Texas.

Plots are presented in Figure 6-11 as well as in Figures 6-8 through 6-10 for the (1) total observed storage contents, (2) portion of the observed storage contents that is contained in active Texas conservation pools, and (3) active conservation capacity controlled by water managers for use by water users in Texas. The July 1933 through December 2023 time series plots reflect storage in reservoirs that existed in 2023, but the reservoir projects were constructed at different times. The majority were constructed during the 1960s to 1980s.

The 1940s to 1980s construction era of water resources development is apparent in the reservoir storage plots. The best sites for constructing reservoir projects were developed early. Economic feasibility (benefits exceeding costs), financial feasibility (funding availability), and environmental impacts of constructing additional new projects became more constraining over time. Population and economic growth resulted in continually increasing demands for the services provided by reservoir projects but also increased the economic and environmental costs of committing additional land and other resources to developing new projects.

Construction of Cooper Dam and Jim Chapman Lake, originally called Cooper Dam and Lake, on the Sulphur River by USACE FWD began in 1987 and was completed in September 1991. As of January 2024, this is the last large dam and reservoir project in Texas with construction totally completed since 1991. With conservation and flood control storage capacities of 298,930 and 130,405 ac-ft which sum to a total capacity of 429,335 ac-ft, Jim Chapman Lake is not included in the 28 reservoirs with capacity exceeding 500,000 listed in Tables 1-4 and 1-5.

Impoundment of water was initiated in April 2021 at the Bois d'Arc Reservoir project on Bois d'Arc Creek in the Red River Basin. However, as of January 2024, construction of some components is still underway for this project owned by the North Texas Municipal Water District. As discussed in Chapter 1, other new reservoir projects were also under construction in 2023. The Texas Water Plan includes several proposed new reservoirs for development in the future (TWDB 2022). Future projects are expected to be smaller and, in many cases, to be off-channel storage facilities rather than dams constructed across major rivers.

The large volumes of storage capacity discussed here were developed in the past to provide for needs projected to occur during a several-decade distant future as well as the then current demands for water services. Population, economic development, and water needs have grown continually over the last many decades and are expected to continue to grow in the future.

Reservoir Storage Contents as a Drought Index

As discussed in earlier chapters, the 1950 to 1957 drought is the most hydrologically severe drought for much if not the majority of Texas and is at least one of the three most hydrologically severe droughts for essentially all of Texas since before 1900. This drought began gradually in 1950 and ended in April 1957 with one of the greatest floods on record with large rainfall depths across the state. The 1950 to 1957 drought and April–May 1957 flood motivated creation of TWDB, statewide adjudication of water rights, aggressive development of additional reservoir projects, and other water management endeavors to deal with future droughts and floods.

Hydrologically severe drought refers to abnormally low rainfall and streamflow. Other more recent droughts have been more economically costly than the 1950 to 1957 drought due to population and economic growth since the 1950s. The historic Dust Bowl drought of the 1930s in Texas, Oklahoma, New Mexico, Colorado, and Kansas predates most of the reservoir development and population growth and was not as widespread within Texas as the 1950 to 1957 drought.

The volume of water in storage is a basic measure of water availability that can also be viewed as a drought index. The volume of water stored in one or multiple reservoirs reflects recent and past water use, reservoir evaporation, interactions between surface and groundwater, and various aspects of weather and hydrology including rainfall and watershed evapotranspiration. Reservoir storage volume provides a comprehensive water availability metric reflecting water resources development and use as well as weather and hydrology. However, the governing shortcoming of using historical observed storage contents as an index for comparing past and current drought conditions is that the number of reservoirs and associated storage capacities and water demands have grown dramatically over time. For example, water use, number of reservoirs, and total reservoir storage capacity during the 1950 to 1957 drought were much less than during the 2011 to 2014 drought.

Storage simulated in the WRAP/WAM computer modeling system for a defined stationary condition of water resources development and use, combined with historical natural hydrology, is adopted as a drought index in Chapter 7 and actual water resources planning and management, rather than actual observed storage. The WRAP/WAM modeling strategy deals with the issue of water development and use growing historically over time. A defined constant scenario of water development and use is combined with computed reasonably stationary hydrology. The modeling system is introduced in Chapter 4 and employed later in Chapter 6, as well as in Chapters 5 and 7.

Storage Contents during 26 Years of Nearly Constant Capacity

The following discussion focuses on the most recent 26 years because the number and storage capacity of major reservoirs in Texas have remained almost stable during this period. The population of Texas grew greatly from 9.6 million people in 1960 to 20.9 million in 2000 and 29.7 million in 2020 resulting in increased water demands. However, reservoir storage capacity has remained essentially constant over the last 26 years. Although other factors affecting reservoir storage such as population and economic growth, urbanization and other watershed land use changes, and water resources management and use have changed significantly, at least the total storage capacity contained in major reservoirs has remained reasonably constant during the period from January 1998 through December 2023 as addressed in the following discussion.

Many different metrics may be employed for measuring drought conditions including those tabulated in Table 6-3. Statewide annual precipitation and reservoir evaporation rates provide measures of meteorological conditions and may be viewed as indicators of meteorological drought. The statewide average annual precipitation and reservoir evaporation depths in inches for January through December of each year from 1998 to 2022 in Table 6-3 are subsets of the 1940 to 2022 precipitation and 1954 to 2022 evaporation time series datasets discussed in Chapter 5. The 2023 statewide annual precipitation and evaporation depths were still unknown when this book went to press.

The summation of conservation storage contents expressed as a percentage of total active conservation storage capacity for the 122 water supply reservoirs is also tabulated in Table 6-3. The volumes of water stored in the 122 major reservoirs at the end of the months of March, June, September, and December of each year of the period 1998 to 2023 are tabulated. The storage was at 84.5% of capacity at the end of December 1997 and 74.2% capacity on February 22, 2024.

The mean, minimum, and maximum of the annual quantities are shown immediately below the headings in Table 6-3. The annual precipitation and evaporation depths and reservoir storage content as a percentage of capacity for each year are listed below the summary statistics. The conservation storage volumes at the end of March, June, September, and December of each year are tabulated as a percentage of the corresponding conservation storage capacity.

The statewide average 1940 to 2022 annual precipitation and 1954 to 2022 annual reservoir evaporation depths plotted in Figure 5-5 have means of 28.03 and 59.43 in. Subperiods are compared as follows. Statewide average 58-year mean 1940 to 1997 annual precipitation is 28.32 in. The 44-year mean 1954 to 1997 annual reservoir evaporation mean is 58.52 in. The 25-year 1998 to 2022 means for annual precipitation and evaporation depths are 27.36 and 61.04 in.

The 10 smallest values for statewide annual precipitation in inches during 1940 to 2022 are 13.6 (2011), 16.7 (1956), 19.1 (1954), 19.9 (1963), 21.8 (1999), 21.9 (1948), 22.3 (2005), 22.4 (1943), and 22.64 (1988 and 2022). The 10 largest values in inches for statewide annual precipitation during 1940 to 2022 are 40.6 (1941), 39.9 (2015), 39.1 (2004), 37.1 (1991), 36.3 (1973), 35.8 (1957), 35.4 (2007), 34.4 (1997), 33.9 (2018), and 33.8 (1949). The highest annual reservoir evaporation rates in inches are 73.0 (2011), 69.2 (1999), 68.6 (1956), 67.6 (1993), and 66.2 (2006).

As discussed in Chapter 5, annual precipitation appears stationary during 1940 to 2022, including 1998 to 2022. If long-term trends or permanent changes have occurred, they are hidden by the great variability of rainfall. Evaporation rates appear to have possibly increased, which would be consistent with global warming. However, continuous variability makes long-term trends and permanent changes difficult to detect and measure. From the perspective of storage in major reservoirs, major droughts have been defined by sequences of multiple abnormally dry years rather than a single dry year or season.

Reservoir Storage Volumes during Droughts

Individual water managers are interested in storage in their specific reservoir or multiple-reservoir system, which tends to fluctuate more than the summation of storage volumes in many reservoirs located throughout the state. The timing of stream inflows, water supply diversions and releases, and resulting reservoir storage drawdowns differ between reservoirs. Different reservoirs are drawn down and refilled at different times. Summations of storage in 122 reservoirs average out the fluctuations of storage in the individual reservoirs as illustrated in the preceding storage time series plots. Fluctuations in statewide totals of reservoir storage contents are significant but not as dramatic as many of the individual reservoirs. However, broader basin-wide and statewide perspectives also provide interesting and relevant insight into hydrology and water management both in Texas and nationwide.

Streamflows at the gauge sites shown in the map of Figure 5-11 are extremely variable as discussed in Chapter 5. Reservoir storage greatly attenuates streamflow fluctuations. Although seasonal and short-term fluctuations in streamflow, water use, and reservoir storage are important in reservoir operation in Texas, the governing concern is preparation for severe multiple-year droughts. Large volumes of water must be stored to meet water needs during droughts extending over several years such as 1950 to 1957 or 2011 to 2014.

The largest drawdowns of reservoir storage during 1998 to 2023 occurred during 2011 to 2014. The lowest minimum storage at the end of March, June, September, and December occurring during any year since 1998 shown in Table 6-3 were in March 2014 (64.6% of capacity), June 2013 (65.6%), September 2013 (60.0%), and December 2011 (60.8% of capacity), respectively. The storage plots in Figures 6-1 through 6-10 illustrate the differences in storage drawdowns at different locations throughout the state.

Table 6-3 includes 104 end-of-month storage volumes expressed as a percent of capacity at 3-month intervals extending from March 1998 through December 2023. The six lowest storage levels in this compilation occurred in September 2013 (60.0%), September 2011 (60.7%), December 2011 (60.8%), December 2014 (62.6%), September 2014 (63.7%), and December 2013 (64.3%).

The minimum statewide average annual rainfall and maximum statewide average annual reservoir evaporation rate during the 1998 to 2022 period of analysis were 13.6 and 73.0 in., respectively, both in 2011. The annual precipitation of 13.6 in. in 2011 is the minimum annual precipitation during 1940 to 2022 as well as 1998 to 2022. The annual evaporation of 73.0 in. in 2011 is also the maximum annual evaporation during 1954 to 2022.

The second lowest annual precipitation and second largest annual evaporation rate in Table 6-3 are 21.9 and 69.2 in., respectively, both in 1999. However, reservoir drawdowns in 1999 and years immediately preceding and following 1999 are not as great as during 2011 to 2014.

General Observations

Reservoirs significantly but only partially reduce losses resulting from droughts. Large areas of forests, pastures, croplands, and urban landscapes subject to adverse effects of drought have no access to water stored in surface reservoirs or groundwater aquifers. Direct rainfall is important even if supplemented by irrigation. Water supply capabilities during drought may be limited by water conveyance, treatment, and distribution capacities rather than reservoir storage. Water managers deal with diverse aspects of drought that include emergency demand management decisions regarding short-term actions to assure water users continue to have water that week and during the next several weeks.

The large volumes of water stored in the reservoirs discussed in this chapter are necessitated by extreme hydrologic variability and increasing water needs associated with population and economic growth. Large storage capacities in reservoirs throughout the state are a key component of managing hydrologic variability and uncertainties regarding future water availability and water needs. Operational modifications such as storage reallocations to optimize available reservoir storage capacity and an array of strategies for integration of reservoir operations with long-term and emergency water demand management measures and groundwater use are also important components of drought management.

Observed reservoir storage content during 1940 to 1990 is highly nonstationary due largely to construction of many reservoir projects at different times. Although reservoir storage capacity has been fairly constant since 1990, storage content from 1940 continuing into the future is highly nonstationary due to continual population and economic growth and associated water use. Global warming and associated climate change add to the already great future uncertainty.

Severe drawdowns are generally in response to a series of multiple very dry years rather than the dry season of a single year. Storage in individual reservoirs typically fluctuates much more than the aggregated total storage content of many reservoirs.

Partially Understood Recent and Unknown Future Conditions

When this book went to press in early 2024, the year 2023 had been the hottest on record for most of Texas and other regions of the nation and planet. The year 2023 was also extremely dry with abnormally low rainfall in many regions of the state. The preceding year 2022 was also generally hotter and drier than normal throughout Texas, though some regions experienced near-average rainfall in some months of both 2022 and 2023. Rainfall varies much more than temperature both spatially and temporally.

The statewide average annual precipitation and reservoir evaporation rate during 2022 was 22.6 in. and 65.1 in. The 2022 precipitation is the fourth lowest, and the 2022 evaporation is the fourth highest during the 1998 to 2022 period of analysis of Table 6-3. The end of March 2023,

June 2023, September 2023, and December 2023 combined storage contents of 122 reservoirs of 74.6%, 76.5%, 65.9%, and 68.8% of active conservation storage capacity are the fifth, seventh, fifth, and fifth, respectively, lowest storage levels for those days of the year since 1998. The storage was at 84.5% of capacity at the end of December 1997. For comparison, the storage contents for these four days of the year during the period December 1997 through December 2023 varied within the range of 91.2% of capacity in March 2008 to 60.85% of capacity in December 2011.

The future is uncertain. Any day without rain could be the first day of the next multiple-year drought that is more hydrologically severe and economically costly than the 1950 to 1957 and 2011 to 2014 droughts. In January 2024, Texas water users and water managers know that the hot and dry 2022 to 2023 could be the beginning of the next extreme drought. The years 2024 and 2025 may be even hotter and drier than 2023. Conversely, the 2022 to 2023 hot and dry conditions, like similar occurrences in the past, may end either quickly and dramatically with major floods or gradually over several years of close-to-average rainfall. Local thunderstorms may release large amounts of intense rainfall throughout any year in any river basin, even occasionally in the desert of West Texas. Prolonged, widespread rain is also common throughout the state.

Droughts may end dramatically with major floods resulting from tropical storms or other types of extreme rainfall events as well as, more commonly, ending more gradually. Hurricane Idalia crossed the Gulf of Mexico reaching Florida on August 30, 2023, causing great damage in Northern Florida and Southeast Georgia. A summer of unusually hot and dry conditions in Florida and Georgia quickly switched to extreme damage from high winds and intense flooding. Hurricane Idalia missed Texas. However, Texas has a long history of massive flooding resulting from similar tropical storms, the most recent being Hurricane Harvey in 2017 and Tropical Storm Imelda in 2019. Hurricane Harvey, with rainfall depths exceeding 60 in. in some locations, flooded coastal areas of Texas during August 25–29, 2017, causing the most economic damage of any hurricane in the history of the state.

Flood Control Storage

Reservoir operations for flood control versus conservation purposes in multipurpose federal reservoirs are organized based on designating a top of conservation pool elevation which serves as the bottom of the flood control pool as illustrated by Figure 3-4 and explained in Chapter 3. Flood control operations are activated by the water surface encroaching into the flood control pool. Otherwise, operations focus on the use of the conservation pool for water supply, hydropower, recreation, and other conservation pool purposes. However, reservoirs serving only conservation purposes and the conservation pool storage in federal multipurpose reservoirs incidentally also contribute to flood control. To a lesser extent, flood control pool operations may incidentally increase the volume of water stored in conservation pools.

Flood control operating rules and procedures are described in Chapter 3. Unless the flood control pool is overtopped into surcharge storage, storage and releases from the flood control pool of a reservoir or the flood control pools of a system of multiple reservoirs are based on making no releases that would contribute to flows at the dams and multiple specified downstream gauge sites exceeding specified allowable nondamaging streamflow levels. In some cases, the allowable flows at downstream gauge sites may vary with storage levels in the flood control pools.

Thirty-Six Flood Control Reservoirs

Thirty-three of the 195 major reservoirs in the appendixes contain both flood control capacities of at least 5,000 ac-ft and conservation capacities of at least 5,000 ac-ft. Three other reservoirs (Addicks, Barker, and Olmos) have flood control capacities greater than 5,000 ac-ft but no conservation storage.

The 36 major reservoirs with designated flood control storage capacities of 5,000 ac-ft or more are listed in Tables 6-4 and 6-5, where FC denotes flood control. Eighteen of the 36 flood control reservoirs in Tables 6-4 and 6-5 have total storage capacities exceeding 500,000 ac-ft and thus are also included in the list of the 28 largest reservoirs in Tables 1-4 and 1-5 with locations shown on the maps of Figures 1-9 and 5-11. The flood control capacity quantities in the fourth column of Tables 6-4 and 6-5 reflect only officially designated flood control pools. The flood control storage capacities are the total storage volumes between the elevations of the top and bottom of designated flood control pools. These 36 reservoirs have conservation, flood control, and total storage capacities that sum to 19,166,752, 19,297,292, and 38,464,044 ac-ft.

Table 6-4. US Army Corps of Engineers Fort Worth District Multiple-Purpose Reservoirs.

Reservoir	Storage Began	Conservation Storage	Flood Control	Height FC Pool	Peak Height	Date of Peak	Times FC Exceeded
		(acre-feet)	(acre-feet)	(feet)	(feet)		
Sulphur River and South Sulphur River							
Wright Patman	6/1956	97,927	2,556,048	32.0	27.1	5/10/2016	0
Jim Chapman	9/1991	298,930	130,405	6.2	8.6	3/2/2018	5
Cypress Creek							
Lake O' the Pines	8/1957	241,363	600,737	21.0	17.4	5/5/1966	0
Angelina and Neches Rivers							
Sam Rayburn	3/1965	2,876,033	1,121,567	8.6	11.0	6/3/2021	5
B.A. Steinhagen	4/1951	66,972	57,728	2.0	87.8	1/3/1992	many
Trinity River and tributaries							
Ray Roberts	6/1987	788,490	276,110	8.0	11.9	5/3/1990	3
Lewisville	11/1954	598,902	382,861	10.0	14.9	5/31/2015	9
Lavon	9/1953	409,300	338,900	11.5	12.9	5/3/1990	2
Benbrook	9/1952	85,648	172,982	16.0	23.5	5/3/1990	7
Bardwell	11/1965	54,900	85,100	18.0	19.7	6/18/2015	1
Navarro Mills	3/1963	49,827	149,403	18.5	18.8	10/31/2015	1
Grapevine	7/1952	163,064	235,136	25.0	28.5	1[illegible]/1/1981	5
Joe Pool	1/1986	176,900	127,100	14.0	14.0	5/30/2015	1
Brazos River and Tributaries							
Whitney	12/1951	1,999,500	554,203	38.0	37.3	5/29/1957	1

(Continued)

Table 6-4. US Army Corps of Engineers Fort Worth District Multiple-Purpose Reservoirs. (Continued)

Reservoir	Storage Began	Conservation Storage	Flood Control	Height FC Pool	Peak Height	Date of Peak	Times FC Exceeded
Belton	3/1954	435,225	662,375	37.0	40.4	3/6/1992	1
Waco	2/1965	189,773	536,587	38.0	26.5	12/24/1991	0
Stillhouse Hollow	2/1968	227,825	402,575	44.0	46.0	3/5/1992	1
Somerville	1/1967	150,293	347,400	20.0	21.8	5/29/2016	1
Aquilla	4/1983	43,279	102,721	27.0	14.4	12/24/1983	0
Proctor	9/1963	54,762	319,438	35.0	35.7	6/4/2016	2
Georgetown	3/1980	36,904	93,896	43.0	44.9	3/4/1992	1
Granger	1/1980	51,882	192,378	24.0	26.1	3/5/1992	1
Concho River and Hords Creek in Colorado River Basin							
O. C. Fisher	2/1952	115,743	276,944	30.5	8.5	10/14/1957	0
Hords Creek	4/1948	7,959	16,678	20.0	7.7	6/23/1997	0
Guadalupe River							
Canyon	6/1964	378,852	362,048	34.0	41.2	7/6/2002	2

Table 6-5. Flood Control Reservoirs in Texas Not Included in Table 6-4.

Reservoir	Storage Began	Conservation Storage	Flood Control	Height FC Pool	Peak Height	Date of Peak	Times FC Exceeded
		(acre-feet)	(acre-feet)	(feet)	(feet)		
USACE Galveston District Projects on Buffalo Bayou and Mayde Creek in San Jacinto Basin							
Addicks	1/1948	none	199,614	42.0	43.1	8/30/2017	1
Barker	1/1945	none	209,600	35.5	33.0	8/30/2017	0
USACE Tulsa District Reservoir Projects in Red River Basin							
Texoma	10/1943	2,516,000	2,877,000	23.3	28.7	5/7/1990	3
Kemp	10/1972	245,430	322,470	12.0	12.0+	–	many
Pat Mayse	8/1967	189,100	71,256	9.5	9.5+	12/12/1971	2
Constructed by USBR and Operated by Nonfederal Sponsors (LCRA, San Angelo, CRMWD)							
Travis	9/1940	1,134,956	786,775	33.0	23.1	2/8/1992	0
Twin Buttes	12/1962	186,000	446,000	45.2	0.2	2/15/1975	0
Meredith	1/1965	817,970	589,630	68.4	-0-	–	0
International Boundary and Water Commission (IBWC) Reservoirs on the Rio Grande							
Amistad	5/1968	3,275,530	1,734,400	23.4	18.6	9/23/1974	0
Falcon	8/1953	2,646,810	501,330	5.5	6.8	10/19/1958	0
Reservoir on Olmos Creek Owned and Operated by San Antonio River Authority (SARA)							
Olmos	1926	none	12,600	54	–	–	0

Operation of Wright Patman Reservoir follows a seasonal rule curve with the top of conservation pool elevation set at 220.6 ft above mean sea level during November through March and raised to 227.5 ft from April through October in an annual cycle. The capacity at elevation 220.6 ft is 97,927 ac-ft (Table 6-4).

The designated top of conservation pool elevation of Lake Texoma is optionally varied seasonally between 615 and 619 ft above mean sea level. The lake has 2,516,000 ac-ft conservation capacity encompassing a surface area of 74,690 ac at elevation of 617 ft (Table 6-5). The total storage capacity is 5,393,000 ac-ft at top of flood control pool of elevation 640 ft. The top of dam is at elevation of 670 ft, and the uncontrolled spillway crest is at elevation of 640 ft. The maximum design water surface is 666.4 ft. Figure 3-1 in Chapter 3 is a photograph of the spillway and hydroelectric plant at Denison Dam.

The 25 reservoirs in Table 6-4 are owned and operated by FWD. All 25 are multipurpose reservoirs containing both large conservation and flood control pools. Conservation, flood control, and total storage capacities sum to 8,154,956, 11,546,617, and 19,701,573 ac-ft. The information tabulated in Table 6-4 was obtained from the USACE FWD website listed in Table 4-1. The website provides an array of information about USACE FWD reservoirs. The storage capacities in Table 6-4 are from pertinent data sheets compiled by USACE FWD for each reservoir. The pertinent data sheets include information regarding the configuration and dimensions of dam and outlet structures and history of the project.

Data in the last three columns of Table 6-4 are from historical storage observations for each reservoir that can be accessed at the USACE FWD website (Table 4-1) as either time series plots or tabular data. Observed historical time series plots for each of the 25 reservoirs accessible at the USACE FWD website include annotations of the timing and magnitude of encroachments into the flood control pools since project construction.

The 11 flood control reservoirs other than the 25 USACE FWD projects are listed in Table 6-5. The data in Table 6-5 were obtained from the Addicks and Barker Water Control Manual (USACE Galveston District 2019), Lake Texoma Master Plan (USACE Tulsa District 2017), TWDB database discussed earlier in this chapter, and other sources.

The 25 multipurpose reservoirs in Figure 6-4 owned and operated by the USACE FWD are located in the Brazos, Trinity, Colorado, Neches, Sulphur, and Guadalupe River Basins. Addicks and Barker Reservoirs owned and operated by the USACE Galveston District are located in the city of Houston upstream of downtown in the Buffalo Bayou subbasin of the San Jacinto River Basin. The land within the flood control pools of Addicks and Barker Reservoirs is used for recreation and sports between flood events, but the projects provide no water supply storage.

Lake Texoma on the Red River and Pat Mayse Reservoir on Sanders Creek are multipurpose reservoirs owned and operated by the USACE Tulsa District. The USACE Tulsa District has oversight responsible for flood control operations of Lake Kemp on the Wichita River owned and maintained by the Wichita County Water Improvement District. Construction of the original smaller nonfederal Lake Kemp was completed in 1922. USACE Tulsa District completed a major reconstruction project in 1974 that added a large flood control pool funded by the federal government.

Lakes Meredith, Twin Buttes, and Travis are owned and operated by the CRMWD, city of San Angelo, and LCRA, respectively, but were originally constructed by the USBR with flood control pools included at federal expense. The USACE FWD has oversight responsibility for operation of the flood control pools of these three reservoirs.

The San Antonio River Authority owns and operates Olmos Reservoir on Olmos Creek in the San Antonio River Basin for flood control and recreation. Olmos Reservoir with construction completed in 1926 is the only nonfederal flood control reservoir in Texas with at least 5,000 ac-ft of flood control storage controlled by human-operated gated outlet structures. Olmos Dam is located about four miles upstream of San Antonio. The watershed area above the dam is 32 sq. mi.

IBWC maintains and operates International Amistad and Falcon Reservoirs pursuant to a treaty and other agreements between the two countries. Storage contents of these two large reservoirs on the Rio Grande are plotted in Figure 6-1.

Storage in Sam Rayburn and Ray Roberts Reservoirs operated by USACE FWD is plotted in Figures 6-3 and 6-6. The plots of total daily storage in Figures 6-8 through 6-10 for the 19, 28, and 24 reservoirs in the Colorado, Brazos, and Trinity River Basins include storage in five, nine, and eight flood control pools.

Peak Flood Levels

The conservation storage capacity in the third columns of Tables 6-4 and 6-5 is the storage volume below the designated top of conservation pool elevation. The total storage volume below the designated top of flood control pool consists of the summation of the conservation and flood control storage volumes tabulated in the third and fourth columns of Tables 6-4 and 6-5. The vertical depth of the flood control pool in feet is tabulated in the fifth column of Tables 6-4 and 6-5. This is the vertical height of the top of flood control pool elevation above the top of conservation pool elevation.

As explained in Chapter 3, flood control operations are based on making no releases contributing to downstream flooding as long as the storage level does not exceed the top of flood control elevation. If the flood control pool fills, releases through emergency spillways and other outlet structures are designed to protect the integrity of the dam, even though releases may contribute to flooding downstream.

The last three columns in the two tables describe the maximum water surface heights that have occurred during the life of each of the reservoir projects. The maximum height in feet above the bottom of flood control pool to which water has ever risen is recorded in the sixth column of Tables 6-4 and 6-5. The date of this peak flood event occurring during the life of the project is shown in the next-to-last column. A count of the number of floods that have overtopped the flood control pool is tabulated in the last column. The water surface rises into uncontrolled temporary surcharge storage whenever the flood control pool storage capacity is exceeded by an extreme flood event.

Wright Patman and Jim Chapman Reservoirs owned and operated by USACE FWD in the Sulphur River Basin of Northeast Texas are the first two reservoirs listed in Table 6-4. The top of flood control pool of Wright Patman Reservoir is 32.0 ft above the top of conservation pool. The highest water surface height reached since initial impoundment in June 1956 was 27.1 ft above the top of conservation pool during the flood of May 2016. The encroachment of 27.1 ft into the flood control pool did not reach the pool depth of 32.0 ft. The flood control pool of Wright Patman Reservoir has never been overtopped.

The flood control pool in Jim Chapman Reservoir has a vertical height of 6.2 ft. The peak water surface elevation occurring since initial impoundment in September 1991 occurred on March 2, 2018, with a depth above the top of conservation pool of 8.6 ft. Thus, this flood encroached 2.4 ft above the top of flood control pool into surcharge storage. The flood control pool of Jim Chapman Reservoir has been overtopped by five different flood events since completion of construction in 1991.

Tables 6-4 and 6-5 indicate that, as of January 2024, the flood control pools of 13 of the 36 reservoirs have never been overtopped. The flood control capacity of 10 of the reservoirs have each been exceeded once. For the other 13 reservoirs, multiple floods have encroached above the top of flood control pool into surcharge storage.

Several previously noted multipurpose reservoirs in western areas of the state have conservation pools that essentially never fill, and thus water never or almost never encroaches into their flood control pools. However, most multipurpose reservoirs have water in lower portions of flood control pools fairly frequently whenever the conservation pools are full and high inflows are spilling through outlet structures. Most of the occasional major flood events are contained at various levels in the flood control pools without rising near the top of the flood control pools.

Storage Reallocations

As discussed in Chapter 3, storage capacity can be reallocated between flood control and conservation pools by raising or lowering the designated elevation defining the top of the conservation pool, which of course is also the bottom of the flood control pool. Storage within the conservation pool can also be reallocated between water supply and hydroelectric power generation. The bottom of an active hydropower pool may be set as a designated top of inactive pool elevation. The following modifications in reservoir operations are reallocations of flood control storage capacity to water supply and for Whitney Reservoir also to hydropower.

The designated top of conservation pool (bottom of flood control pool) of Belton Reservoir was raised from 569.0 ft above msl to 594.0 ft msl in August 1971. The top of conservation pool of Lavon Reservoir was raised from 472.0 ft above msl to 492.0 ft in December 1975. The Lake Lewisville conservation pool was raised from 515.0 to 522.0 ft in November 1988 in coordination with completion of construction of Lake Ray Roberts. The Sam Rayburn conservation pool was raised from 164.0 to 164.4 ft in April 1972. As noted in Chapter 3, the Lake Waco conservation pool was raised from 455.0 to 462.0 ft in September 2003, converting 47,500 ac-ft of the flood control pool to additional water supply storage for the city of Waco. The top of conservation pool of Whitney Reservoir has raised three times: from 520.0 to 522.0 ft above msl in March 1968, from 522.0 to 523.0 in March 1969, and from 523.0 to 533.0 in May 1972.

Storage of Flood Waters in Conservation Pools

Lake Meredith is owned and operated by CRMWD as discussed in Chapter 7. Since the project was originally constructed by USBR with a federally funded flood control pool, USACE has oversight responsibility for flood control pool operations. However, a flood would have to fill a large volume of conservation storage capacity before the reservoir water surface would rise into the flood control pool. This has never happened.

As implied by Figure 6-1, operations of International Amistad and Falcon Reservoirs rely heavily on infrequent major floods to refill drawn-down conservation pool capacity to supply water needs. Depending on the timing and severity of long multiple-year depletions of conservation pool storage, the conservation pools may be a major contributor to storage of flood flows during infrequent flood events. Lakes Meredith, Amistad, and Falcon illustrate the extremes of long severe conservation pool drawdowns, but conservation storage capacity in other reservoirs in West Texas is also severely depleted much of the time, thus providing incidental flood storage.

Toledo Bend Reservoir was constructed and is operated solely with nonfederal funds and thus has no designated flood control pool. The conservation pool is full or almost full most of the time. However, the elevation of the top of the spillway gates is one foot above the designated top of conservation pool, providing some operating flexibility during high flows. Toledo Bend has an extremely large surface area, which means that even a foot of additional storage translates to a significant volume. This storage serves to attenuate flood flows.

Droughts accompanied by multiple-year severe reservoir storage depletions have been ended in the past by major floods that refill the reservoirs. Drawn-down conservation pools often incidentally contribute to flood control as noted in preceding paragraphs. Conversely, flood waters temporarily stored in flood control pools conceptually could contribute to maintaining higher storage levels in conservation pools. However, simulation modeling studies have shown that flood control pool operations for most multipurpose reservoirs result in only minimal if any improvements in water supply capabilities (Wurbs 2019a, b, 2020, 2022). Flood control pools are emptied as quickly as practical without contributing to downstream flooding and are usually completely emptied while river flows are still much higher than normal.

Ungated Flood-Retarding Dams and Flood Detention Facilities

This book focuses on major dams with gated outlet structures operated to regulate flows of major streams. However, thousands of small ungated flood-retarding structures located throughout Texas attenuate flood flows from small watersheds. Numerous cities have stormwater management programs that include the use of detention facilities to mitigate the increases in peak flows resulting from urban development. For many cities, many detention facilities are located within each city. The NRCS has constructed 2,040 flood-retarding dams in rural watersheds of Texas that are maintained by nonfederal entities. In some cases, the rural watersheds in which NRCS flood-retarding structures were constructed many years ago have since urbanized. Although temporary storage of urban and rural runoff may also increase infiltration and evapotranspiration, the primary purpose of temporary storage is to attenuate hydrographs to reduce peak flow rates. Flow attenuation or retardation primarily changes the timing rather than total volume of storm runoff. Effects of the smaller storm runoff retarding structures on the volume of inflows to major reservoirs and streams are typically minimal.

Authorized Conservation Storage and Diversions

The statewide water rights system administered by TCEQ for allocating surface water resources is introduced in Chapter 2 and further discussed in Chapter 3. Streamflow and water stored in reservoirs are owned by the state of Texas. A water right authorizes its holder to store and divert specified amounts of water for specified uses following rules outlined in the permit or other relevant water right documentation. Water rights typically authorize storage up to a specified reservoir storage capacity and a maximum annual diversion amount along with relevant standard requirements and special provisions.

The surface water rights system in Texas like most western states in the United States is based on the prior appropriation doctrine. As explained in Chapters 2 and 3, priorities or seniorities are assigned to each water right. Senior rights are protected from having their supplies diminished by junior rights. Except for the lower Rio Grande, priorities are based on dates that streamflow was first appropriated for beneficial use or water right permit applications were administratively approved, which dates to the late 1800s and early 1900s for the most senior rights. For the Rio Grande below Fort Quitman, priorities are based on the three water use classifications of municipal, agricultural Class A, and agricultural Class B. The allocation system is explained in Chapter 3.

A water right that authorizes withdrawal of water from a stream without reservoir storage is called run-of-river. Most of the larger water rights and more recently granted (junior) water rights depend on large volumes of reservoir storage capacity.

Various forms of water supply contracts also govern reservoir operations. Access by nonfederal sponsors to conservation storage capacity in USACE reservoirs is authorized by water supply contracts that include provisions for repayment to the federal government of all costs allocated to water supply, including payments with interest on the original construction costs and continuing operation and maintenance costs. The river authorities, water districts, and cities that contract with USACE are paying for storage capacity, without regard to the quantity of water that is withdrawn from the reservoir. Nonfederal sponsors, not USACE, are responsible for obtaining and adhering to TCEQ-administered water rights authorizing storage and use of water. River authorities and water districts manage their own reservoirs and contracted storage in USACE reservoirs to supply cities, industries, and farmers wholesale water under various contractual agreements.

Explanations of TCEQ procedures for submitting water use permit applications are available online. Information for each active water right is publicly accessible at the TCEQ Water Rights Viewer website listed in Table 4-1. TCEQ also maintains a web page with information for pending water right applications currently in the review process. Each water right holder is required to submit an annual report of water use. The reported annual water use data is also publicly available online.

Information is compiled in Table 6-6 as totals for each river basin. The contributing drainage area in square miles of each river basin and the drainage area within Texas shown in the second and third columns are from TWDB information. The other columns in Table 6-6 reflect water right information available from TCEQ.

River basin totals of authorized annual diversion and storage are given in the fourth and fifth columns of Table 6-6. These are approximate totals of annual diversion amounts in acre-feet/year and reservoir storage capacities in acre-feet authorized by the individual water rights with storage and diversion sites located in each river basin.

The authorized diversion amount for each water right is the maximum amount of water that the right holder can divert from the river and/or reservoir each year subject to water availability. Likewise, authorized reservoir storage capacities in the fifth column are summations for all water rights as specified in the certificates of adjudication and water use permits.

The counts of reservoirs in the last two columns of Table 6-6 were compiled by the author from the TCEQ WAMs introduced in Chapter 4. The last two columns show the total number of reservoirs and number of major reservoirs with conservation storage capacity of at least 5,000 ac-ft. The water rights include reservoirs that have conservation storage capacities of at least 200 ac-ft. Flood control storage does not normally require a water right permit and is not included in Table 6-6. The 3,464 authorized reservoirs include several projects with approved water rights that have not yet been constructed. The thousands of smaller storage and flood-retarding facilities excluded from water right licensing requirements are not reflected in the data in Table 6-6. In some cases, storage in a single large reservoir may be treated as multiple authorizations held by different water right holders.

The Red River Basin including its Sulphur River and Cypress Creek subbasins and the Sabine Basin encompass portions of neighboring states. The waters of the Rio Grande are shared by the two nations. The authorized diversion and storage amounts in Table 6-6 include only the portions of the conservation storage capacity in the international and interstate reservoirs that are authorized by Texas water rights administered by TCEQ to supply water use in Texas. The storage volumes in the appendixes are totals regardless of the international or interstate allocations.

The 210 reservoirs with authorized storage capacities of 5,000 ac-ft or greater contain about 98% of the storage capacity of the 3,464 storage facilities authorized by water rights. Eighty-one reservoirs with conservation storage capacities of at least 50,000 ac-ft contain about 92% of the authorized storage capacity of the 3,464 reservoirs.

As of late 2023, water right holders in Texas held a total of 6,235 active water rights documented by 4,892 certificates of adjudication and 1,343 water use permits. Another 125 pending applications for new or amended water rights were in the TCEQ review process. Existing water rights were formalized through an adjudication process during the 1970s to 1980s pursuant to the Water Rights Adjudication Act of 1967 as discussed in Chapters 1, 2, and 3. These water rights are recorded as certificates of adjudication. After completion of the adjudication process

Table 6-6. Watershed Drainage Areas and Authorized Water Supply Diversions and Storage.

River Basin	Drainage Area Total (square miles)	Drainage Area Texas (square miles)	Supply Diversion (acre-feet/year)	Storage Capacity (acre-feet)	Number of Reservoirs Total Authorized	Number of Reservoirs Major ≥5,000 ac-ft
International and Interstate River Basins						
Rio Grande	182,215	49,387	6,910,758	3,023,656	80	8
Canadian River	47,705	12,865	165,329	1,478,892	47	3
Red River	93,450	24,297	1,049,348	4,568,916	236	25
Sabine River	9,756	7,570	2,640,348	6,393,413	212	13
Sulphur River and Cypress Creek Basins						
Sulphur River	3,767	3,580	599,341	962,528	57	4
Cypress Creek	3,552	2,929	567,507	949,139	86	13
Major River Basins Contained Totally or Almost Totally within Texas						
Nueces River	16,700	all	530,384	1,047,017	121	2
San Antonio River	4,180	all	225,397	367,028	122	2
Guadalupe River	5,953	all	6,440,632	451,818	222	4
Lavaca River	2,309	all	217,647	265,247	22	2
Colorado River	42,318	39,428	4,836,182	5,270,565	484	31
Brazos River	45,573	42,865	6,439,815	5,050,716	673	49
San Jacinto River	3,936	all	1,454,511	637,190	114	4
Trinity River	17,913	all	7,494,232	7,815,297	683	32
Neches River	9,937	all	4,346,960	3,904,100	180	13

(Continued)

Table 6-6. Watershed Drainage Areas and Authorized Water Supply Diversions and Storage. (Continued)

River Basin	Drainage Area (square miles) Total	Drainage Area (square miles) Texas	Supply Diversion (acre-feet/year)	Storage Capacity (acre-feet)	Number of Reservoirs Total Authorized	Number of Reservoirs Major ≥5,000 ac-ft
Coastal Basins between Lower Reaches of Major River Basins						
Nueces-Rio Grande	10,442	all	1,897,831	146,131	64	4
San Antonio-Nueces	2,652	all	81,547	3,548	9	0
Lavaca-Guadalupe	998	all	4,600	0	0	0
Colorado-Lavaca	939	all	985,444	9,666	8	0
Brazos-Colorado	1,850	all	96,310	30,959	0	0
San Jacinto-Brazos	1,440	all	4,483,282	51,042	0	0
Trinity-San Jacinto	247	all	44,474	18,633	13	0
Neches-Trinity	769	all	425,377	64,481	31	1
Totals for 15 Major River Basins and Eight Coastal Basins						
Totals	508,601	263,186	51,937,254	42,509,984	3,464	210

in the 1990s, additional water rights were granted in the form of water use permits.

With changing conditions and water needs; water right holders sometimes request changes to their water rights. Existing water rights are revised as water use permit amendments through the TCEQ water right application submittal and evaluation process. Periodically updated and TCEQ approved water conservation plans are required for most major water rights. The more complex water rights also include system operations water management plans that are periodically updated by the water right holder and submitted to TCEQ for approval.

The quantities of Table 6-6 include both permanent and short-term water right permits. Regular water rights are in effect for perpetuity if the right holders continue to use water as specified by the water rights. Many water rights for municipal use include predicted increased water needs in the future. The city will need the full amount of the authorized diversion amount in the future though less than the full amount is currently needed. TCEQ also issues term permits to deal with emergencies or special situations. A term water right permit is valid only for a fixed period of up to several years. The authorized annual diversion amounts in Table 6-6 sum to 51,937,254 ac-ft/year, and the total storage amount is 42,509,984 ac-ft. This includes authorized annual diversion amounts of 176,700 ac-ft/year for temporary term permits.

The preceding discussion focuses on the Texas water rights system that regulates the quantities of streamflow that are stored in reservoirs and withdrawn for use for beneficial purposes or maintained as instream flows. TCEQ also administers regulatory programs for protecting water quality and public health. Various aspects of hydroelectric power generation are regulated by the Federal Energy Regulatory Commission.

Water Use and Reuse

The most common types of use of water diverted from lakes and streams are agricultural irrigation, municipal water supply, thermal-electric energy production, manufacturing, mining including oil and gas production, rural domestic, and raising livestock. The most common instream and in-lake water uses in Texas are recreation, preservation and enhancement of fish and wildlife and ecosystems, and hydroelectric power generation.

Return flows and water reuse are important aspects of water management. Return flows to streams originate from water supplied from both surface water and groundwater sources. Return flows from municipal water use are comprised mainly of wastewater treatment plant discharges. Municipal return flows typically vary seasonally with different mixes of lawn watering and indoor water use. Return flows from irrigated agriculture consist of drainage from fields to streams, with return flow amounts varying with different soil characteristics, crops, and irrigation practices.

Recent and projected future annual water demands compiled by TWDB are tabulated in Table 1-1 and discussed in Chapter 1. These estimated quantities represent water needs assuming efficient water use practices during drought conditions comparable to the most hydrologically severe past drought. Hydrologists and water managers know that droughts including those as severe or worse hydrologically and economically than past droughts will occur in future, but the timing of the droughts is not known. The future year 2030 or 2050 may be extremely wet, extremely dry, or somewhere in between. Precipitation and resulting streamflow and reservoir storage have varied dramatically from month to month, year to year, and decade to decade in the past. The extreme hydrologic variability is expected to continue in the future. Water management also deals with uncertainties regarding future population and economic growth, global warming and climate change, and other factors affecting future water use.

The annual water demands in Table 1-1 are from TWDB-administered planning studies. Annual diversion amounts authorized by water rights in Table 6-6 are from the TCEQ-administered water allocation process. The two datasets are fundamentally similar but differ in respect to an array of complexities. The two datasets handle the following types of water use differently: Water for creating steam and condensing (cooling) steam in thermal-electric power generation processes, hydroelectric power generation, and various instream flow requirements.

Water for thermal-electric power generation plants comprise the water use sector that diverts the largest volume of water from lakes in Texas and throughout the United States. Most of this water used for producing and condensing steam is discharged back into lakes and streams after circulating through the power plants. Return flows are almost as large as diversions. Except for steam-electric power plant use, the water demand quantities in Table 1-1 and TWDB planning reports are diversion amounts without consideration of the portions of the diversions returned to a lake or stream. The steam-electric demands in Table 1-1 are limited to consumptive evaporative losses excluding return flows. Most of the water rights for steam-electric power plants include total diversion amounts, though some steam-electric water rights include only consumptive use.

The water demand quantities in Table 1-1 do not include hydroelectric power, recreation, and instream flow requirements. Water right permits for hydroelectric power generation include annual diversion amounts representing water that flows through the hydropower plants. Water right permits for hydropower are not structured the same for all hydropower plants. In some cases, hydropower releases are viewed as diversions with 100.0% return flow. In other cases, hydropower is treated as incidental to releases for other water supply purposes. Water rights may also be specified in terms of electric energy generated. Water rights for recreational use of streamflow in some cases include diversion amounts as well as reservoir storage. Some rights also include authorized diversions to supply instream flow requirements.

Most types of water use involve significant return flows. In the past, most water rights have not included specifications regarding return flows. Return flow specifications have been included in more recently granted or amended water right permits.

Bed and banks water right permits have received increasingly more attention in recent years. A city or other water right holder has the right to use the water withdrawn from a stream system until the water is discharged back into the stream system. Upon discharge into the stream, wastewater treatment plant effluent becomes part of the streamflow owned by the state. Water reuse becomes increasingly important as water demands exceed supply. Several cities have pursued water management strategies that include reusing wastewater treatment plant discharges to irrigate parklands or industrial uses that involve reuse locations at various distances downstream of wastewater treatment plants. Bed and banks water right permits allow cities to use the stream as a channel to convey their treated wastewater discharge to downstream diversion locations for reuse without losing their rights to use the wastewater effluent.

Reservoir Surface Evaporation

Friedrich et al. (2018) explore the science and management challenges of reservoir surface evaporation in the western United States. They also provide a lengthy list of publications dealing with reservoir evaporation.

Wurbs and Zhang (2014) and Wurbs and Ayala (2014) investigated the volume of water evaporated from the reservoirs of Texas employing WRAP simulations with the full authorization scenario WAMs for all river basins of the state. The long-term annual mean of reservoir evaporation from more than 3,400 reservoirs authorized by water rights was estimated to be 6.10 million ac-ft/year. Evaporation varies with varying reservoir storage levels and corresponding water surface areas and with seasonally and annually varying evaporation rates. Evaporation rates are highest and storage levels are lowest during droughts or other hot dry periods. The estimated statewide annual evaporation volumes with exceedance frequencies or probabilities of 75%, 50%, and 25% are 5.73, 6.06, and 6.44 million ac-ft/year, respectively.

Referring to Table 1-1 for comparison, the total annual municipal water use by all cities in Texas from all supply sources was 5.22 million ac-ft in 2020. Reservoir evaporation varies seasonally and from year to year. However, the estimated long-term mean and median (50% exceedance) annual evaporation volumes of 6.10 and 6.06 million ac-ft are larger than the total statewide 2020 annual municipal water use from all sources of 5.22 million ac-ft.

The quantities in the preceding paragraphs are for gross evaporation from reservoir surfaces without consideration of precipitation. Rainfall falling on reservoir water surfaces provides reservoir inflows that partially offset losses due to evaporation.

Net evaporation refers to evaporation less precipitation. Precipitation fell on the land covered by each reservoir before the reservoirs were constructed. Without a reservoir, precipitation transformed to infiltration into the ground, evapotranspiration, and runoff to the stream. All precipitation falling on the water surface of a reservoir is inflow to the reservoir. Likewise, evapotranspiration would occur at a reservoir site even if the reservoir had not been constructed. Construction of a reservoir increases both evaporation and capture of precipitation.

Strategies have been investigated over many decades in Texas and worldwide to reduce evaporation from reservoirs. Wurbs and Ayala (2014) include a literature review of reservoir evaporation reduction strategies and methods. Measures that have been implemented or proposed for reducing evaporation from reservoirs include monomolecular films, floating devices, suspended floating covers, and wind retarding devices. Barnes (2008) provides a thorough literature review of the technology and issues associated with the use of monomolecular films in reducing reservoir surface evaporation. Prime et al. (2012) investigated reservoir evaporation suppression technologies that could contribute to water management in Australia.

Monomolecular films, floating devices, and other techniques have been applied successfully to reduce evaporation from swimming pools, small ponds, and small constructed water storage facilities. However, successful implementation of these types of strategies for major reservoirs has proven to be challenging. Monomolecular films have been promoted more than other evaporation suppression techniques for major reservoirs, but constraints have been encountered. Monomolecular films spread across large water surfaces tend to be broken up by wind and wave action and washed to the shore.

Reservoir surface evaporation is a consideration in conjunctive management of supplies from reservoirs with high evaporation rates and groundwater aquifers with low or negligible evapotranspiration. Evaporation is a function of reservoir surface area which decreases with storage drawdowns due to water use. Reservoir evaporation is decreased by supplying water demands from reservoir storage. Water supply reliabilities can be significantly increased by a backup source such as groundwater even if water is supplied from reservoir storage most of the time. Shifts between surface and groundwater can be based on reservoir storage contents.

Water Availability and Supply Reliability

The water management community of Texas maintains and operates an extensive inventory of dams and reservoirs, conveyance facilities, and other constructed infrastructure to manage extremely variable streamflow to meet the diverse water-related needs of the people of Texas. The future is highly uncertain. Droughts more hydrologically and economically severe than those experienced during the last 100 years along with other less severe dry periods are anticipated at unpredictable times in the future.

Reliability Assessment

Water availability is assessed in terms of likelihood, frequency, risk, or reliability. Estimates of water supply reliability metrics are employed to probabilistically quantify capabilities for supplying water for specific users and types of use at specific locations during the uncertain future.

Assessment of water availability and supply reliability employing the WRAP/WAM modeling system is discussed later in this chapter. The generalized WRAP modeling system was implemented with WAM datasets during 1998 to 2003 pursuant to the 1997 SB1. The resulting modeling and analysis capabilities greatly improved assessments of water supply capabilities. However, most water rights already existed prior to the 1990s. The reliability criteria based on WRAP/WAM analyses currently applied in evaluating water right permit applications were not applied in the adjudication of pre-existing water rights.

TCEQ has employed WRAP/WAM-based analyses since about 2001 to ensure that approval of new water rights or amendments to existing water rights for agricultural irrigation meet the criterion of supplying at least 75% of the proposed new water demand at least 75% of the time. Likewise, new municipal water rights are limited to firm yield diversion amounts that can be supplied 100% of the time during a WAM simulation. These reliability criteria for new permit applicants must be achieved without adversely affecting reliabilities of existing water rights. However, the reliabilities of many existing water rights are already much lower than current reliability criteria.

Water Supply Capabilities

Authorized water supply amounts are supplied in both actual reality and the WRAP/WAM simulation model only to the extent that water is available. Supply reliabilities are less than 100% for many water rights, in many cases much less than 100%. During 2000 to 2003 as the initial WAMs for each river basin were completed, TCEQ sent letters to each water right holder providing reliability metrics for their individual water rights. Water right holders were informed that the water supply diversion amounts authorized by their certificates of adjudication or water use permits would not necessarily be fully available to them during drought or dry conditions.

Reliabilities for supplying water differ significantly between water right holders. Reliabilities depend on hydroclimatic characteristics of the diverse regions of Texas, competition and cooperation between numerous water users for limited water resources, water right priorities, and access to reservoir storage. Surface water users in the Neches and Sabine River Basins have high reliabilities. Surface water users in the Rio Grande Basin and West Texas have much lower reliabilities.

Approval of new run-of-river water right diversions is not feasible for most locations in Texas. Unappropriated flows occur during floods or periods of extremely high precipitation. However, large volumes of reservoir storage capacity are necessary to store these flows for later beneficial use during dry periods. Institutional, economic, and environmental feasibility of construction of new reservoir projects is difficult to achieve. Water management has been relying increasingly in recent decades on more efficiently managing existing infrastructure and water resources to supply growing water needs.

Water right holders can sell their rights to other water users subject to TCEQ approval of transactions through the water right application evaluation process. Most water transfers to date have been from farmers to cities, mostly in the Lower Rio Grande Valley.

Environmental Flow Standards

Environmental flow standards for the river systems in Table 3-1 have been established following procedures created by the 2007 SB3 as discussed in Chapter 3. EFS are being incorporated into WAMs. The SB3 process includes periodic reevaluation and improvement of EFS. Establishment of SB3 EFS for other stream sites in the future is anticipated.

The SB3 EFS priority dates listed in Table 3-1 for the different river systems are November 2009, March 2011, October 2011, and March 2012. Water rights with priority dates senior to those of EFS are not required to curtail streamflow appropriations to protect EFS. However, EFS constrain unappropriated flows available for new water right applicants or amendments to existing water rights. Remaining unappropriated flows for most river reaches in Texas were limited before establishment of SB3 EFS. SB3 EFS represent a recently established appropriation of streamflow that places even greater constraints on quantities of unappropriated flows available for future water right permit applicants.

Streamflows are insufficient to fully satisfy all EFS all the time. Formulation and implementation of strategies for supplying water needs for preserving ecosystems, fish and wildlife, freshwater inflows to bays and estuaries, and other environmental resources reflected in SB3 EFS are important current and future endeavors.

WRAP/WAM Modeling System

The WRAP modeling system and WRAP simulation input datasets called WAMs from the TCEQ water availability modeling system are introduced in Chapter 4 and employed in Chapters 5 and 7 and the last section of this chapter. Publications referenced in Chapter 4 and accessible at the WRAP website included in Table 4-1 document the modeling system and its applications in detail.

Application of the Modeling System

The WRAP/WAM modeling system is used by TCEQ staff and water right permit applicants or their consultants to prepare and evaluate water right permit applications; TWDB staff and regional planning groups or their consultants in regional and statewide planning; river authorities and other entities in operational or project planning; and by researchers in investigating various water management issues. The WRAP/WAM modeling system is employed in this book to explore statewide river/reservoir system water management. Tables 5-5 and 6-7, and Figures 5-39, 5-40, 7-3, 7-5, 7-8, 7-12, 7-16, and 7-17 are derived from information generated by the modeling system.

A WRAP/WAM simulation combines historical natural hydrology with a specified scenario of water resources development, management, and use. The simulations referenced in Chapters 6 and 7 are based on modeling premises of all water right holders appropriating the full amounts of streamflow to which they are entitled, subject to water availability, during a repetition of past natural hydrology.

TCEQ applies the current use scenario version of WAMs in evaluating applications for term water right permits and the full authorization scenario for evaluating applications for regular water right permits. The full authorization WAMs do not include term permits. All diversion amounts and reservoir storage capacities in a current use WAM represent current or recent conditions rather than the premise of all water rights using their full authorized amounts.

Full authorization scenario WAMs are used to support discussions in Chapters 6 and 7. The full authorization scenario simulates the situation in which all present water right holders continuously store and divert the full amounts of streamflow to which they are legally entitled subject to water availability. Return flows are included only to the extent specified in the water rights. Authorized annual diversion amounts are distributed over the months of the year in the simulation based on past observed seasonal water use patterns. In some cases, water supply diversion targets vary with reservoir storage contents and hydrologic conditions.

The plots of actual historical conservation storage capacities and contents in Figures 6-8 through 6-11 reflect water use and available storage capacities that have changed dramatically over time. Likewise, the past sequences of observed total storage levels of each of the reservoirs reflected in Figures 6-1 through 6-7 extend over lengthy periods characterized by growing population and water demands and changing conditions that include construction of more new reservoirs and other infrastructure. WRAP/WAM simulations allow a specified stationary scenario of water development and use to be combined with a long stationary sequence of natural hydrology.

Streamflow Quantities

Time series of streamflows at a location of interest are expressed as flow rates in ft^3/s or flow volumes in acre-feet during time steps of days, months, or years. The following types of time series sequences of streamflow quantities are fundamental to WRAP/WAM modeling studies:

- Observed flows are quantities compiled from USGS gauge measurements accessible at the National Water Information System website.
- Naturalized flows at about 470 gauge sites in the WAM datasets have been developed and periodically extended forward in time (updated) by TCEQ and TWDB staff and contractors by adjusting observed flows to remove the effects of water resources development and use that have occurred during the hydrologic period-of-analysis. Naturalized flows represent natural conditions without human water resources development and use. Naturalized flows at more than 14,000 other relevant locations are generated within the simulation model based on the naturalized flows input for 470 gauged sites and input datasets of watershed parameters.
- Simulated regulated flows are computed in WRAP simulations with WAM input datasets. Regulated flows in the simulation

results represent the flows that would occur with the simulated water development and use scenario during a repetition of past natural hydrology.

- Unappropriated flows at a stream location computed in a WRAP simulation represent streamflow remaining after all water rights have appropriated their authorized amounts.

Simulated regulated flows at a location may include combinations of both appropriated and unappropriated flows. Unappropriated flows in a particular month or day of a monthly or daily simulation may be less than the regulated flow due to environmental instream flow requirements at the site or requirements to pass flows for downstream more senior water rights. Instream flow requirements result in streamflows at a site being appropriated, reducing unappropriated flows, but remaining in the stream and thus not reducing regulated flows.

Naturalized streamflows in WAMs reflect a significant expenditure of time and effort by TCEQ and TWDB staff and consulting firm contractors. The basic approach to developing hydrologic period-of-analysis monthly naturalized flow sequences at USGS gauge sites consists of adjusting historical measured flows to remove the effects of relevant upstream reservoir operations, water supply diversions, return flows, and stream channel conveyance losses (Wurbs 2006, 2019c, 2024). Other factors are incorporated into the adjustments for certain locations. Various approaches have been adopted for periodically updating the naturalized flow sequences as the length of the gauge periods of record and corresponding simulation hydrologic periods of record increase with time.

Probabilistic Expressions of Likelihood, Reliability, or Frequency

Water resources planning and development address future rather than past needs and conditions. However, historical past natural hydrology over a period of many years, such as 1940 to 2022, without water development and use is adopted in the modeling system as being statistically representative of future hydrology (Wurbs 2006, 2019c, 2021, 2024). Statistical analyses of the results of WAM simulations provide metrics for assessing water supply reliabilities, storage frequencies, and streamflow frequencies. Several probability-based metrics routinely employed within the WRAP modeling system are defined as follows. These likelihood metrics routinely employed with WRAP are applied by other modeling systems as well.

For brevity, the words *probability, frequency,* and *reliability* are used here for the expressions *estimated probability, estimated frequency,* and *estimated reliability.* The future is highly uncertain, and many factors contribute to modeling and analysis uncertainties. The topic of quantifying uncertainty in statistical analyses is addressed extensively in the literature. However, uncertainty descriptors designed to quantify the uncertainties reflected in estimates of probabilities are themselves also approximate estimates.

The frequency and reliability estimates discussed subsequently reflect all premises incorporated in the WRAP/WAM model, including the hydrologic period of analysis. A particular municipal water right is said to have a simulated reliability of 100% of being fully supplied even though water managers know that a future drought more severe than any drought experienced during 1940 to 2022 or other hydrologic period of analysis may result in failures to supply water demands. Water managers prepare for the future, not the past. Reliability and frequency metrics provide meaningful estimates of the likelihood of supplying water needs in the future under various water management scenarios.

The computed metric "period reliability" routinely used in planning and water allocation is the percentage of months of a WRAP simulation that at least a specified percentage of a water supply target is supplied. Water availability modeling analyses are employed in the water rights permitting process to estimate whether at least 75% of a proposed increase in agricultural water use can be supplied at least 75% of the time in the future without affecting the reliabilities of senior water rights. The maximum annual municipal water demand supplied 100% of the time without adversely affecting reliabilities of other more senior water rights is often of interest in WAM applications.

Volume reliability is routinely computed along with period reliability. Volume reliability is the percentage of water demand volume that is supplied during a simulation. Another similar metric employed with WRAP simulations is the annual hydroelectric energy supplied as a percentage of the annual energy target.

Firm or safe yields are routinely computed in planning studies performed by TWDB staff, SB1 regional planning teams, consultants, and others. The conventional firm yield is the maximum annual water supply or hydroelectric energy demand, which varies seasonally through each year, that can be supplied continuously without shortage during the hydrologic period of analysis of a long-term, such as 1940 to 2022, simulation. The firm yield just empties the reservoir or multiple-reservoir system during the most severe drought of the simulation. WRAP also includes a safe yield variation in which a specified minimum storage reserve is maintained throughout the simulation.

The likelihood of reservoir storage or streamflow levels falling below or above specified levels of interest under specified conditions is also estimated based on long-term simulations. Frequency analyses for a variety of other simulation results variables may also be performed.

The preceding paragraphs address interests in future conditions over the long term. Discussions earlier in this chapter addressed the issue of predicting drought or non-drought conditions over the next several months or years, given that reservoir storage contents today are known. WRAP/WAM short-term CRM facilitates estimates of storage contents and other relevant variables occurring one or several months, a year, or perhaps further into the future associated with specified ranges of probability. The probability estimates associated with storage contents or other quantities occurring at specified times in the future conditioned upon known present storage are computed with WRAP in CRM simulation mode based on the same WAM datasets of historical natural hydrology and specified water management and use scenario that are regularly employed in long-term planning and water right permitting applications.

The last column of Tables 6-4 and 6-5 shows the number of times that each of 36 flood control pools has been exceeded. The daily computational time step version of the WRAP simulation model includes flood control operations of reservoir systems following the procedures outlined in Chapter 3. WRAP includes a feature based optionally on either the log-normal or log-Pearson Type III probability distributions for computing the annual probability of overtopping flood control pools based on frequency analyses of simulation results. These same frequency analysis options can be applied to either peak annual streamflows or peak annual storage. The basic log-normal and log-Pearson Type III frequency analysis methods are found in many other modeling systems as well as WRAP and explained in many textbooks (Wurbs and James 2002).

The array of USACE HEC software discussed in Chapter 4 provides capabilities for detailed modeling and analysis of flooding and flood control aspects of river and reservoir system development and management. HEC simulation models are extensively applied in flood plain management, stormwater management, design of hydraulic structures, and real-time flood control operations of reservoir systems throughout Texas and nationwide.

Applications of WRAP and more typically applications of HEC modeling systems for flood analyses often include determinations of peak flow rates and stream water surface elevations associated with an annual exceedance probability or its reciprocal, the annual recurrence interval. For example, floodplain regulation in conjunction with the National Flood Insurance Program (Chapter 2) is commonly based on floodplain delineations associated with an annual exceedance probability of 0.010 (1.0% exceedance frequency), which is also called the 100-year recurrence interval. Design of hydraulic structures such as bridges and culverts and sizing of stormwater detention facilities and reservoir flood control pools are commonly based on simulated design storms with a specified annual exceedance probability (Wurbs and James 2002).

Effects of Water Development on River Flows

Daily, monthly, and annual observed flows of the Brazos River at the USGS gauge at the city of Waco in Central Texas are plotted in Figures 1-5, 1-6, 5-37, and 5-38 as an illustrative example of flow characteristics representative of many locations throughout Texas. Monthly and annual means of the naturalized and observed flows of the Brazos River at this gauge site shown in the basin map of Figure 7-2 are plotted for comparison in Figures 5-39 and 5-40.

These and the other flow plots in Chapter 5 illustrate the great variability characteristic of river flows throughout Texas. Large volumes of reservoir storage are required to reduce flooding, develop reliable water supplies, and otherwise deal with extreme streamflow variability.

The flow plots for the Brazos River at Waco also illustrate varying effects of reservoirs on downstream river flows. The attenuation effects of reservoirs on downstream river flows vary depending on the averaging time interval. Flood flows are affected differently than low flows. Effects on downstream flows of temporarily storing flood flows in upstream reservoirs may be pronounced in daily flows but dissipated in the process of averaging daily mean flows to obtain monthly or annual flow quantities. Effects of reservoirs are pronounced immediately downstream of a dam but dissipated as the flow continues downstream.

Analyses of long-term historical sequences of observed flows at multiple gauge sites indicate that long-term trends of decreases in certain flow metrics are evident at some locations, flow increases have occurred in other cases, and some gauge sites exhibit no evident long-term trends or flow changes.

Flood control reservoir storage decreases high flows but has only minimal effect on long-term mean flows. Reservoir evaporation reduces storage levels, which results in future long-term reductions in downstream flows. Reservoir drawdowns for beneficial use incidentally reduce evaporation. Municipal, industrial, and agricultural water use generally result in decreases in long-term mean flows. Reservoir conservation storage and associated water use may either decrease or increase streamflow at downstream locations during low flow periods. Water supply diversions decrease downstream flows. Return flows from water supply diversions supplied by reservoirs and groundwater sources increase downstream flows. Hydropower releases increase downstream flows. Releases from a dam may be diverted from the river for water supply at pumping stations located significant distances below the dam. Streamflow between a dam and downstream diversion sites is increased by water supply operations. Thus, reservoir operations to reduce flooding and supply human water needs can both adversely and/or beneficially affect environmental instream flow requirements.

River flows at the outlets of the rivers listed in Table 6-7 for natural versus developed conditions are compared employing the WRAP modeling system with the full authorization scenario WAM datasets. These same naturalized flows are included in the comparisons of Table 5-5 of Chapter 5. The Texas border serves as the outlets for the Sulphur River and Cypress Creek. The other rivers in Tables 5-5 and 6-7 discharge directly into the Gulf of Mexico within Texas. The Guadalupe-San Antonio WAM covers the entire Guadalupe Basin, which includes the San Antonio River subbasin of the Guadalupe River Basin.

The flow metrics presented in Tables 5-5 and 6-7 demonstrate the long-term reduction in streamflows that result from river and reservoir system development. Monthly naturalized and simulated regulated monthly flows at the outlets of the rivers listed in the first column of Table 6-7 were determined for each month of the WAM hydrologic period of analysis shown in the second column. The mean and median of each of the 22 series (naturalized and regulated flows at 11 sites) of monthly flows were computed. The mean is the average of the period of analysis monthly flow rates. The median is the flow rate exceeded during

Table 6-7. WAM Naturalized and Simulated Regulated Flow at River Outlets.

River Basin	Analysis Period	Naturalized Flow		Regulated Flow in ft^3/s and % of Naturalized			
		Mean	Median	Mean	Median	Mean	Median
		(ft^3/s)	(ft^3/s)	(ft^3/s)	(ft^3/s)	(%)	(%)
Nueces	1934–2021	695.5	141.2	168.9	0.000	24.3	0.000
Guadalupe	1934–2021	3,086	1,718	2,506	1,157	81.2	67.3
Lavaca	1940–2021	1,366	221.2	1,151	143.1	84.3	64.7
Colorado	1940–2016	3,369	1,501	1,545	333.1	45.9	22.2
Brazos	1940–2017	8,445	3,293	6,304	1,340	74.6	40.7
San Jacinto	1940–2021	3,547	1,710	2,823	1,038	79.6	60.7
Trinity	1940–2018	9,114	3,712	6,204	1,891	68.1	50.9
Neches	1940–2019	8,503	3,906	6,165	1,226	72.5	31.4
Sabine	1940–2021	9,465	5,980	6,146	3,236	64.9	54.1
Cypress	1948–2021	2,442	1,194	1,879	686.7	76.9	57.5
Sulphur	1940–2021	3,370	1,252	2,694	685.1	79.9	54.7
Totals		53,402	24,628	37,586	11,736	46.1	31.2

50% of the months in the hydrologic period-of-analysis of the WRAP/WAM simulations.

Graphical comparisons of time series plots of monthly or daily naturalized flows and the corresponding simulated regulated or observed flows and statistical frequency metrics for specific locations of interest may also be employed to explore effects of development on flow variability and the differences between the effects on low flows versus high flows.

The mean and median of the simulated regulated flows are tabulated in the last two columns of Table 6-7 as a percentage of the corresponding naturalized flows. The flows are highly variable but have long-term averages and 50% exceedance frequency quantities shown in Table 6-7. The median is much smaller than the mean since the frequent flows below the median reflect low flow rates but the less frequent flood flows include extremely high flow rates. Flood flows increase the mean (average) more than the median (50% exceedance).

Totals for the 11 rivers are shown as the last line of Table 6-7. Under natural conditions with no human water resources development and use, the long-term mean and median of the flow of these 11 rivers at their outlets are 53,402 and 24,628 ft^3/s, respectively. If all water right holders appropriate their full authorized amounts of water subject to water availability, the simulated regulated flows from the river basins have a mean and median of 37,586 and 11,736 ft^3/s, respectively, which are 46.1% and 31.2% of the mean and median of naturalized flows. Water resources development represented by the WAM full authorization scenario affects median flows more than mean flows because low flows are affected more than high flows.

Means of 53,402 and 37,586 ft^3/s for the total naturalized and regulated flows of the 11 rivers in Table 6-7 are equivalent to 38,688,000 and 27,230,000 ac-ft/year, respectively. The total authorized annual diversion amount in Table 6-7 for these 11 river basins is 35,793,000 ac-ft/year.

Flow differences between natural and developed conditions are affected by a variety of complexities discussed throughout this book. For example, water is supplied to the city of Houston in the San Jacinto River Basin by pipeline from Lake Livingston on the Trinity River. Flows of the San Jacinto River at its outlet depend on return flows from municipal water use that includes groundwater and Trinity River sources as well as San Jacinto River Basin reservoirs.

The simulation results in Table 6-7 indicate that with the full authorization scenario, the long-term mean flow of the Nueces River at its outlet is an estimated 24.3% of the mean flow under natural undeveloped conditions and that there would be no flow at the outlet over half of the time. The mean of 168.9 ft^3/s and median of zero for the simulated flow for the Nueces is much less than the quantities for the other rivers. The small flows at the Nueces River outlet are due to flow loss to groundwater recharge upstream as the Nueces River and its tributaries flow over the recharge zones of the Edwards, Carrizo–Wilcox, Gulf Coast, and other aquifers as discussed in Chapters 5 and 7.

Table 6-7 is a continuation of Table 5-5 and accompanying discussion in Chapter 5. Table 5-5 compares the precipitation falling on the 11 river basins and natural streamflow leaving the river basins at their outlets. Table 6-7 focuses on the effects of water development on streamflow leaving the 11 river basins. Table 5-5 deals with natural conditions with no water development and use by people. Table 6-7 adds the effects of water development and use. The summation of long-term mean volumes of simulated natural flows of the 11 rivers at their outlets is 13.4% of the estimated long-term mean volume of the precipitation falling on the watersheds above the 11 river basin outlets (Table 5-5). The long-term mean simulated regulated flow is 46.1% of the mean natural flow. Thus, based on all the previously discussed modeling and analysis premises and approximations, with water development and use, the volume of the river flows leaving these 11 river basins averaged over many years is estimated to be about 6.2% of the volume of the precipitation falling on the river basin watersheds.

Most of the precipitation falling on the river basins is infiltrated into the ground and/or transported back to the atmosphere through transpiration from plants and evaporation from land and water surfaces. Much or most of the rainfall that infiltrates into the ground is later transported back to the atmosphere through evapotranspiration. Water resources development and use by people also greatly affects both long-term quantities and variability characteristics of river flows.

References

Barnes, G. T. 2008. "The potential for monolayers to reduce the evaporation of water from large water storages." *J. Agric. Water Manag.* 95 (4): 339–353.

Friedrich, K., R. L. Grossman, J. Huntington, P. D. Blanken, et al. 2018. "Reservoir evaporation in the western United States: Current science, challenges, and future needs." *Bull. Am. Meteorol. Soc.* 2018 (99): 167–187.

Prime, E., A. Leung, D. Tran, G. Harnam, et al. 2012. *New technology to reduce evaporation from large storages.* Waterlines Report Series No. 80. Canberra, AU: Australian National Water Commission.

TWDB (Texas Water Development Board). 2022. *Water for Texas: 2022 state water plan.* Austin, TX: Texas Water Development Board.

USACE (US Army Corps of Engineers) Galveston District. 2019. *Water Control manual: Addicks and Barker Reservoirs, Buffalo Bayou and Tributaries, San Jacinto River basin.* Galveston, TX: US Army Corps of Engineers.

USACE Tulsa District. 2017. *Lake Texoma master plan, Red River basin.* Tulsa, OK: US Army Corps of Engineers.

Wurbs, R. A. 2006. "Methods for developing naturalized monthly flows at gaged and ungaged sites." *J. Hydrol. Eng.* 11 (1): 55–64.

Wurbs, R. A. 2019a. *Daily water availability model for the Brazos River basin and San Jacinto-Brazos Coastal basin.* TR-513. College Station, TX: TWRI.

Wurbs, R. A. 2019b. *Daily water availability model for the Trinity River basin.* Contract No. 582-18-80410. Austin, TX: TCEQ.

Wurbs, R. A. 2019c. *Water rights analysis package river system hydrology.* TR-431. College Station, TX: TWRI.

Wurbs, R. A. 2020. *Daily water availability model for the Neches River basin*. Contract No. 582-18-80410. Austin, TX; TCEQ.

Wurbs, R. A. 2021. "Monthly river flows in Texas for natural and developed conditions." *Water Cycle J.* 2 (2021): 1–14.

Wurbs, R. A. 2022. *Daily water availability model for the Colorado River basin*. Contract No. 582-18-80410. Austin, TX: TCEQ.

Wurbs, R. A. 2024. *Incorporation of environmental flow standards and other water management complexities in water availability modeling*. TR-552. College Station, TX: TWRI.

Wurbs, R. A., and R. A. Ayala. 2014. "Reservoir evaporation in Texas." *J. Hydrol.* 510 (2014): 1–9.

Wurbs, R. A., and W. P. James. 2002. *Water resources engineering*. Upper Saddle River, NJ: Prentice Hall.

Wurbs, R. A., and Y. Zhang. 2014. *River system hydrology in Texas*. TR-461. College Station, TX: TWRI.

Chapter 7

Water Management in Each River Basin

Many reservoir projects and related constructed facilities are operated throughout Texas. The water resources and facilities that are most important to each person are those that supply that person's own personal water needs, recreational opportunities, and other services. Likewise, for water managers, the most important projects are those for which they are directly responsible. Detailed descriptions of every individual water management project in Texas in a single book is not feasible. Chapter 7 highlights selected projects that play central roles in water management in each of the river basins. The 15 major river basins of the state are discussed individually in the different sections of this chapter. The eight coastal basins are covered as a single section.

River Basins and Reservoirs

The 268,600 sq. mi area of Texas comprising the 15 major river basins and eight coastal basins delineated in Figure 1-4. The 15 major river basins are also shown in Figure 7-1. The coastal basins are the areas adjacent to the Gulf of Mexico between the lower reaches of the major river basins. The Canadian and Red Rivers are interstate tributaries of the Mississippi River as shown in Figure 1-1. Although the Sulphur River and Cypress Creek are tributaries of the Red River, the portions of their upper basins located in Texas are treated as separate Texas river basins. Except for the Canadian and Red Rivers and their tributaries, the

Figure 7-1. Major river basins of Texas.

major river basins and coastal basins discharge directly into the Gulf of Mexico in Texas.

Each river has many tributaries each of which has many tributaries that also have tributaries and so forth composing numerous subwatersheds that can be delineated at any scale of interest. Many water resources management endeavors are at the scale of a particular small watershed. This book has a statewide scope focused on major rivers.

The 15 major river basins and eight coastal basins are listed in Table 7-1. The total contributing drainage area and portion of this area in Texas are tabulated in the second and third columns. The Rio Grande includes an additional large noncontributing area, and several other basins encompass additional small noncontributing areas not included in these watershed areas. The watershed areas in Table 7-1 are also included in Table 6-6.

Major Reservoirs

Appendix A lists the 195 existing dam and reservoir projects in Texas with storage capacities of at least 5,000 ac-ft grouped by river basin. Five of the eight coastal basins have no major reservoirs. Water use in the coastal basins is supplied from reservoirs located in the major river basins. Water is also transported through pipelines between major river basins. The fourth and fifth columns of Table 7-1 are the number of major reservoirs located in each basin that have conservation storage and flood

Table 7-1. Major River Basins and Coastal Basins of Texas.

	Watershed Area		Number of Reservoirs			Storage Capacity	
			195 Major				
Major River Basin or Coastal Basin	**Total (square miles)**	**Texas (square miles)**	**Cons.**	**FC**	**Author-rized**	**Authorized (acre-feet)**	**Total FC (acre-feet)**
Major River Basins							
Brazos River	45,573	42,865	42	9	673	5,050,710	4,102,667
Trinity River	17,913	all	31	8	683	7,815,297	1,767,592
San Jacinto River	3,936	all	4	2	114	637,190	409,214
Colorado River	42,318	39,428	28	4	484	5,270,565	1,526,397
Neches River	9,937	all	10	2	180	3,904,100	1,179,295
Sabine River	9,756	7,570	12	0	212	6,393,413	0
Cypress Creek	3,552	2,929	10	1	86	949,139	600,737
Sulphur River	3,767	3,580	4	2	57	962,528	2,686,453
Red River	93,450	24,297	24	3	236	4,568,916	3,270,726
Canadian River	47,705	12,865	2	1	47	1,478,892	589,630
San Antonio River	4,180	all	3	1	122	367,028	12,600
Guadalupe River	5,953	all	5	1	222	451,818	362,048
Lavaca River	2,309	all	1	0	22	265,247	0
Nueces River	16,700	all	3	0	121	1,047,017	0
Rio Grande	182,215	49,387	7	2	80	3,023,656	2,235,730
Coastal Basins							
Nueces-Rio Grande	10,442	all	4	0	64	146,131	0
San Antonio-Nueces	2,652	all	0	0	9	3,548	0
Lavaca-Guadalupe	998	all	0	0	0	0	0
Colorado-Lavaca	939	all	0	0	8	9,666	0
Brazos-Colorado	1,850	all	0	0	0	30,959	0
San Jacinto-Brazos	1,440	all	1	0	0	51,042	0
Trinity-San Jacinto	247	all	1	0	13	18,633	0
Neches-Trinity	769	all	0	0	31	64,481	0
Totals	508,601	263,186	192	36	3,464	42,509,984	18,743,089

Note: Cons. = Conservation; FC = Flood control.

control storage, respectively. The three river basins with the largest number of major reservoirs described in the appendixes are the Brazos, Trinity, and Colorado River Basins with 42, 31, and 28 major reservoirs.

As indicated by Table 1-3 of Chapter 1, the 195 major reservoirs have storage capacities totaling approximately 58,872,709 ac-ft comprised of about 40,129,620 ac-ft of conservation storage capacity in 192 reservoirs and 18,743,089 ac-ft of flood control storage capacity in 36 reservoirs. Thirty-three of the 195 major reservoirs contain both conservation and flood control storage. Addicks and Barker Reservoirs in the San Jacinto River Basin and Olmos Reservoir in the San Antonio River Basin have only flood control capacity.

The 36 reservoirs with flood control storage pools are listed in Tables 6-4 and 6-5 of Chapter 6 grouped by river basin. The total flood control storage capacities of these 36 reservoirs are tabulated by river basin in the last column of Table 7-1.

Water Rights

The quantities in the sixth and seventh columns of Table 7-1 are also found in Table 6-6. The number of reservoirs included in the water rights system administered by TCEQ is shown in the sixth column of Table 7-1. The seventh column is the total reservoir storage capacity authorized by TCEQ-administered Texas water rights. Table 6-6 also includes the annual volumes of water that can be legally appropriated for beneficial use from streamflow or reservoir storage.

Water right permits are required prior to construction of a new reservoir project. Authorized reservoir storage capacities tabulated in Tables 6-6 and 7-1 include reservoirs permitted but not yet constructed. Water rights are not required for flood control. Authorized storage does not include flood control pool capacity. In general, all reservoirs with conservation storage capacities of at least 200 ac-ft require water right authorization. In some cases, storage capacities controlled by different water right permit holders for storage in the same reservoir are viewed as multiple reservoirs. The 3,464 storage capacities of 200 ac-ft or greater authorized by water rights documented by certificates of adjudication or water right permits include 210 reservoirs or reservoir components with conservation storage capacities of at least 5,000 ac-ft.

Simulated Reservoir Storage

The WRAP/WAM modeling system is described in Chapters 4 and 6. The mean and median of naturalized and simulated regulated flows at the outlets of eight major river basins are presented in Table 6-3. The modeling system generated the flows of Table 6-3 and the reservoir storage sequences discussed in the next paragraph.

The reservoir storage sequences plotted as Figures 7-3, 7-5, 7-8, 7-12, 7-16, and 7-17 were computed in WRAP daily simulations with the full authorization WAM datasets for the Brazos, Trinity, Neches, Colorado, Lavaca, and Nueces River Basins (Wurbs 2019a, b; 2020; 2022; 2023a, b; 2024). These plots are summations of daily simulated reservoir storage volumes in all reservoirs authorized by water rights in the Brazos (673 reservoirs), Trinity (683 reservoirs), and Neches (180 reservoirs) WAMs; the six Highland Lakes on the Colorado River; the one major reservoir in the Lavaca River Basin, and the two major reservoirs in the Nueces River Basin.

The storage plots illustrate the basin-wide water supply capabilities of these river basins. The hydrologic period-of-analysis simulated reservoir storage sequences also serve as a meaningful basin-wide drought index reflecting the combined effects of natural hydrology and authorized water use. Similarities and differences between the river basins are demonstrated.

Brazos River Basin

The 45,570 sq. mi Brazos River Basin is a little larger than the state of Tennessee and much larger than several other states in the United States and many countries in the world. Many local, state, and federal entities contribute to water management in the basin. The FWD of USACE and BRA own, maintain, and operate a multiple-purpose, 12-reservoir system that is central to water management in the Brazos River Basin.

BRA was created by the Texas legislature in 1929 as the first government entity in the United States responsible for developing and managing the water resources of an entire large river basin. A staff of about 250 people work in the BRA headquarters in Waco and at project facilities. The agency is self-supporting, except for occasional governmental grants to help pay the costs of specific projects. Funding is through revenues from customers for services provided.

BRA is responsible for developing and managing water supplies, water and wastewater treatment, monitoring water quality, and pursuing water conservation through public education programs. BRA leases water supply storage capacity in nine USACE reservoirs and owns and operates three other water supply reservoirs, two pipeline conveyance systems, regional water treatment systems, and several wastewater treatment plants located throughout the basin.

Dams and Reservoirs

Forty-two reservoirs in the Brazos River Basin are described in Appendixes A and B. Whitney, Belton, Waco, and Possum Kingdom are included in Tables 1-4 and 1-5 as the 7th, 12th, 20th, and 28th largest reservoirs in Texas. The 14 largest in the Brazos Basin, which includes the 12 USACE/BRA reservoirs, containing about 74% of the authorized conservation capacity and all the flood control storage capacity are included in the basin map of Figure 7-2.

Environmental flow standards created following procedures established by the 2007 SB3 are described in Chapter 3. The Brazos River and its tributaries and other river systems for which EFS have been

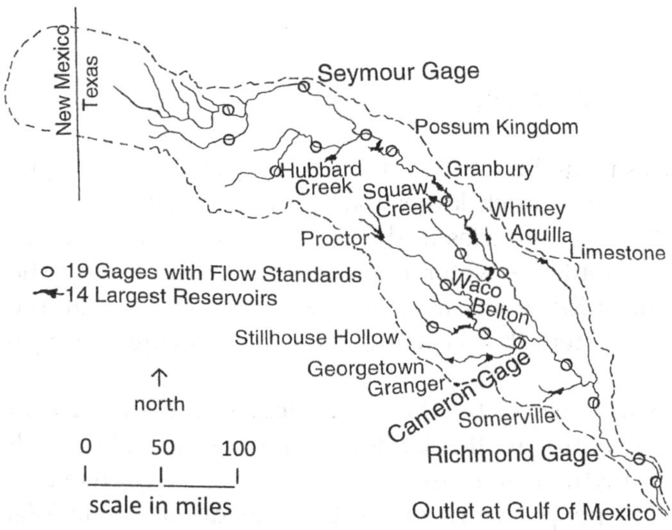

Figure 7-2. Brazos River Basin.

established are listed in Table 3-1. The locations of the 19 USGS gauge sites in the Brazos River Basin with EFS are shown in Figure 7-2.

Flows of the Brazos River at USGS gauges at Waco and Richmond and the Navasota River at Easterly are plotted in Figures 1-5 and 5-35 through 5-40. The historical daily storage contents of Possum Kingdom Reservoir are plotted in Figure 6-7. The summation of the observed daily reservoir storage contents of 28 reservoirs on the Brazos River and its tributaries from initial impoundment through 2023 are plotted in Figure 6-9.

The summation of 1940 to 2022 daily (end-of-day) storage contents of the 673 reservoirs in the full authorization scenario Brazos WAM are plotted in Figure 7-3. This simulation includes conservation storage capacities totaling 4,694,850 ac-ft in 673 reservoirs and flood control pool capacities totaling 3,945,830 ac-ft in the nine USACE multipurpose reservoirs.

Storage depletions resulting from 1950 to 1957 and 2010 to 2015 droughts and other less severe droughts are evident from Figure 7-3. Wurbs (2019a) presents storage plots for each of the 12 individual USACE/BRA reservoirs along with other detailed results from a simulation study. Storage content tends to fluctuate more in individual reservoirs than in the summation of storage in many reservoirs. Summing the storage contents of multiple reservoirs smooths or averages out variability. The effects on water supply capabilities of balancing storage are

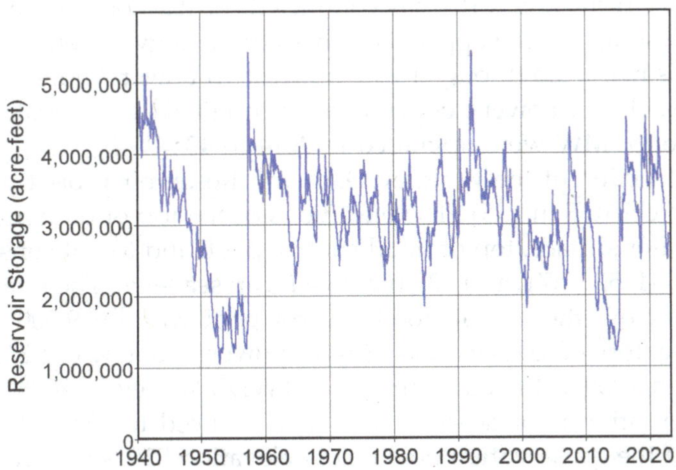

Figure 7-3. Simulated total storage contents of 673 reservoirs in Brazos WAM.

reflected in the system operation permit and associated water management plan discussed later in this section.

Squaw Creek Reservoir provides water for operating the Comanche Peak Nuclear Power Plant owned by Luminant Generation. A pipeline from Lake Granbury to Squaw Creek Reservoir facilitates purchase of water from BRA to maintain a constant level in the cooling water lake. The West Central Municipal Water District operates Hubbard Creek Reservoir and associated pumping and water delivery systems to supply its member cities of Abilene, Albany, Anson, and Breckenridge. BRA owns and operates Lakes Possum Kingdom, Granbury, and Limestone.

USACE FWD owns and operates the other nine reservoirs in Figure 7-2. The Corps of Engineers is responsible for flood control operations of the nine multipurpose federal projects. BRA has contracted for a portion of the conservation pool storage capacity of Lake Whitney and most of the conservation storage capacity of the eight other USACE reservoirs.

Flood control operations following procedures explained in Chapter 3 are based on maximum allowable non-damaging flow rates at the dams and 10 other USGS gauge sites on the Brazos River and its tributaries. The maximum allowable flow rate of the Brazos River at Waco is 25,000 ft^3/s. The allowable flow at all gauges on the Brazos River downstream of Waco including the Richmond gauge is 60,000 ft^3/s. The allowable flow of the Little River at Cameron is 10,000 ft^3/s.

The front cover of this book has a photograph of Whitney Dam and Reservoir. USACE FWD operates Lake Whitney for flood control, hydropower, and recreation. BRA has contracted for 22% of the conservation pool for use as water supply. Hydroelectric energy generated at Lake Whitney is marketed through the Southwestern Power Administration to the Brazos Electric Power Cooperative. The original two generators with a capacity of 30 MW were upgraded in 2014 to 43 MW.

Construction of the Whitney Dam and Reservoir project was completed by USACE in 1951. The elevations of the top of dam, maximum design water surface, top of flood control pool, and top of conservation pool are 584, 573, 571, and 533 ft above mean sea level. The total storage capacity below the top of flood control pool was 1,999,500 ac-ft and below the top of conservation (hydropower) pool was 627,100 ac-ft based on a 1959 volumetric survey. A TWDB survey in 2005 indicates the conservation pool capacity had been reduced to 554,200 ac-ft. The portion of the conservation pool below elevation 520 is inactive storage providing head for hydropower but no releases.

BRA holds a water right permit from TCEQ and a contract with USACE FWD for 50,000 ac-ft of the active conservation pool acquired in 1982 through a reallocation from hydropower to water supply. University researchers (Wurbs and Carriere 1988) and more recently the SB3 Region G Planning Group and others have explored options for additional future reallocations of hydropower or flood control storage capacity in Whitney Reservoir to water supply and reallocations of flood storage to water supply in other reservoirs in the basin.

Lake Waco is a component of the nine-reservoir system operated by USACE for flood control. The conservation pool is committed to supplying water solely for the city of Waco and adjacent smaller communities. Lake Waco was constructed by USACE during 1958 to 1965, completely submerging the much smaller original Lake Waco constructed by the city of Waco in 1928 to 1929. At the request of BRA and the city of Waco, USACE raised the designated top of conservation pool 7 ft in 2003 converting a portion of the flood control pool to water supply.

Allen's Creek Reservoir is authorized by a water right permit but has not been constructed and thus is not included in the 42 reservoirs described in the appendixes. When the project will be constructed and its size are uncertain. The proposed off-channel storage project on Allen's Creek is to be filled with water pumped from the Brazos River near the Richmond gauge. The BRA holds a water right permit to construct the project to supply water for Houston.

Natural Salt Pollution

Natural salt pollution in the upper Brazos River Basin and adjoining basins is described in the last section of Chapter 5. USACE, USGS, and USBR performed salinity control studies during the 1960s to 1990s for the upper Brazos, Red, Canadian, Colorado, and Pecos River Basins (Rought 1984, Wurbs 2002). Salt control projects were proposed for all these basins and implemented in some. USACE recommended construction of three salt control dams on the Salt Fork of the Brazos River and its tributaries that were never funded and thus never constructed (USACE Fort Worth District 1976, McCrory 1984, Wurbs 2002). These three proposed brine control dams like others actually constructed in the Colorado and Red River Basins are designed to store highly saline runoff from small primary salt source watersheds. The water evaporates, leaving the salt to accumulate behind the dams.

Statistics from salinity concentration measurements collected by USGS for USACE salt pollution studies are presented in Tables 5-6 and 5-7. Salt enters the upper Brazos River from geologic formations in the Permian Basin Region. Concentrations are diluted downstream by inflows from low-salinity tributaries. Referring to Table 5-6, estimated long-term mean TDS concentrations decrease from 56,960 mg/L in Salt Croton Creek near Aspermont to 12,400 mg/L in the Salt Fork of the Brazos River near Aspermont to 1,530 mg/L in the Brazos River below Morris Sheppard Dam at Possum Kingdom Reservoir to 928 mg/L below Whitney Dam to 339 mg/L at the USGS gauge on the Brazos River at Richmond.

Salinity affects water supply operations of the three reservoirs on the Brazos River and multiple-reservoir system operations that include the tributary reservoirs. Brazos salinity may also be relevant to Houston and other water users in the San Jacinto River Basin and coastal areas in comparing strategies for acquiring water from the Brazos River, Trinity River, or other alternative sources.

The city of Granbury operates a desalination treatment plant that allows use of Lake Granbury for municipal supply. BRA transports water from Lake Granbury by pipeline to Squaw Creek Reservoir for use as cooling water for the Comanche Peak Nuclear Power Plant. Releases from Possum Kingdom and Whitney Reservoirs diluted with tributary inflows partially controlled by tributary reservoirs supply water needs in the lower basin and adjoining coastal basins.

Salinity in the lower reach of the Brazos River, like other Texas rivers, is also affected by saltwater intrusion from the Gulf of Mexico. Seawater propagates further upstream in rivers during low flows. Water supply pumping plants for industrial water use located on the Brazos River about 25 and 60 mi upstream of its outlet are affected by reservoir operations during periods of low flows from the dual perspectives of stopping seawater intrusion and dilution of salinity from the upper basin.

System Operations

A BRA system operation permit and associated water management plan approved by TCEQ in 2016 significantly increase water supply capabilities. BRA acquired water rights for the individual reservoir projects historically as the different projects were constructed over several decades. Authorized diversion amounts were based on the supply capabilities of each constructed project operated individually.

The system operations permit recognizes that BRA operates a multiple-reservoir system. The water management plan required for the water right permit documents system operations which supply much more water than the individual reservoirs operated independently.

The system operation permit and water management plan are publicly accessible at the BRA website along with detailed technical reports documenting their development. Simulation and reliability analysis capabilities provided by the WRAP/WAM modeling system played a fundamental role in developing the system permit and management plan.

USACE FWD responsibilities for flood control and hydropower operations are separate from BRA water supply operations. The system water right permit and associated management plan address BRA water supply operations. The conservation pool of Waco Reservoir is dedicated to supplying water to the city of Waco and neighboring towns. The conservation pools of the other eleven reservoirs are operated as a system to supply diversions both lakeside and at various distances downstream of the dams. The BRA system supplies water to cities, industries, and other customers at certain locations from specific reservoirs. However, much of the water use occurs in the lower Brazos River Basin and adjoining coastal area supplied by releases from multiple upstream reservoirs.

The system operation permit and water management plan increase water supply capabilities through several strategies that include multiple-reservoir risk sharing, combining regulated and unregulated flows, employing both firm and interruptible water supply commitments, and downstream reuse of return flows. The 11 reservoirs scattered throughout the large river basin experience different levels of drought conditions at different times. Multiple-reservoir system operation allows balancing of storage depletions between reservoirs. The precipitation runoff from the large watershed between the dams and downstream diversion sites results in extremely variable but often significantly high flows which are combined with reservoir releases to achieve acceptable levels of reliability.

The concept of varying levels of supply reliability is another key enhancement. The amount of water available for contractual commitment to BRA customers was limited to firm or very high reliability supply in the past. The amount of water provided by the BRA system can be increased substantially by accepting a higher risk of shortages in supplying target amounts. Certain customers prefer a larger amount of

BRA water most of the time even at the risk of occasional shortages during drought conditions. Municipal use typically requires a higher reliability than agricultural irrigation. Interruptible supply curtailment may be triggered by reservoir storage falling below specified operating levels. Temporary conversions to groundwater sources or demand management strategies can be activated during infrequent BRA supply curtailments. Options are provided by BRA to water customers for either firm (high reliability) or interruptible (lower reliable) water supply commitments.

The BRA system operation permit and water management plan include the proposed Allen's Creek off-channel reservoir in the lower basin that has not yet been constructed. The water right permit and management plan include alternative versions of operating strategies for without versus with construction of the Allen's Creek project.

Trinity River Basin

About 26% of the residents of Texas live in the upper Trinity River Basin. Most of the population of the Trinity River Basin reside in the Dallas and Fort Worth (DFW) metropolitan area in the upper basin, which had a 2020 census population of 7.6 million people, making it the most populous metropolitan area in Texas and the fourth largest in the United States. Dallas and Fort Worth have populations of 1,290,000 and 935,000 people, respectively. Seventy other cities in the DFW metropolitan area have populations exceeding 10,000. This is one of fastest growing large metropolitan areas in the United States.

Dams and Reservoirs

Figure 7-4 is a map of the Trinity River Basin. Daily observed flows of the Trinity River at USGS gauges at Dallas, Oakwood, and Romayor plotted in Figures 5-24 through 5-26 illustrate the tremendous variability of streamflow in this basin and throughout Texas. The daily storage levels from initial impoundment through September 2023 of Lakes Livingston, Richland-Chambers, and Ray Roberts are plotted in Figures 6-4 through 6-6. The summation of the observed daily reservoir storage of 24 reservoirs in the Trinity River Basin are plotted in Figure 6-10.

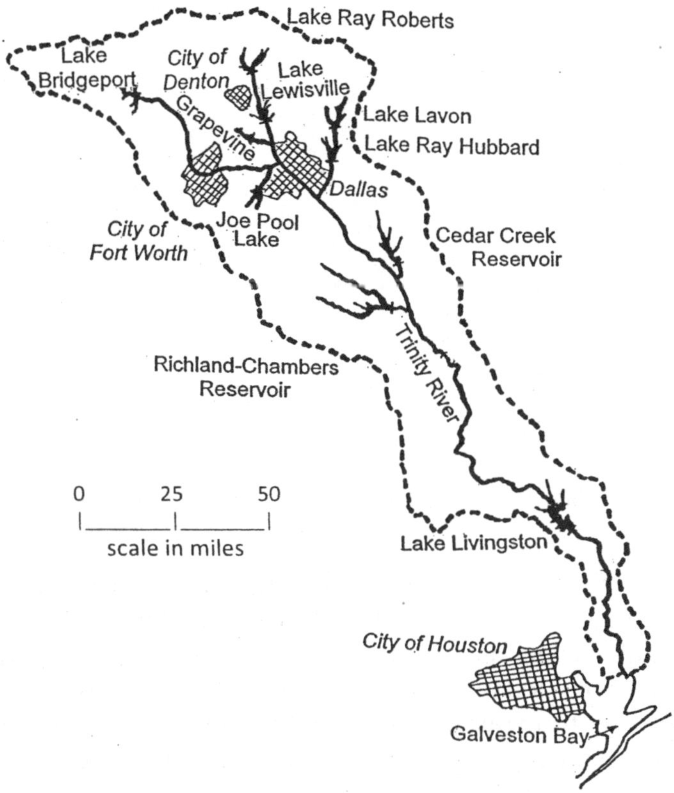

Figure 7-4. Trinity River Basin.

Statistics for naturalized and regulated flows of the Trinity River at its outlet generated by the WRAP/WAM simulation model are included in Table 6-3.

Summations of end-of-day storage levels of the 683 reservoirs included in the Trinity full authorization WAM for January 1940 through December 2022 hydrology are plotted in Figure 7-5. Thirty-one major reservoirs contain 98.0% of the total conservation storage capacity of 7,596,675 ac-ft in the 683 reservoirs. The daily simulation includes flood control operations of eight USACE reservoirs with flood control pool capacities totaling 1,579,420 ac-ft.

The WRAP/WAM storage plot of Figure 7-5 indicates that the 1950 to 1957 drought is the hydrologically worst drought since before 1940 in this river basin. The WRAP/WAM simulation combines presently authorized water demands and water development infrastructure with

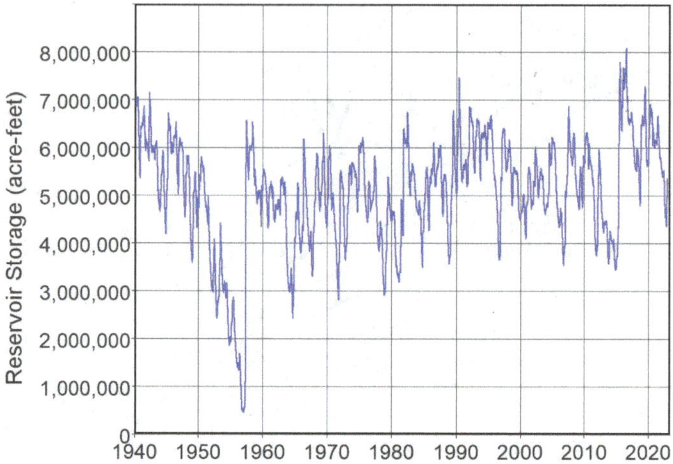

Figure 7-5. Simulated storage contents of 683 reservoirs in Trinity WAM.

1940 to 2022 natural hydrology. Population and water demands have greatly increased since the 1950 to 1957 drought. Drought conditions as hydrologically severe as 1950 to 1957 have never been experienced in combination with today's population and associated water demands. Economic losses from hydrologic drought conditions like those of 1950 to 1957 would be much greater today even though more reservoir storage capacity is now available.

It is also interesting to note that the 1950 to 1957 drought ended with the April–May 1957 flood, which is one of the most widespread and devastating floods on record throughout Texas as well as in the DFW metroplex. Four of the eight USACE FWD flood control reservoirs in the upper Trinity River Basin were constructed before and the other four constructed after the 1957 flood. Likewise, the levee and channel improvements discussed later include both projects completed before and subsequent to the 1957 flood.

USACE FWD owns and operates eight of the 14 largest reservoirs in the basin. The eight multiple-purpose reservoirs are operated by USACE for flood control. Nonfederal sponsors hold contracts for the water supply storage that were executed in accordance with the provisions of the Water Supply Act of 1958 discussed in Chapter 2. The nonfederal water supply sponsors for the eight federal reservoirs include the TRA, Tarrant Regional Water District (TRWD), North Texas Municipal Water District (NTMWD), and cities of Dallas, Fort Worth, Grapevine, and Denton.

Thirty-one major reservoirs with permitted storage capacities exceeding 5,000 ac-ft account for 98.0% of the total authorized storage capacity of the 683 reservoirs included in WAM. Lakes Livingston, Richland-Chambers, Ray Roberts, and Lavon have storage capacities exceeding 500,000 ac-ft and thus are included in Tables 1-4 and 1-5 and Figure 1-9.

Dallas Water Utilities (DWU) of the city of Dallas supplies water to about 30 cities in addition to Dallas residents from Lakes Ray Roberts, Lewisville, Grapevine, and Lavon owned by USACE FWD and Lake Ray Hubbard and White Rock Lake owned by the city of Dallas. NTMWD supplies its customers from Lake Lavon under a water supply storage contract with USACE FWD. The other major reservoirs in the Trinity River Basin are owned by various cities and electric power companies. Recreation is very popular at most of the lakes in the basin.

The cities of Dallas and Denton have contracted with USACE FWD for the conservation storage capacity of both Lakes Lewisville and Ray Roberts. Dallas holds a water supply contract with the Corps of Engineers for 74.0% of the conservation storage capacity in both Lakes Ray Roberts and Lewisville. Denton holds a contract for the remaining 26.0% of the conservation storage capacity of both reservoirs. A photograph of the outlet conduit intake structure at Lake Ray Roberts is presented as Figure 7-6. Two 6 ft by 13 ft gates near the bottom of the tower are employed to control outflows through a 13 ft diameter conduit through the dam that discharges into the Elm Fork of the Trinity River downstream of the dam. Lake Lewisville is located immediately downstream of Ray Roberts Dam.

The author was a consultant providing computer modeling support during a 2003 to 2006 collaboration between the Lake Lewisville Association, DWU, and USACE FWD that illustrates common reservoir management issues. The investigation of operational modifications involving allocating storage drawdowns between Lakes Lewisville and Ray Roberts was instigated by the Lake Lewisville Association, which represents private marina owners operating under leases with USACE FWD. Reservoir storage fluctuations are disruptive to marina boat storage and other aspects of lake recreation. From a water supply perspective, storing water upstream in Ray Roberts rather than Lewisville improves operating flexibility. Water can be released from Lake Ray Roberts downstream to Lake Lewisville through the outlet structure in Figure 7-6 any time by gravity flow, but water cannot be conveniently transported upstream from Lake Lewisville to Lake Ray Roberts. A plan

Figure 7-6. Outlet conduit intake structure at Ray Roberts Dam.
Source: Courtesy of USACE.

was formulated with USACE FWD implementing small reallocations of flood control capacity to water supply seasonally under certain conditions that would mitigate adverse effects on water supply reliability related to DWU balancing storage between the two reservoirs in a manner more favorable to recreation interests.

Lake Lewisville is the only reservoir in the Trinity River Basin used for hydroelectric energy generation. A low-head run-of-river hydropower unit located in the river below Lewisville Dam operates using water supply releases through the dam. A similar hydropower facility at Ray Roberts Dam was installed in 1991 but discontinued and decommissioned in 2003.

Richland-Chambers, Cedar Creek, Bridgeport, and Eagle Mountain, which are ranked among the nine largest water supply reservoirs in the basin, are owned and operated by TRWD to supply water to Fort Worth and other cities. Lakes Bridgeport and Eagle Mountain are operated by TRWD as a system, along with Lake Worth, which is operated by the City of Fort Worth for recreation and water supply. TRWD also supplies water to the cities of Fort Worth, Weatherford, and Benbrook from Lake Benbrook which is owned by USACE.

TRWD, DWU, and other water management entities in the upper Trinity River Basin import water from the Sabine, Neches, and Red River

Basins through pipelines to supplement water supplies for the many cities in the DFW metropolitan area. These interbasin transfers are discussed in the later Neches, Sabine, and Red River Basin sections of this chapter.

Lake Livingston owned and operated by TRA under contract with the city of Houston and located on the lower Trinity River is the largest reservoir in the basin and ninth largest in Texas. Water is transported by pipeline from Lake Livingston through a regional water supply system to Houston in the adjoining San Jacinto River Basin and water users in the lower Trinity Basin. TRA supplies its customers in the upper and middle Trinity Basin from Lakes Bardwell, Navarro Mills, and Joe Pool, owned by the USACE.

Flood Control Levees and Channel Improvements

Multiple flood control levee and channel improvement projects on the upper Trinity River and tributaries have been constructed, maintained, extended, and rehabilitated since the 1940s by the Corps of Engineers and local water management agencies working both individually and collaboratively. Additional projects are proposed for future implementation.

Federal levee and channel projects are constructed by USACE with nonfederal cost-sharing and maintained by nonfederal sponsors. For local floodway projects, real estate and relocation costs are nonfederal, and construction costs are federal. TRWD maintains 27 mi of USACE-constructed levees along with its other flood control programs on the Tarrant County side of the metroplex. The Flood Control Division of DWU maintains the USACE-constructed floodway in Dallas along with its other stormwater management responsibilities. Other regional and local agencies also construct and maintain flood control projects in the upper Trinity River system. Construction, maintenance, and enlargement of the structural flood control projects overlap floodplain regulation endeavors by local entities and FEMA in collaboration with the National Flood Insurance Program.

The River and Harbor Act of 1945 authorized USACE construction of several reservoir and floodway projects in the Trinity River Basin (USACE 1992). A basin-wide plan of improvement authorized by the River and Harbor Act of 1965 included additional reservoir and floodway projects.

The Fort Worth Floodway constructed by the Corps of Engineers during the 1950s and extended during the 1960s comprises levees and channel improvements on the West Fork and Clear Fork of the Trinity River. The city of Dallas constructed levees and channel improvements during 1928 to 1932. The Dallas Floodway Project constructed by the Corps of Engineers during 1953 to 1960 consisted of strengthening and lengthening the existing city improvements. Later improvements included construction of levees and enlarged channels on other reaches of the Trinity River and tributaries between Fort Worth and Dallas. TRWD, DWU, and other local sponsors are responsible for maintenance and repair of the floodway system.

The city of Fort Worth, TRWD, and USACE are partners on the Central City Flood Control Project under development to enhance the economic growth of downtown Fort Worth. The project consists of a bypass channel, levee system, and associated improvements to divert flood flows around a portion of the existing floodway system. The ongoing endeavor also includes proposed oxbow ecosystem restoration of about 1,060 ac severed by the channel realignment.

The Water Resources Development Act of 2007 authorized modifications to the Dallas Floodway that address improved capabilities for managing flood risk along with recreation, ecosystem preservation, and other concerns. The multiple-faceted endeavor involves a collaborative partnership of USACE and multiple other federal, state, and local entities.

San Jacinto River Basin

Houston is the most populous city in Texas and fourth most populous in the United States. The US census April 2010 population of the city of Houston of 2,099,450 increased 9.77% to an April 2020 census population of 2,304,580. The 2020 population of the nine-county Houston, Woodlands, Sugar Land metropolitan statistical area was 7,210,000, which was 24.0% of the population of Texas. The city of Houston is spread primarily over the Buffalo Bayou subwatershed of the San Jacinto River Basin. The San Jacinto River and Buffalo Bayou merge into the Houston Ship Channel of the Port of Houston before flowing into Galveston Bay.

The six major reservoirs in the 3,936 sq. mi San Jacinto River Basin are described in Appendixes A and B. Addicks and Barker Reservoirs contain flood control storage capacities totaling 409,210 ac-ft but have no conservation storage. Shelton, Houston, Lewis Creek, and Conroe Reservoirs have conservation storage capacities of 5,420, 16,400, 140,520, and 429,890 ac-ft, respectively, but no flood control storage capacity. Lake Houston is located on the San Jacinto River upstream of the Houston Ship Channel. Lake Conroe is further upstream on the West Fork of the San Jacinto River adjacent to the City of Conroe. The population of the City of Conroe has grown from 56,210 people in April 2010 to 89,890 in April 2020 and 101,410 by July 2022.

Houston Ship Channel

The Houston Ship Channel shown in the photograph of Figure 7-7 is a 52 mi long deep-draft waterway through Galveston Bay, the San Jacinto River, and Buffalo Bayou to a turning basin about 4 mi east of downtown Houston. The ship channel is directly linked to numerous transportation facilities, industrial plants, and other enterprises that use the channel to ship products to markets throughout the world. The USACE Galveston

Figure 7-7. Houston ship channel.
Source: Courtesy of Port of Houston Authority.

District constructed components of the Houston Ship Channel over several decades as authorized by the 1958 and 1965 River and Harbor Acts and 1986 Water Resources Development Act and continues to maintain the project along with other navigation improvements along the Texas coast.

Buffalo Bayou was originally a tributary of the San Jacinto River. Their confluence is now in the enlarged ship channel. Prior to development of the ship channel, Buffalo Bayou was swampy, marshy, and overgrown with dense vegetation. Steamboats and shallow draft boats were the only vessels able to navigate its complicated channel during the 1800s and early 1900s before creation of the improved Houston Ship Channel.

The ship channel was created and is maintained by dredging. The channel has been widened to 530 ft and deepened to 45 ft to accommodate large ships. Islands have been formed in the ship channel from soil pulled up by dredging. Islands and salt marches and associated bird habitat are included in the environmental mitigation responsibilities of the Houston Port Authority. The ship channel has five vehicle crossings that include a tunnel and four bridges.

The Port of Houston is one of the busiest seaports in the world. The ship channel is the conduit for ocean-going vessels between Houston-area terminals directly to the Gulf of Mexico and also serves traffic from the Gulf Intracoastal Waterway. The ship channel has many public and private terminals and berthing locations along Buffalo Bayou and Galveston Bay. Major types of cargo shipped include petroleum, petrochemicals, iron and steel, fertilizers, minerals, grains, wood, and wood products.

The Gulf Intracoastal Waterway extends from Apalachee Bay, Florida, for a length of 1,100 mi to the Mexican border at Brownsville, Texas. The federally developed and maintained navigable waterway consists of a dredged and otherwise improved channel paralleling the coast behind barrier islands. A 150 ft wide, 12 ft deep channel is maintained mainly between barrier island beaches and the mainland, though some stretches are not protected by islands. The waterway provides a route for barges that accesses several ports for ocean-going vessels. The Port of Houston with its ship channel is one of the ports connected to the Gulf Intracoastal Waterway.

Water Supply and Land Subsidence

Before the 1970s, Houston and other nearby cities relied almost entirely on groundwater from the Gulf Coast Aquifer for water supply. Several decades of groundwater pumping exceeding recharge resulted in major subsidence problems. The land surface subsided several feet in downtown Houston and even more in other areas. Consequently, Houston and neighboring cities and industries have been shifting to surface water over the last several decades.

Lakes Houston, Conroe, and Livingston supply water to multiple cities and industries in the area around Houston as well as to the city of Houston. The 1,742,000 ac-ft Lake Livingston on the Trinity River was constructed by TRA with impoundment beginning in 1969. The 429,890 ac-ft Lake Conroe owned and operated by the San Jacinto River Authority began impounding water in 1973. The older and much smaller Lake Houston owned by the city of Houston also provides municipal and industrial water supply.

Excessive groundwater use causes land subsistence, which results in increased flooding. The Harris–Galveston Subsidence District (HGSD) is a special-purpose district created by the Texas legislature in 1975 to manage groundwater withdrawal throughout Harris and Galveston counties for the purpose of preventing land subsidence. HGSD controls subsidence by regulating the use of groundwater resources and working with surface water suppliers to convert the region to a greater reliance on surface water. A basic policy objective is to enforce regulatory disincentives for those who rely too heavily on groundwater. HGSD also conducts water conservation, education, and outreach programs.

The Brownwood Subdivision in the city of Baytown is a classic example of the effects of land subsidence in the region. The ground level in this area near the Houston Ship Channel was close to sea level even before subsidence. Decades of overpumping of groundwater resulted in lowering the neighborhood of about 400 homes by as much as 10 ft. Hurricane Alicia destroyed the neighborhood in 1983. The remaining houses were demolished and removed. The Baytown Nature Center now occupies a portion of the site.

Flood Control

Frequent flooding and stormwater drainage problems are prevalent in the low-lying flat coastal plain of Houston and vicinity. Intense rainstorms are common. Urbanization has greatly increased watershed runoff to the bayous and creeks.

Addicks and Barker Reservoirs on South Mayde Creek and Buffalo Bayou upstream of downtown Houston are owned and operated by the USACE Galveston District. The two flood control projects are the only USACE reservoirs in Texas with no conservation pool. Numerous flood control reservoirs with no conservation pool have been constructed and are operated by the Corps of Engineers in other states. Construction of the Addicks and Barker Dams was completed in 1948 and 1945. Additional improvements to both dams were performed during 1986 to 1991 to upgrade to updated engineering standards.

As of January 2024, Hurricane Harvey in August 2017 is the only flood to produce flows over the emergency spillways of either of the two projects. However, large releases that contribute to damages to residential and commercial properties located downstream of the dams are necessary to protect upstream properties before the water level reaches the emergency spillways. The government-owned land behind the dams is completely inundated, with damage to residential and commercial properties on adjacent private lands at flood levels significantly below the elevations of the emergency spillways. Operations during major floods deal with both upstream and downstream flooding. The other reservoir projects in Texas owned by USACE include government lands behind the dams extending well above the top of the flood control pools.

The two dams have a combined watershed drainage area of 279 sq. mi (178,560 ac). The normally dry federally owned impoundment areas of 13,500 ac behind Addicks Dam and 12,500 ac behind Barker Dam are used extensively for recreation. The city of Houston and Harris County lease reservoir lands from the federal government for use as public parks.

The HCFCD was created in response to devastating floods in 1929 and 1935 and serves as the nonfederal sponsor for Addicks and Barker Reservoirs. The special-purpose district was created by the Texas legislature in 1937 and is governed by the Harris County Commissioners Court. The jurisdictional boundary of HCFCD coincides with the 1,777 sq. mi Harris County, which includes most of the city of Houston as well as surrounding areas. Harris County encompasses all or parts of 33 different municipalities as well as unincorporated areas.

HCFCD works with 23 primary watersheds, each with its own flooding problems. HCFCD constructs and maintains detention basins, channel improvements, and levees and has a program to buy out and demolish houses subject to frequent major flooding. The agency maintains 2,500 mi of improved channels and numerous detention basins. Hurricane Harvey in August 2017 damaged more than 120,000 properties in Harris County and is considered the worst flood dealt with by HCFCD to date.

Colorado River Basin

The Colorado River Basin extends from New Mexico across Texas to Matagorda Bay. The headwaters in New Mexico and northwest Texas are at elevations of about 4,000 ft msl. Observed daily flows of the Colorado River at San Saba, Austin, and Columbus are plotted in Figures 5-29, 5-30, and 5-31 of Chapter 5. Annual flows at these three USGS gauge sites are compared in Figure 5-32. Lake Buchanan is generally considered the upper limit of the Lower Colorado River. The Upper Colorado River Basin is the West Texas portion of the basin above Lake Buchanan.

Reservoir Storage

The 484 reservoirs in the Colorado River Basin licensed by water rights and included in the full authorization WAM dataset include 31 with authorized storage capacities exceeding 5,000 ac-ft. These 31 major reservoirs include 28 existing reservoirs and three others that are not yet constructed. One of the proposed but not yet constructed projects may be modified later to consist of storage in multiple LCRA off-channel reservoirs though now treated as one project. The 28 existing major reservoirs account for 88.3% of the total authorized storage capacity of 5,270,600 ac-ft of the 484 reservoirs included in the full authorization WAM dataset.

The Colorado River Municipal Water District (CRMWD) and LCRA control most of the reservoir storage capacity in the upper and lower basins, respectively. Lakes J. B. Thomas, E. V. Spence, and O. H. Ivie owned and operated by CRMWD have authorized storage capacities that total 1,247,100 ac-ft. The storage capacity authorized by water right permits for the six Highland Lakes operated by LCRA total 2,349,032 ac-ft, which is 50.5% of the authorized capacity of the existing 28 reservoirs with storage capacities exceeding 5,000 ac-ft and 44.6% of the total storage capacity of the 484 reservoirs in the Colorado WAM. The three largest

CRMWD reservoirs and six LCRA Highland Lakes account for 77.3% of the authorized storage capacity of the 28 existing major reservoirs and 68.3% of the total authorized storage capacity of the 484 reservoirs. CRMWD and LCRA also own and operate several smaller reservoirs in addition to these nine larger projects.

Daily time series plots of the summations of total observed storage contents and active conservation storage contents and capacities of 19, 28, and 24 selected large water supply reservoirs in the Colorado, Brazos, and Trinity River Basins are presented in Figures 6-8 through 6-10, respectively, discussed in Chapter 6. The observed reservoir storage contents viewed as a percentage of storage capacity is much lower in the Colorado Basin than in the Brazos and Trinity Basins. Stated another way, the storage drawdowns in the Colorado River Basin are much greater than in the two other river basins. The LCRA reservoirs on the Lower Colorado River and even more so the CRMWD reservoirs and other reservoirs in the upper basin are dependent on inflows from rainfall runoff from dry watersheds in West Texas to fill storage.

Referencing Figure 6-8, the active conservation storage capacity of the 19 larger water supply reservoirs in the Colorado Basin increased steadily between June 1933 and December 1968 as construction of the individual reservoir projects was completed. The active conservation capacity of the 19 reservoirs is shown in Figure 6-8 to be a constant 3,411,600 ac-ft from December 1968 to March 1990 and then vary slightly between 3,911,200 and 4,029,100 ac-ft between September 1990 and early 2024. The maximum volume ever stored in these 19 reservoirs was 3,736,200 ac-ft in June 1997, which was 95.5% of the active conservation storage capacity. The minimum storage level since before 1968 was 1,066,900 ac-ft or 26.7% of their active conservation capacities in September 2013. The active conservation capacity of these 19 reservoirs was 36.1% full on February 22, 2024.

Upper Colorado River Basin

The Colorado River Municipal Water District is the largest reservoir owner, water right holder, and water supplier in the upper Colorado River Basin. CRMWD was created by the Texas legislature in 1949 for the purpose of providing water to its member cities of Odessa, Big Spring, and Snyder. CRMWD also has water supply contracts with the cities of Midland, San Angelo, Stanton, Robert Lee, Grandfalls, Pyote, and

Abilene and the Millersview–Doole Water Supply Corporation. CRMWD owns and operates J. B. Thomas, E. V. Spence, and O. H. Ivie Reservoirs, which have authorized water supply storage capacities of 204,000, 488,760, and 554,340 ac-ft, respectively. CRMWD also operates four groundwater well fields used primarily to supplement surface water sources during the summer months. The surface water reservoirs in the upper basin are badly over-appropriated.

CRMWD owns nine other reservoirs that are used to prevent low-quality, high-salinity water from flowing downstream. Water is permanently impounded or diverted for other uses. These nine salinity control impoundments are Sulphur Draw Reservoir, Red Lake Reservoir, Natural Dam Lake, Barber Reservoir, Mitchell County Reservoir, Red Draw Reservoir, Beals Creek Sump, Three Mile Lake, and Four Mile Lake.

USACE FWD owns and operates Hords Creek Dam and Reservoir and O. C. Fisher Dam and Reservoir for flood control, water supply, and recreation. Hords Creek is by far the smallest, and O. C. Fisher is smaller than most of the other USACE reservoirs in Texas. The Central Colorado River Authority contracted for the water supply storage of Hords Creek Reservoir, which is used to supply the city of Coleman. The Upper Colorado River Authority contracted for the water supply storage of O. C. Fisher Reservoir. USACE FWD is also responsible for oversight of operations of the flood control pools of Lake Travis owned by LCRA and Twin Buttes Reservoir owned by the city of San Angelo that were constructed by USBR.

Lower Colorado River Basin

The Colorado River flows through Austin and serves as the primary water supply source for Austin and neighboring cities as well as major farming operations in the lower basin and other water users throughout the basin. Austin, with a 2020 census population of 962,900, is the largest city in the Colorado River Basin, fourth largest city in Texas, and one of the fastest growing cities in the nation. The five-county Austin, Round Rock, and San Marcos metropolitan area had a 2020 population of 2,352,000. Austin both holds its own water rights and contracts with LCRA for water supplied under LCRA water rights.

LCRA created by the Texas legislature in 1934, like the other river authorities, has no taxing authority and operates solely on utility revenues and fees generated from supplying electrical energy, water, and community services. It has about 2,100 employees. LCRA supplies wholesale electrical power to 43 city-owned utilities and electric cooperatives that serve more than a million people in Central Texas. The agency owns and operates three gas-fired electric power plants, one coal-fired power plant, and six hydroelectric plants and purchases electricity from three wind farms. LCRA manages more than 16,000 ac of recreational lands along the Colorado River and administers programs supporting community and economic development. The agency operates the off-channel Lakes Bastrop and Fayette County to provide cooling water for thermal-electric power plants, as well as operating the six multiple-purpose Highland Lakes.

LCRA owns five and operates all six of the Highland Lakes on the Colorado River. The lakes are listed in upstream-to-downstream order in Table 7-2. The authorized storage capacities in the third column are from the water right certificates of adjudication or permits. The information in the last four columns is provided at the LCRA website. Capacities and water surface areas for Lakes Buchanan, LBJ, Marble Falls, and Travis reflect volumetric surveys performed during 2020 to 2021.

Summations of WRAP/WAM full authorization scenario simulated daily storage levels of the six Highland Lakes on the Colorado River operated by LCRA are plotted in Figure 7-8. Simulated storage depletions in Figure 7-8 are severe for both the 1950 to 1957 and 2010 to 2015 droughts. The Highland Lakes are almost emptied during 2014 to 2015 under this simulation scenario. Likewise, in actual real-world water management, water use was greatly curtailed during this period, particularly for agricultural irrigators with interruptible water supply commitments.

Table 7-2. Highland Lakes on the Colorado River Operated by LCRA.

Dam	Lake	Authorized Capacity (acre-feet)	Updated Capacity (acre-feet)	Surface Area (acres)	Elevation at Top of Normal Pool (feet msl)	Elevation at Top of Dam (feet msl)
Buchanan	Buchanan	992,475	880,356	22,452	1,020	1,025.5
Inks	Inks	17,545	13,668	777	888	922
Wirtz	LBJ	138,500	131,618	6,432	825	838.5
Starke	Marble Falls	8,760	7,597	613	737	761.5
Mansfield	Travis	1,170,752	1,115,076	19,044	681	750
Tom Miller	Austin	21,000	24,644	1,830	492.8	517

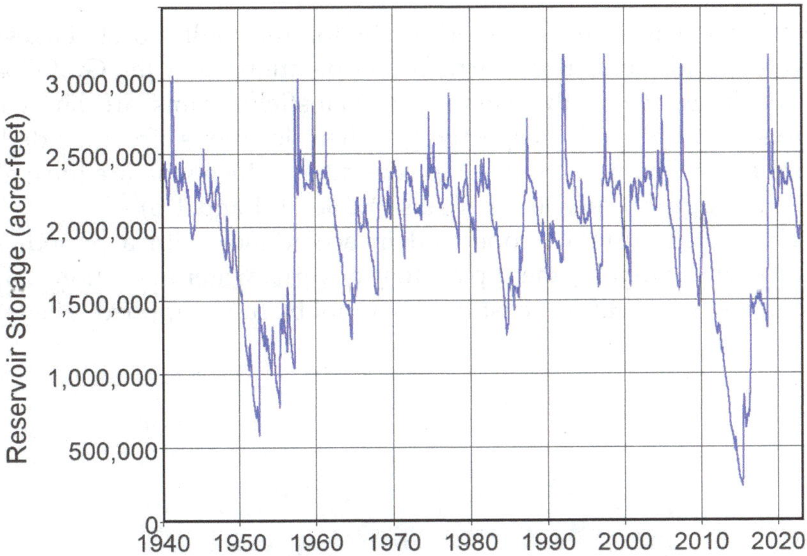

Figure 7-8. Simulated storage in the six Highland Lakes on the Colorado River.

The full authorization scenario simulation of Figure 7-8 assumes that all present water right holders store and/or divert their full permitted volumes of streamflow each year during a repetition of 1940 to 2022 historical natural hydrology. The minimum simulated storage levels reached by the 1946 to 1952 and 2010 to 2015 drawdowns are 24.7% and 10.1% of the total conservation storage capacity of 2,349,032 ac-ft, occurring on September 9, 1952 and May 10, 2015, respectively. The peak flood storage level of 3,168,000 ac-ft occurs in February 1992, June 1997, and November 2018 of the hydrologic period of analysis. The end-of-day storage volume is 1,982,840 ac-ft on December 31, 2022, the last day of 30,316-day simulation.

Tom Miller Dam and Lake Austin located in Austin are owned by the city of Austin and operated by LCRA. The five other lakes are owned by LCRA and located upstream of Austin. Hydroelectric power plants at each of the six dams are operated to use water supply releases for downstream diverters to help meet peak electric power demands. Releases for only hydroelectric energy generation occur only during energy-related emergencies. Lake Travis has a flood control pool. Lake LBJ provides cooling water for an LCRA thermal-electric power plant. Lakes Buchanan and Travis contain water supply storage used primarily to supply municipal and industrial users in Austin and vicinity and agricultural irrigation needs primarily for the Gulf Coast, Lakeside, Garwood, and Pierce Ranch irrigation operations near the Gulf Coast.

The large scenic Buchanan and Mansfield Dams are shown in Figures 7-9 and 7-10, respectively. Their locations are included in Figure 1-9 of Chapter 1. Lake Travis and Lake Buchanan are the eighth and 15th largest reservoirs in Texas (Tables 1-4 and 1-5).

The LCRA system is operated in accordance with a periodically updated water management plan that governs water allocation during droughts when all LCRA customers cannot be fully supplied. The latest

Figure 7-9. Buchanan Dam and Lake on the Colorado River.
Source: Courtesy of TWDB.

Figure 7-10. Mansfield Dam and Lake Travis on the Colorado River.
Source: Courtesy of TWDB.

updated plan was submitted by LCRA to TCEQ in 2019 and approved in 2020. Water is released from Lakes Buchanan and Travis whenever flows in the lower river are inadequate to meet downstream needs, including environmental instream river flows and freshwater inflows to Matagorda Bay. The water management plan divides supply between firm and interruptible based on storage level triggers in Lakes Buchanan and Travis. Firm water is available even during a severe drought. During water shortages, interruptible water, which is used primarily for agricultural irrigation, is curtailed as necessary to protect firm water supply commitments for primarily municipal, industrial, and thermal-electric cooling uses.

Neches River Basin

Ten existing major reservoirs and three other proposed projects with capacities exceeding 5,000 ac-ft account for 98.7% of the total authorized storage capacity of the 180 reservoirs in the Neches River Basin licensed by water rights. The conservation pools of the major reservoirs in the 9,937 sq. mi Neches River Basin of water-abundant East Texas have been full or comfortably close to full continuously in the past. The minimum combined storage level for the major reservoirs in the basin since initial filling in the late 1960s was 56.9% of capacity in November 2011. The Neches Basin is a source of water supply for the adjoining upper Trinity River Basin and coastal areas as well as for residents of the sparsely populated Neches Basin. Exports of water westward to the adjoining Trinity River Basin and Neches-Trinity coastal basin will likely increase in the future.

Major reservoirs are shown in the basin map of Figure 7-11. The total storage contents in the 180 reservoirs computed in a full authorization WAM simulation are plotted in Figure 7-12. The daily simulation includes operations of the large flood control pool of Sam Rayburn Reservoir as well as the 3,904,100 ac-ft of conservation storage capacity in the 180 reservoirs.

Sam Rayburn Reservoir on the Angelina River and Lake Palestine on the upper Neches River are the largest reservoirs in the Neches River Basin. These two reservoirs with authorized storage capacities of 2,876,000 and 411,800 ac-ft, respectively, contain 84.8% of the authorized capacity of the 180 storage facilities. The proposed Lake Columbia,

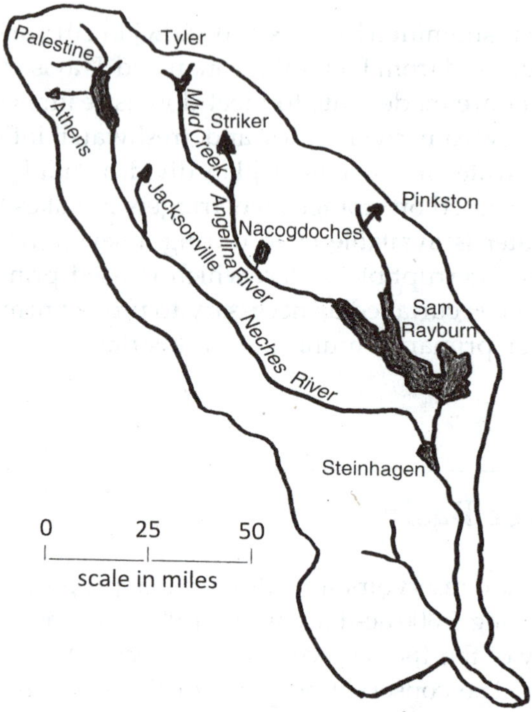

Figure 7-11. Major reservoirs in the Neches River Basin.

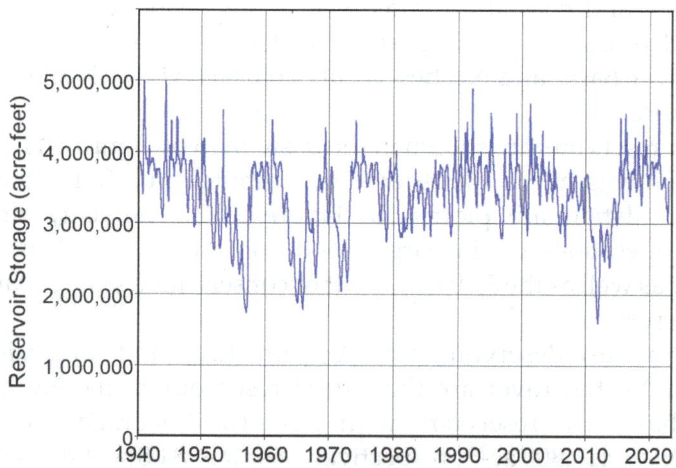

Figure 7-12. Simulated total storage in 180 reservoirs in the Neches WAM.

which is not yet constructed, is the third largest licensed project with 195,500 ac-ft or 5.0% of the authorized storage. If and when the project will be constructed is uncertain.

The 4 million ac-ft Sam Rayburn Reservoir in Figure 7-13 is the largest reservoir located wholly within Texas. The 94,200 ac-ft B. A. Steinhagen Reservoir impounded by Town Bluff Dam is downstream of Sam Rayburn Dam. The two-reservoir system is owned and operated by USACE for flood control, hydroelectric power, water supply, and recreation.

The purposes of B. A. Steinhagen Reservoir are to reregulate the intermittent hydropower releases from Sam Rayburn Dam, provide head for hydroelectric power and diversion into a water supply canal, and provide some flood control storage. Hydropower releases from Sam Rayburn Dam are stored in B. A. Steinhagen Reservoir for release as needed for water supply diversions from the lower Neches River for municipal, industrial, and agricultural use in the Neches-Trinity coastal basin and lower Neches River Basin.

Water released through hydropower turbines at the two dams is diverted from the Neches River downstream of Town Bluff Dam for

Figure 7-13. Sam Rayburn Dam, outflow structures, and hydropower facilities. *Source*: Courtesy of USACE.

water supply. The Lower Neches Valley Authority is the primary nonfederal water supply sponsor. The city of Lufkin has contracted for a small amount of the conservation storage capacity of Sam Rayburn Reservoir. The Southwestern Power Administration markets the hydroelectric energy to the Sam Rayburn Municipal Power Agency for distribution to customers in Jasper, Liberty, and Livingston, Texas and Vinton, Louisiana.

Figure 5-21 of Chapter 5 is a plot of observed 1904 to 2023 daily flows at a USGS gauge on the Neches River 20 mi upstream of the confluence of the Angelina River. Daily, monthly, and annual flows at a USGS gauge on the Neches River at Evadale 25 mi upstream of the City of Beaumont are plotted in Figures 5-22 and 5-23. This gauge is downstream of Sam Rayburn and B. A. Steinhagen Reservoirs. As discussed in Chapter 5, the effects of the two reservoirs include increases in low flows and decreases in high flows at the gauge site.

Capabilities have been recently implemented for transporting water from the upper Neches River Basin to the upper Trinity Basin. Lake Palestine, shown in the basin map of Figure 7-11, on the Neches River is owned by the Upper Neches River Authority. The Integrated Pipeline Project constructed during 2014 to 2022 is a 150 mi long pipeline with multiple pumping stations that transports water from Lake Palestine to Lake Benbrook on the Clear Fork of the Trinity River west of Fort Worth. Other existing pipelines transport water from Lake Benbrook to Richland-Chambers and Cedar Creek Reservoirs also in the Trinity Basin. The Integrated Pipeline Project is being implemented as a partnership of TRWD and DWU, partly financed through the SWIFT administered by TWDB.

Sabine River Basin

The Sabine River Basin is about 300 mi long and has a maximum width of 48 mi. Its total drainage area is 9,756 sq. mi, with 7,570 sq. mi (77.6%) in Texas and the remainder in Louisiana. The Sabine River along with Toledo Bend Reservoir serve as a 265 mi segment of the state border. Major tributaries include Cow Bayou, Anacoco Bayou, Toro Bayou, Tenaha Creek, Martin Creek, Murvaul Bayou, Big Sandy Creek, and Lake Fork Creek. The Sabine River flows into Sabine Lake, which is a 14 mi long, 90,000 ac estuary and bay formed by the confluence of the Neches and Sabine Rivers. Sabine Pass is the natural outlet of Sabine Lake into the Gulf of Mexico. The largest city in the Sabine River Basin is Longview located in the upper basin with a 2020 census population of 81,640.

Figure 1-10 is a photograph of Toledo Bend Dam and Reservoir. The reservoir project and the Sabine River Compact allocating the water resources of the basin between the two states are described in Chapter 2. Observed 1924 to 2023 daily flows of the Sabine River at the USGS gauge at Ruliff upstream of Sabine Lake are plotted in Figure 5-20. The observed storage contents of Toledo Bend Reservoir since its initial impoundment in 1969 is plotted in Figure 6-2. The lowest storage level experienced in Toledo Bend Reservoir since initially filling in 1966 has been 59% of capacity occurring in 2011.

The 13 major reservoirs with storage capacities of 5,000 ac-ft or greater located in Texas account for about 99% of the total storage capacity of 6,393,413 ac-ft in 212 reservoirs authorized by Texas water rights administered by TCEQ. Toledo Bend Reservoir is shared by Louisiana and Texas. The other 12 major reservoirs licensed by Texas water rights are located wholly in Texas. Lakes Vernon and Anacoco on Anacoco Bayou are the only major reservoirs in the Sabine River Basin located in Louisiana and are not included in the Texas water rights system. Lakes Vernon and Anacoco have conservation capacities of 57,000 and 24,000 ac-ft and drainage areas of 112 and 209 sq. mi and are used only for recreation.

The 4,477,000 ac-ft storage capacity of Toledo Bend Reservoir is the largest conservation storage capacity of any reservoir in Texas or the southern United States. The lake covers an area of 181,600 ac (684 sq. mi). The reservoir extends 65 mi from the dam upstream to Logansport, Louisiana. The project was constructed and is jointly operated by the Sabine River Authority (SRA) of Texas and SRA of Louisiana. The reservoir was created primarily for water supply with secondary purposes of hydroelectric power generation and recreation. Most of the water supplied from Toledo Bend for municipal, industrial, and agricultural water supply is released through the hydropower turbines. The water supply and hydroelectric energy provided by the project are shared equally by Texas and Louisiana.

The hydroelectric plant is located in the dam abutment and consists of two vertical units of equal size with Kaplan turbines that generate about 207,000 MW hours annually. The Gulf States Utilities Company, Central Louisiana Electric Company, and Louisiana Power and Light contract with the two river authorities for purchase of the hydroelectric energy.

None of the reservoirs in the Sabine River Basin include a formally designated flood control pool. However, the top of emergency spillway gates at Toledo Bend Reservoir is one foot above the normal operating level providing operational flexibility for attenuation of flood flows.

Lake Tawakoni and Lake Fork Reservoir, the two other reservoirs owned and operated by SRA of Texas, have storage capacities of 872,000 and 637,000 ac-ft and initial impoundment dates of 1960 and 1979. These are the second and third largest reservoirs in the Sabine River Basin. Lake Tawakoni and Lake Fork Reservoir are located about 50 and 70 mi, respectively, east of Dallas in the upper Trinity River Basin. Dallas Water Utilities contracts with SRA for water supply from Tawakoni and Lake Fork Reservoirs. Use of water from reservoirs in the Trinity River Basin is more economical for DWU, but the capabilities for supplemental supply delivered by pipeline from Lake Tawakoni and Lake Fork Reservoir increases supply reliability. Other cities and electric utilities in the Sabine, Trinity, and Sulphur River Basins also contract with SRA for water supply from the two reservoirs.

Other reservoirs in the Sabine River Basin also provide municipal and industrial water supply, recreation, and incidental flood mitigation. Lake Gladewater is owned and operated by the city of Gladewater for municipal and industrial water supply. Lakes Quitman, Holbrook, Winnsboro, and Hawkins are owned and operated by the Wood County Fresh Water Supply District for recreation and flood control. Although there is no controlled flood control storage, the limited spillway discharge capacity provides surcharge storage. Lake Murvaul owned by the Panola County Water Supply District provides municipal and industrial supply for the city of Carthage and other customers in Panola County. Lake Cherokee owned by the Cherokee Water Company provides water for a Southwestern Electric Power Company plant and municipal and industrial water for the cities of Kilgore, Longview, and White Oak. Martin Lake is owned and operated by Texas Utilities Generating Company to provide water for steam-electric power plants.

Cypress Creek Basin

The ecologically unique Caddo Lake pictured in the photograph of Figure 1-7 is described in the "Interstate Reservoirs" section of Chapter 2. The 129,000 ac-ft lake on Cypress Creek extends across the border between Texas and Louisiana.

Lake O' the Pines (Figure 3-3) is owned and operated by USACE FWD for flood control, municipal and industrial water supply, and recreation. The Northeast Texas Municipal Water District has contracted for the conservation capacity and supplies water to the cities of Daingerfield, Lone Star, and Hughes Springs and several industries.

Welsh and Johnson Creek Reservoirs owned by Southwestern Electric Power Company provide cooling water for operation of steam-electric power plants located at the reservoirs and are maintained at constant operating levels by diversions from Lake O' the Pines. Ellison Creek Reservoir is owned by Lone Star Steel Company and supplies water for its steel mill. Monticello Lake is owned and operated by Texas Utilities Generating Company to supply water for steam-electric power plants located near the reservoir and is maintained at a constant operating level by diversions from Lake Bob Sandlin. Titus County Fresh Water Supply District No. 1 owns and operates Lake Bob Sandlin which supplies the water needs of the city of Mount Pleasant and provides water to Texas Utilities Generating Company for steam-electric power plant steam and cooling. Lake Cypress Springs owned by the Franklin County Water District supplies water to the city of Mount Vernon, rural areas in Franklin County, and Texas Utilities Generating Company.

The Cypress Bayou Operating Agreement is a cooperative arrangement between reservoir owners designed to optimize water supplies in the basin through coordinated operations. Northeast Texas Municipal Water District, Franklin County Water District, Titus County Fresh Water Supply District No. 1, and Lone Star Steel entered into this operating agreement approved by the Texas Water Rights Commission (TCEQ predecessor) in 1973 that includes rules for the operation of their reservoirs and accounting for the waters held in storage. The agreement provides that Lake Cypress Springs and Lake Bob Sandlin can store water previously appropriated to the downstream Lake O' the Pines and Ellison Creek Reservoir subject to releases from upstream storage to satisfy downstream requirements upon demand. Storage accounts were established such that the basin waters are divided through exchange of storage in accordance with existing water rights.

Sulphur River Basin

Lakes Wright Patman, Jim Chapman, Sulphur Springs, and River Crest are the four major reservoirs located in the Texas portion of the Sulphur River Basin. Construction began in 2021 on Lake Ralph Hall and is scheduled for completion in 2026.

The 2.3 mi long Leon Hurse Dam impounding the 181,000 ac-ft Lake Ralph Hall on the North Sulphur River is being constructed by Upper Trinity Regional Water District (UTRWD) to supply water for use in about 29 communities in the upper Trinity River Basin. The project also includes 32 mi of 54 to 60 in. diameter pipeline, a pump station, and a balancing reservoir. The untreated water pipeline will join an existing pipeline that now delivers water from Jim Chapman Lake. The existing pipeline has enough available capacity to carry water from Lake Ralph Hall to Lewisville Lake on the Elm Fork of the Trinity River for treatment and distribution. UTRWD holds a water right permit for the project issued by TCEQ in December 2013 and a Clean Water Act Section 404 permit from the USACE issued in January 2020. Construction of the project is being completed during 2021 to 2026.

Lakes Wright Patman and Jim Chapman with initial impoundment in 1956 and 1991, respectively, are owned and operated by USACE Fort Worth District for flood control, municipal and industrial water supply, and recreation. Jim Chapman Reservoir shown in Figure 7-14 is the most recently constructed USACE reservoir project in Texas.

Figure 7-14. Jim Chapman Dam and Lake on the South Sulphur River.
Source: Courtesy of USACE.

Most of the storage capacity of the 2,654,900 ac-ft Wright Patman Reservoir is allocated to flood control, but the City of Texarkana contracts with the Corps of Engineers to provide a conservation pool for water supply purposes. The Corps of Engineers varies the operating rules seasonally to increase the supply of water available each summer.

Jim Chapman Reservoir has conservation and flood control capacities of 298,900 and 131,400 ac-ft. The North Texas Municipal Water District and the Sulphur River Municipal Water District are the non-federal sponsors for the USACE project. Water is supplied to cities and industries in both the Sulphur and Trinity River Basins.

Lake Sulphur Springs is owned and operated by the city of Sulphur Springs and provides municipal and industrial water for the city and surrounding area. River Crest Reservoir owned by Texas Power and Light Company is an off-channel storage facility providing water for steam-electric power plant cooling.

Red River Basin

The Red River is a tributary of the Atchafalaya River, which is a distributary of the Mississippi River that flows separately into the Gulf of Mexico at New Orleans. The Red River Basin has a contributing drainage area of 93,450 sq. mi that encompasses 24,300 sq. mi in Texas and portions of the states of Oklahoma, New Mexico, Arkansas, and Louisiana. The western part of the basin is flat and arid with large areas contributing no runoff to streams. The Red River and Lake Texoma impounded by Denison Dam serve as a major length of the border between Oklahoma and Texas. The Red River Compact allocating water resources between the four states is described in Chapter 2. Observed 1924 to 2023 daily flows of the Red River at the USGS gauges just below Denison Dam and further downstream on the Texas border near the border between Oklahoma and Arkansas are plotted in Figures 5-18 and 5-19.

Dams and Reservoirs

The 24 major reservoirs located in the Texas portion of the Red River Basin are listed in Appendix A. As the largest reservoir located wholly or partially in Texas, Lake Texoma is included in Tables 1-4 and 1-5 and Figure 1-9. Lake Texoma impounded by Denison Dam is the oldest USACE reservoir project in Texas. Completion of construction and initial impoundment of streamflow occurred in 1943.

The Denison Dam and Lake Texoma project was constructed and is operated by USACE Tulsa District for flood control, hydroelectric power, water supply, and recreation. The conservation pool is used primarily for hydroelectric energy generation and recreation. The Water Resources Development Act of 1986 (Public Law 99-662) authorized USACE to reallocate conservation storage capacity in Lake Texoma from hydropower to water supply in increments as needed to meet growing water supply needs. Although water supply capabilities are constrained by salinity, municipal and industrial demands for water from Lake Texoma are growing.

The Red River Authority, city of Denison, Texas Power and Light Company, Atlantic Richfield Company, and Texaco Company have contracted for use of the conservation storage capacity of Lake Texoma. Lake Randall, owned by the city of Denison for a municipal water supply, is also used for regulating diversions from Lake Texoma. Valley Lake, owned and operated by Texas Power and Light Company, is also supplemented by diversions from Lake Texoma to maintain a constant operating level for steam-electric power plant operations. The city of Dallas and NTMWD have pumped water through a pipeline from Lake Texoma to Lake Lavon in the Trinity River Basin in the past. Dilution of Red River water with Trinity River water has been viewed primarily as a supplemental drought emergency backup.

Pat Mayse Reservoir was constructed and is operated by the Corps of Engineers for flood control, municipal and industrial water supply, and recreation. The city of Paris has contracted for the conservation storage. Coffee Mill Creek Lake is owned and operated by the US Forest Service for recreation.

The city of Wichita Falls owns and operates Wichita, Kickapoo, and Arrowhead lakes and is co-owner with the Wichita County Water Improvement District No. 2 of Lake Diversion and Lake Kemp. Lakes Kickapoo and Arrowhead are the principal sources of surface-water supply for the Wichita Falls area. Lake Diversion supplies agricultural irrigation as well as municipal uses. In 1974, the USACE Tulsa District completed reconstruction of the Lake Kemp Dam for improved safety and to provide a designated allocation of storage capacity for flood control. The city and water improvement district reimbursed the federal government for the reconstruction cost allocated to conservation storage. The federal government reimburses these nonfederal sponsors annually for operation and maintenance cost allocated to flood control.

Bivins Lake, owned by the city of Amarillo, is used for aquifer recharge. Electra, Baylor Creek, and Crook Reservoirs are owned by and provide water for the cities of Electra, Childress, and Paris, respectively. Santa Rosa Reservoir is owned by the W. T. Waggoner Estate and has been used for livestock watering and oil and gas secondary recovery operations. Lake Bonham owned by the Bonham Municipal Water Authority supplies municipal and industrial water to the city of Bonham. North Fork Buffalo Creek Reservoir is owned by Wichita County Water Control and Improvement District No. 3 and supplies most of the water used by the city of Iowa Park. Hubert Moss Lake is owned by the city of Gainesville in the Trinity River Basin. Farmers Creek Reservoir owned and operated by the North Montague County Water Supply District supplies the city of Nocona and other areas of Montague County. Mackenzie Reservoir is owned by the Mackenzie Municipal Water Authority and serves several member cities. Likewise, Greenbelt Reservoir is owned by the Greenbelt Municipal and Industrial Water Authority and serves several member cities.

Natural Salt Pollution Control

Natural salt pollution from geologic formations in the Permian Basin Region underlying portions of the upper watersheds of the Pecos, Colorado, Brazos, Red, and Canadian River Basins in Texas, New Mexico, Oklahoma, and Kansas is described in the last section of Chapter 5. Lake Texoma and other reservoirs in the Red River Basin are among the many reservoirs with water supply capabilities constrained by salinity from salt source areas in their upper watersheds. The effects of salinity are magnified during droughts with decreases in water volumes and corresponding increases in salt concentrations.

The USACE Tulsa District (1976, 1995) conducted a series of chloride studies dating back to the 1960s. Construction of salt control facilities at eight sites recommended by USACE was authorized by Congress in the Water Resources Development Act of 1970 (Public Law 91-611) and preceding legislation. Projects were subsequently implemented at two of the sites, the Estelline Spring ring dike and the Truscott brine reservoir and associated brine collection system. Both involved prevention of brine from small high-concentration salt source watersheds from entering major streams. Funding, environmental, and other concerns have prevented implementation of other proposed projects (Rought 1984, Wurbs 2002).

Estelline Spring is in the floodplain of the Prairie Dog Fork of the Red River in Oklahoma. The implemented project consists of a ring dike constructed around the spring to stop the saline flow. The Truscott Brine Dam and Lake is on a tributary of the South Fork of the Wichita River above Lake Kemp in Texas. Brine is collected and pumped into the 107,000 ac-ft Truscott Reservoir to prevent its flow into Lake Kemp. The water evaporates leaving the salt impounded. The project is designed for a 100 year accumulation of salt and sediment.

The Water Resources Development Act of 1986 (Public Law 99-662) authorized development of salt control projects for other primary salt source areas pursuant to further feasibility studies. Studies continued for many years focused mainly on reducing salinity concentrations in Lake Kemp and Lake Texoma. Officials of the US Fish and Wildlife Service, Oklahoma Wildlife Department, and environmental organizations expressed opposition to salt control projects based on concerns that reduction in salinity would damage ecosystems. The striped bass population of Lake Texoma, which is very popular with sport anglers, could be severely affected. Proposed brine storage reservoirs may also be a potential hazard to resident and migratory wildlife. Conversely, project proponents emphasized the benefits of proposed salt control projects to municipal, industrial, and agricultural water supply.

According to an online TWDB desalination plant database, Texas has 53 municipal desalination facilities that have a total capacity of 176,000 ac-ft/year. Sixteen of the facilities treat brackish surface water. Thirty-six treat brackish groundwater. One facility treats reclaimed wastewater. None treat seawater. The two largest desalination plants in Texas that treat surface water are owned and operated by the city of Sherman to treat water from Lake Texoma and the city of Granbury to treat water from Lake Granbury on the Brazos River.

Canadian River Basin

The Canadian River is a tributary of the Arkansas River which is a tributary of the Mississippi River. The Canadian River Basin has a watershed area of 47,700 sq. mi extending over portions of Colorado,

Oklahoma, and Texas that includes 12,860 sq. mi in Texas. Observed 1930 to 2023 daily flows of the Canadian River at USGS gauge sites upstream and downstream of Lake Meredith are plotted in Figures 5-16 and 5-17. Lake Meredith is included in Tables 1-3 and 1-4 and Figure 1-9 as the 10th largest reservoir in Texas.

Lakes Meredith and Rita Blanca are the only two major reservoirs in the Texas portion of the Canadian River Basin. Rita Blanca Lake was constructed by the Soil Conservation Service (renamed NRCS) and is maintained by the city of Dalhart for recreation. Sanford Dam and Lake Meredith were constructed by USBR and are now owned and operated by the Canadian River Municipal Water District. USACE is responsible for flood control operations, though as noted in Chapter 6, water has never to date risen into the flood control pool.

The Canadian River Municipal Water Authority (CRMWA) was created by the Texas legislature in 1953 to provide a source of municipal and industrial water for its eleven member cities: Amarillo, Brownfield, Borger, Lamesa, Levelland, Lubbock, O'Donnell, Pampa, Plainview, Slaton, and Tahoka. The 11 member cities with populations totaling more than 500,000 people are located across the Texas Panhandle and South Plains. CRMWA headquarters are located at Sanford Dam (Lake Meredith) about 37 mi northeast of Amarillo.

CRMWA supplies raw water through a 358 mi aqueduct system that includes multiple pumping stations. The current source of water is Lake Meredith and the John C. Williams Aqueduct and Wellfield in the Texas Panhandle. Lake Meredith was initially the sole source of water supply. However, hydrologic conditions have resulted in insufficient river inflow to fill Lake Meredith. Development of a 50 sq. mi well field with 27 groundwater wells and an extensive pipeline network connecting the wells was necessary to supply the water needs of the member cities.

Salinity of water in Lake Meredith has been a concern since the initial water supply deliveries from the lake in 1968. The main salt source was identified as brine from a shallow aquifer discharging into the Canadian River in New Mexico downstream of Ute Dam. USBR and CRMWA completed and has operated a brine control project since 2001 comprised of seven shallow wells that intercept the saline water and a deep well for disposing of the brine.

Guadalupe and San Antonio River Basin

The San Antonio River is a tributary of the Guadalupe River. The Guadalupe River flows into Guadalupe Estuary a short distance downstream of the San Antonio River confluence and then into San Antonio Bay. The complete Guadalupe River Basin including its San Antonio River subbasin encompasses an area of 10,130 sq. mi. The San Antonio River Basin has a watershed area of 4,180 sq. mi. The Guadalupe River is about 230 mi long, and the San Antonio River is about 240 mi long. More than 30% of the total consumptive water use authorized by water rights from the two rivers and their tributaries is diverted below their confluence. Flows of the San Antonio and Guadalupe Rivers and tributaries are significantly affected by the Edwards Aquifer through both spring flow contributions to streamflow and streamflow losses to groundwater recharge through the aquifer outcrop as discussed in Chapter 5.

San Antonio River Basin

The city of San Antonio in the upper San Antonio River Basin, with a 2020 census population of 1,434,625, is the second most populous city in Texas and seventh most populous city in the United States. It has been one of the fastest growing large cities in the United States over the last several decades. San Antonio is the largest city in the United States that relies exclusively on groundwater for municipal water supply. The city is supplied by the Edwards Aquifer. Limitations on the quantities of water that can be pumped from the Edwards Aquifer and the resulting need for supplemental surface water sources are long-recognized and extensively investigated issues.

Medina, Calaveras, Victor Braunig, and Olmos are the only reservoirs in the San Antonio River Basin with storage capacities of 5,000 ac-ft or greater. The 254,000 ac-ft Medina Lake on the Medina River about 40 mi northwest of San Antonio is operated by the Bexar-Medina-Atascosa County Water Improvement District to supply agricultural irrigation. The 61,800 ac-ft Calaveras Lake on Calaveras Creek and 26,500 ac-ft Victor Braunig Lake on Arroyo Seco Creek owned by the City Public Service Board of San Antonio provide cooling water for steam-electric power plants. Olmos Reservoir on Olmos Creek just north of San Antonio is a 12,600 ac-ft reservoir owned and operated by the San Antonio River Authority (SARA) for flood control.

The San Antonio Channel Improvement Project authorized by the Flood Control Act of 1954 was constructed in multiple units by USACE with SARA as the nonfederal sponsor (USACE 1992). Modifications authorized by the Water Resources Development Act of 1976 included measures to preserve the historic Espada Acequia Aqueduct. The first construction contract was awarded in 1957 and others followed through the 1990s. The Fort Worth District received the Chief of Engineers Award of Merit in 1972 and later other awards honoring work on the project (USACE 1992).

The channel improvements mitigate flooding problems in downtown San Antonio and surrounding areas. The flood control project includes 35 mi of improved channel with widths varying from 12 to 280 ft on the San Antonio River and San Pedro, Apache, Alazan, Martinez, and Six Mile Creeks; concrete and steel piling floodwalls and culverts; and a system of deep tunnels conveying flood waters under downtown San Antonio. Many thousands of visitors enjoy the dining, shopping, and cultural experiences of the popular San Antonio River Walk each year, many perhaps not realizing that they are surrounded by a historic large-scale federal/local flood control system.

Guadalupe River Basin

Canyon Lake shown in Figure 7-15 is owned and operated by USACE FWD. Completion of construction and initial impoundment occurred in 1964. The dam is located on the Guadalupe River about 300 mi upstream of its outlet and has a watershed area of 1,430 sq. mi. GBRA has contracted for use of the 379,000 ac-ft conservation pool. The Corps of Engineers operates the 362,000 ac-ft flood control pool. Canyon Lake and the Guadalupe River downstream are extremely popular for recreation.

GBRA manages the conservation pool of Canyon Reservoir for municipal, industrial, and agricultural water supply and generation of hydroelectric energy. GBRA operates a hydropower plant at Canyon Dam and six hydropower plants on the Guadalupe River downstream that use releases from Canyon Lake to generate electricity. The downstream plants are located at hydropower dams impounding Lakes Dunlap, McQueeney, Nolte, Placid, Gonzales, and Wood. These six

Figure 7-15. Canyon Dam and Lake on the Guadalupe River.

hydropower plants were initially placed in operation during 1928 to 1932 and were purchased by GBRA in 1963. GBRA constructed the hydropower plant at Canyon Dam with operations beginning in 1989. Lakes Dunlap, Gonzales, and McQueeney, with storage capacities of 5,900, 5,200, and 5,000 ac-ft, respectively, are the largest of the six lakes impounded by the six hydropower dams on the Guadalupe River owned by GBRA.

Coleto Creek Reservoir, completed by GBRA in 1980, provides water for producing and cooling steam for a steam-electric power plant. The 35,000 ac-ft reservoir on Coleto Creek and adjacent power plant are located about 15 mi south of the city of Victoria. GBRA maintains recreation facilities at the lake.

Rehabilitation of aging constructed infrastructure, including but not limited to water development projects, is an important concern throughout Texas and the United States. The GBRA hydropower dams on the Guadalupe River constructed in the 1920s are illustrative of more widespread concerns. A spillway gate at Lake Wood broke loose from the dam in 2016, partially draining the lake. A spillway gate at Lake Dunlap similarly separated from the dam in 2019, partially draining

Lake Dunlap. GBRA announced in 2019 a planned systematic drawdown of all six lakes to ensure public safety. This action was subsequently halted by a temporary injunction issued in favor of lakefront property owners interested in preserving the lakes to protect aesthetics and property values. Financing of repairs of the spillway gates and rehabilitation of the dams and appurtenant structures from limited hydropower revenues and/or other funding sources is a challenging issue. The spillway at Lake Dunlap was reconstructed during 2022 to 2023.

Lavaca River Basin

The 2,310 sq. mi Lavaca River Basin encompasses the smallest area of any of the 15 major river basins of Texas. The Navidad River with a watershed area of 1,420 sq. mi is the largest tributary of the Lavaca River. Lake Texana on the Navidad River is the only major reservoir in the Lavaca River Basin. The watershed above Lake Texana Dam has an area of 1,410 sq. mi. The Navidad and Lavaca Rivers confluence downstream of Texana Dam before flowing into Lavaca Bay, which is a secondary bay of the Matagorda Bay system.

Lake Texana was constructed by USBR and is now owned and operated by the Lavaca-Navidad River Authority (LNRA). Construction of the 157,900 ac-ft Lake Texana was completed in 1981. Water is transported from Lake Texana by a 101 mi long, 64 in. pipeline constructed during 1997 to 1998 to supply the city of Corpus Christi located in the Nueces-Rio Grande Coastal Basin near the outlet of the Nueces River. LNRA also supplies other water customers in the lower Lavaca Basin and adjoining coastal area. Most water use within the Lavaca River Basin is supplied from groundwater.

Storage contents of Lake Texana computed in a full authorization daily WRAP/WAM simulation are plotted in Figure 7-16. The simulation is based on the premise that all water right holders appropriate the full amounts of water to which they are entitled during a hypothetical repetition of January 1940 through December 2021 natural streamflow and reservoir evaporation rates. The 1950 to 1957 drought ending with the April to May 1957 flood results in the greatest storage drawdown during the 1940 to 2021 hydrologic period of analysis. Of course, with construction completed in 1981, Lake Texana has never actually experienced a drought as hydrologically severe as 1950 to 1957.

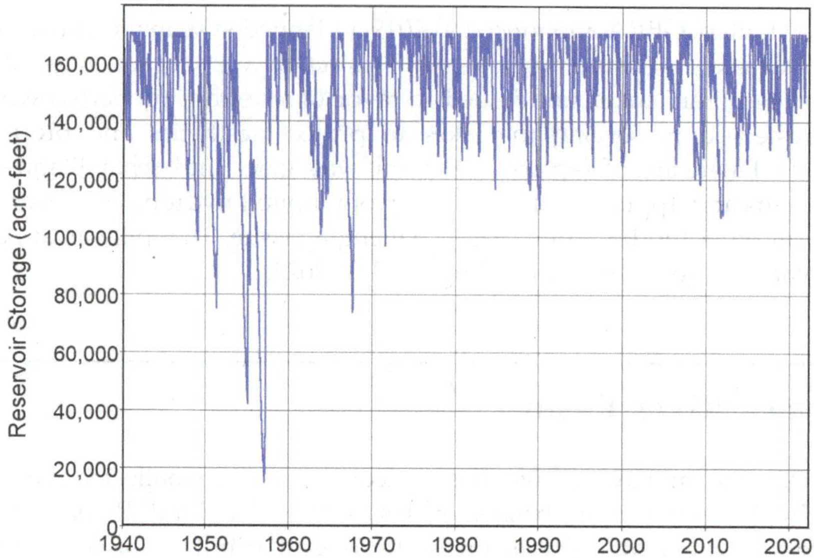

Figure 7-16. Simulated storage in Lake Texana on the Navidad River.

Nueces River Basin

The Nueces River flows into Nueces Bay. Corpus Christi is located in the Nueces-Rio Grande Coastal Basin adjacent to the southwest side of Nueces Bay and the lower Nueces River. Most surface water use supplied from the Nueces River Basin is used by the city of Corpus Christi and its water customers for municipal and industrial use. The majority of water supplied by Corpus Christi from the city-owned and operated Choke Canyon Reservoir and Lake Corpus Christi is diverted downstream at the Calallen diversion dam and saltwater barrier located on the Nueces River about 12 mi upstream of the river outlet into Nueces Bay. Most of the surface water supplied from the Nueces River Basin is used in the two adjoining coastal basins.

Most water use within the Nueces Basin is supplied from groundwater. Uvalde with a 2020 census population of 24,560 is the largest city located within the Nueces Basin. The Edwards Aquifer is the primary source of water supply for the city of Uvalde. Agricultural irrigation in the Winter Garden Region and throughout the Nueces Basin is supplied from groundwater.

The 2020 census population of Corpus Christi is 317,860. The three-county metropolitan area containing most of the water users supplied by

Corpus Christi has a 2020 population of 442,600. Supplies developed during the 1980s and earlier from sources in the Nueces River Basin have been supplemented since the 1990s by interbasin transfer of water by pipeline from the Lavaca and Colorado River Basins. Corpus Christi and vicinity could possibly be the first region of Texas to supplement its water supply with seawater desalination in the future.

Reservoirs in the Nueces River Basin

A total storage capacity of 1,047,000 ac-ft in 121 reservoirs is authorized by water rights and included in the Nueces WAM dataset. Lake Corpus Christi and Choke Canyon Reservoir contain 96.1% of the total authorized storage capacity. One other reservoir has authorized storage of about 5,000 ac-ft. None of the reservoirs including the two largest have designated flood control storage. Choke Canyon Reservoir and Lake Corpus Christi are operated in coordination to provide municipal, industrial, and agricultural water supply and recreation for the Coastal Bend Region.

Choke Canyon Dam and Reservoir on the Frio River was constructed by USBR and is jointly owned by the Nueces River Authority (20%) and city of Corpus Christi (80%). The reservoir began to impound water in 1982. Choke Canyon Reservoir has an authorized storage capacity of 700,000 ac-ft, but a 2012 TWDB volumetric survey indicated that the capacity had been reduced by sedimentation to 662,820 ac-ft. The surface area at full capacity is 25,440 ac. The watershed area above the dam is 4,667 sq. mi.

Lake Corpus Christi and Wesley E. Seale Dam are owned and operated by the city of Corpus Christi for water supply and recreation. Impoundment began in 1958. In addition to supplying its own residents, Corpus Christi sells water to the San Patricio Municipal Water District, Alice Water Authority, and several cities and industries. Lake Corpus Christi has an authorized storage capacity of 300,000 ac-ft and according to a 2012 reservoir site survey by the TWDB a reduced capacity of 254,730 ac-ft with a surface area of 18,700 ac. The watershed area above the dam is 16,660 sq. mi.

Lake Corpus Christi and Wesley E. Seale Dam replaced the smaller Lake Lovenskiold and La Fruta Dam constructed in 1927 to 1929. A partial failure of La Fruta Dam occurred in 1930. Reconstruction of the failed dam at the same site was completed in 1934. Construction of

Wesley E. Seale Dam just downstream of La Fruta Dam began in 1955 and was completed in April 1958.

The third largest reservoir in the basin is the much smaller Upper Nueces Reservoir owned and operated by the Zavala and Dimmit Counties Water Improvement District No. 1 to supply water for agricultural irrigation. Construction was completed in 1948. The authorized storage capacity is 4,010 ac-ft. A volumetric survey by TCEQ in 2012 found the capacity to be 7,590 ac-ft. The reason for this unusual increase in storage capacity is unknown.

Total storage contents of the combined Lakes Choke Canyon and Corpus Christi system computed in a full authorization scenario simulation with a 1934 to 2021 hydrologic period of analysis are plotted in Figure 7-17 to illustrate the extent to which the water resources of the Nueces River Basin have been appropriated (Wurbs 2023b). Storage drawdowns are dramatic. The city of Corpus Christi and other entities hold water rights for much more water than the Nueces River Basin can provide with acceptable levels of reliability. The reservoir storage plot of Figure 7-17 helps explain the motivation for the interbasin water transport projects discussed later. The simulation reflects the modeling premise of setting storage contents at the beginning of January 1934 of the

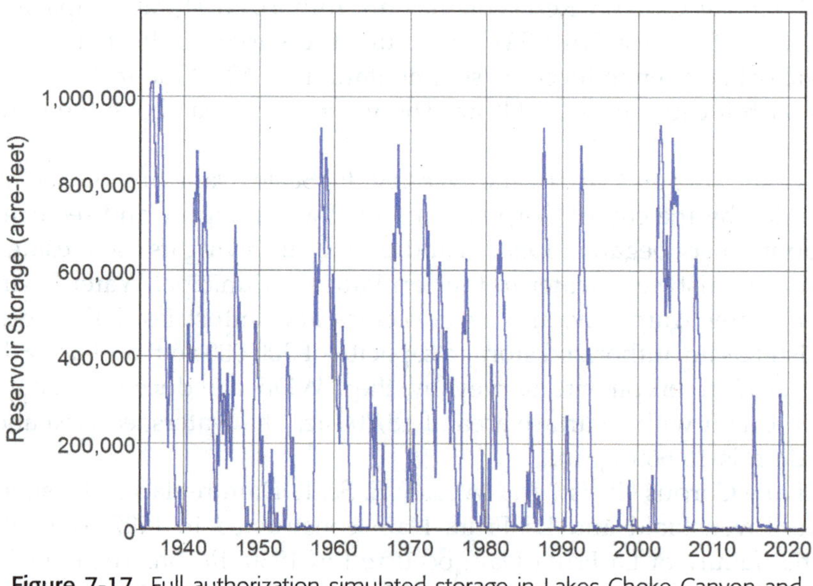

Figure 7-17. Full authorization simulated storage in Lakes Choke Canyon and Corpus Christi.

1934 to 2021 simulation period equal to the contents at the end of December 2021.

Groundwater Recharge in the Nueces River Basin

The hydrology of the Nueces River Basin is greatly affected by interactions between streamflow and groundwater, primarily through streamflow recharge of groundwater systems but also through spring flows to streams. Flows of the Nueces River and its tributaries contribute to the recharge of several aquifers including the Edwards Aquifer and thus the flow of many springs including Leona Springs in the Nueces River Basin and Comal and San Marcos Springs in the Guadalupe River Basin.

Daily flows during 1918 to 2023 at a USGS gauge on the Nueces River just below the Frio River confluence are plotted in Figure 5-34. Long-term mean naturalized flow at the basin outlets of 11 rivers are compared with precipitation falling on the river basins in Table 5-5. The naturalized flow as a percentage of precipitation is much smaller for the Nueces River than the other rivers. Long-term means of simulated regulated flows at the outlets of the 11 rivers are compared with naturalized flows in Table 6-7. The regulated flow as a percentage of naturalized flow is much smaller for the Nueces than the other rivers. Most rainfall runoff reaching streams in the Nueces Basin under either natural or developed conditions is lost to groundwater recharge as discussed in Chapter 5.

The Nueces River and its tributaries including the Frio River, Sabinal River, and other smaller streams flow across aquifer outcrop recharge zones of the Edwards, Carrizo–Wilcox, Gulf Coast, and other groundwater aquifers. The surface streams contribute to recharge of all of these aquifers. However, the Edwards Aquifer recharge zone accounts for the largest volume of streamflow loss to groundwater. The Edwards Aquifer comprises caverns through limestone that are essentially underground streams. The recharge zone extending across the upper portions of the Nueces, San Antonio, and Guadalupe River Basins is a surface outcrop of fractured limestone allowing large quantities of water to flow into the aquifer.

HDR Engineering (1998) estimated that the average annual recharge to the Edwards Aquifer from streams in the Nueces River Basin to be 333,400 ac-ft/year. For comparison, this quantity of streamflow volume recharging the Edwards Aquifer is about 63% as large as the total of all authorized water supply diversions from the Nueces River and tributaries. Other additional streamflow is also lost to recharging other aquifers. The use of streamflow authorized by water rights includes several groundwater recharge enhancement projects.

Interbasin Water Transport

The Mary Rhodes Pipeline Project was completed in two phases. The first phase completed in 1999 transports water to Corpus Christi by pipeline from Lake Texana on the Navidad River in the Lavaca River Basin. The second phase completed in 2016 added water from the Colorado River to the supply transported to Corpus Christi from Lake Texana. The city of Corpus Christi acquired water rights during the 1990s for the Navidad River and Colorado River water.

The first phase of the Mary Rhodes Pipeline Project consists of a 101 mi long, 64 in. diameter pipeline constructed of reinforced concrete in a steel cylinder and two pumping stations that connects Lake Texana and the O. N. Stevens Water Treatment Plant in Corpus Christi. The Texana pipeline was constructed during 1998 to 1999 and has been supplying a major portion of the water used in the Coastal Bend Region since 1999. The second phase of the interbasin water transport system completed in 2016 consists of a 42 mi long pipeline from a pumping station on the Colorado River near Bay City that connects to the Lake Texana pipeline.

The city of Corpus Christi now operates a water supply system for the Coastal Bend Region that obtains water from Choke Canyon Reservoir on the Frio River, Lake Corpus Christi on the Nueces River, Lake Texana on the Navidad River, and the Colorado River. However, with growing water demands, the region will need additional water supply in the future.

Possible Future Seawater Desalination

The city of Corpus Christi, Port of Corpus Christi Authority, and private industrial companies have investigated the feasibility of seawater desalination over the last several decades. As of January 2024, multiple projects for construction of seawater desalination plants in the Corpus Christi area are under consideration. Currently, no seawater desalination plant supplying municipal or agricultural water use is in operation in

Texas. The Coastal Bend Region is perhaps the most likely candidate to become the first region of Texas to supplement its municipal water supplies by construction a seawater desalination plant.

Almost all of the limited seawater desalination that has occurred in the United States to date has been in Florida and California motivated by lack of other feasible options for increasing water supply. Reverse osmosis and electrodialysis technologies are readily available for removing chlorides, sulfates, and other inorganic salts from water. The primary constraint to large-scale desalination of seawater is operation cost. The water treatment processes are energy intensive and expensive. Another constraint is environmental and economic issues associated with disposal of the waste brine generated in the desalination process.

Environmental Protection Agency secondary drinking water standards include a maximum TDS limit of 500 milligrams per liter (mg/L). Brackish water is generally defined as having TDS concentrations between 1,000 and 10,000 mg/L. Saline water has TDS concentrations greater than 10,000 mg/L. Seawater has concentrations of about 35,000 mg/L.

According to an online TWDB desalination plant database, Texas has 36 municipal desalination facilities with a total capacity of 100,769 ac-ft/year that treat brackish groundwater and 16 plants with a total capacity of 72,443 ac-ft/year that treat brackish surface water. Industrial operations mainly in the electric power and semiconductor industries are estimated by TWDB to provide an additional 67,000 to 112,000 ac-ft/year of desalination. A desalination plant in El Paso with a capacity of 30,800 ac-ft/year is the largest municipal desalination plant in Texas. Its supply source is brackish groundwater. The two largest desalination plants for treating brackish surface water are the plant in the city of Sherman that treats water from Lake Texoma and the plant in the city of Granbury that treats water from Lake Granbury.

The city of Corpus Christi and its consultants investigated seawater desalination during the 1990s as an alternative to the Lake Texana pipeline project. However, construction and operation of a seawater desalination plant was found to be much more expensive than the pipeline project that was selected for implementation. Development of each additional increment of water supply from freshwater sources becomes more expensive. The city of Corpus Christi, Port of Corpus Christi Authority, and private companies are considering seawater desalination as an economically competitive alternative for the next addition to the water supply for the region.

Seawater provides an essentially unlimited drought-proof supply of water for the many communities and industries located near the 367 mi long Texas Gulf Coast. The principal constraint to using seawater is the economic and environmental cost of removal and disposal of the salt. Corpus Christi and the Coastal Bend Region could possibly pioneer the use of seawater desalination to supplement conventional freshwater supplies as incremental additions to the use of limited freshwater resources become more expensive.

Coastal Basins

Many cities and towns, industries, and agricultural operations are located in the coastal basins between the lower reaches of the major river basins. A major portion of the total water use in Texas occurs in the coastal basins, supplied largely by reservoirs located in the major river basins. Precipitation runoff drains from the watersheds of the coastal basins through multiple creeks and bayous into the Gulf of Mexico. There are no major rivers in the coastal basins for development of large reservoir projects. However, six of the major reservoirs included in Appendixes A and B are located in three of the eight coastal basins. Four major reservoirs are located in the Nueces-Rio Grande Coastal Basin. The other two reservoirs are located in the Trinity-San Jacinto and San Jacinto-Brazos Coastal Basins.

The 6,600 ac-ft Barney M. Davis Reservoir in the Nueces-Rio Grande Coastal Basin is owned and operated by Central Power and Light Company to store cooling water for the Barney M. Davis steam-electric power plant located adjacent to the Laguna Madre near Corpus Christi. Saline water is withdrawn from Laguna Madre for cooling of the plant condensers. After retention in the Barney Davis cooling pond, the water is returned to Laguna Madre.

Three other major off-channel reservoirs in the Nueces-Rio Grande Coastal Basin with a total capacity of 59,300 ac-ft are used for temporary storage and regulation of water diverted from the Rio Grande. Valley Acres Reservoir is owned and operated by the Valley Acres Water District for municipal and agricultural purposes. Delta Lake is owned and operated by Hidalgo and Willacy Counties Water Control and Improvement District No. 1 for agricultural irrigation. Loma Alta

Reservoir is owned and operated by the Brownsville Navigation District for municipal and agricultural uses.

The 20,000 ac-ft Cedar Bayou Reservoir on Cedar Bayou in the Trinity-San Jacinto Coastal Basin is owned and operated by the Houston Lighting and Power Company. The 7,300 ac-ft Galveston County Industrial Water Reservoir owned by the Galveston County Water Authority is located off-channel from Dickinson Bayou in the San Jacinto-Brazos Coastal Basin.

Rio Grande Basin

Allocation of the waters of the Rio Grande between the United States and Mexico and between states in the United States and operation of International Amistad and Falcon Reservoirs are discussed in Chapter 2. Figure 2-1 is a map of the Rio Grande Basin. Figures 2-2 and 2-3 are photographs of the International Amistad and Falcon Dams and Reservoirs. Chapter 3 explains differences in the Texas water rights system employed for the Rio Grande below Fort Quitman versus the rest of Texas. Daily flows of the Rio Grande observed at IBWC gauges at El Paso, Falcon Dam, and El Paso are plotted in Figures 5-12 through 5-14. As the second and fifth largest reservoirs in Texas, International Amistad and Falcon Reservoirs are included in Tables 1-3 and 1-4 and Figure 1-9. The storage contents of the two reservoirs since their initial impoundment in 1968 and 1953 are plotted in Figure 6-1 along with a discussion of storage capacities and historical storage contents.

The Texas portion of the Rio Grande Basin contains seven major reservoirs including the two large international projects on the Rio Grande and five smaller reservoirs located on tributaries on the US side of the Rio Grande.

International Falcon and Amistad Reservoirs were constructed under the terms of a 1944 treaty between the United States and Mexico. The reservoirs are owned and operated by the United States and Mexico Sections of IBWC. USBR designed and supervised construction of the Falcon project for IBWC. USACE designed and supervised construction of Amistad for IBWC. TCEQ administers the water rights for the US share of the conservation storage in the two projects. The two reservoirs are operated as a system for flood control, water supply, hydroelectric power, and recreation. The US share of storage and releases is used within the state of Texas. The water supply is largely for irrigated agriculture in the Rio Grande Valley below Falcon Reservoir.

Initially before losses to sedimentation, Falcon Reservoir had a total controlled storage capacity of 3,177,600 ac-ft of which 2,267,600 ac-ft was allocated to conservation storage (including 2,800 ac-ft of sediment reserve storage) and 910,000 ac-ft was allocated to flood control. During the winter months, a small volume of flood control capacity may be reallocated for additional temporary conservation storage. The United States has 58.6% and Mexico has 41.4% of the conservation storage capacity.

Initially before losses to sedimentation, Amistad Reservoir had a total controlled capacity of 5,249,700 ac-ft, including 3,505,400 ac-ft of conservation storage (including 8,000 ac-ft of sediment reserve) and 1,744,300 ac-ft of flood control storage. The United States has 56.2% and Mexico 43.8% of the conservation storage.

Much of the water use from the Rio Grande River occurs in the four-county Lower Rio Grande Valley region. Numerous farming operations irrigate crops with water from the Rio Grande. Municipal and manufacturing uses account for much smaller but significant additional use of water in the four-county region. Most of the water use is supplied from releases from Falcon Reservoir, which is refilled by releases from Amistad Reservoir. The TCEQ watermaster office is responsible for administering the water allocation system described in Chapters 2 and 3 and requesting releases by IBWC from the conservation pools of Falcon and Amistad Reservoirs.

Mexico and the United States each have a hydroelectric generating plant on its side of Falcon Dam. Each plant includes three 10,500 kW generators driven by three 14,750 horsepower turbines, with a total capacity in each powerhouse of 31,500 kW. As a part of the construction of Amistad Dam, power penstocks were installed and sites were excavated on each side of the dam for future hydroelectric power plants. In 1983, the United States completed construction of a hydroelectric generating plant consisting of two units of 33,000 kW each.

The flood control pools of Amistad and Falcon Reservoirs are operated by IBWC to reduce flooding in the Rio Grande floodplain below the dams. The storage contents of the two reservoirs plotted in Figures 6-1 show large depletions of storage in the conservation pools. Infrequent floods are relied upon to fill the conservation pools.

Amistad and Falcon Reservoirs are used extensively for recreation. Recreation facilities on the US side of Falcon Reservoir have been developed by the Texas Parks and Wildlife Department, Starr County, and private interests. The National Park Service is responsible for recreation facilities on the US side at Amistad Reservoir. Mexican authorities supervise recreation on the Mexican side of the two reservoirs.

Red Bluff Reservoir is located on the Pecos River and backs water into New Mexico. The reservoir is owned and operated by the Red Bluff Water Power Control District to provide water supplies for irrigation and hydroelectric power generation. Due to inflows of highly saline water, the actual use for irrigation has been much less than originally intended. As discussed in Chapter 2, the Pecos River Compact provides that New Mexico does not allow human activities to reduce the flow of the Pecos River below specified levels.

Lake Balmorhea is owned and operated for irrigation by the Reeves County Water Improvement District No. 1. Much of the inflow to the reservoir is from springs. San Esteban Lake is owned by a private estate. It is heavily silted and dry most of the time due to insufficient local runoff. Casa Blanco Lake is owned and operated by Webb County for recreation and irrigation of a golf course. Imperial Reservoir is owned and operated by Pecos County Water Control and Improvement District No. 2 for irrigation.

References

HDR Engineering. 1998. *Trans-Texas Water Program, West Central Study Area, Phase II: Edwards aquifer recharge analyses*. Austin, TX: San Antonio River Authority, San Antonio Water System, Guadalupe-Blanco River Authority, Lower Colorado River Authority, Nueces River Authority, Texas Natural Resource Conservation Commission, Texas Water Development Board.

McCrory, J. A. 1984. "Natural salt pollution control, Brazos River basin, Texas." In *Salinity in watercourses and reservoirs*, edited by R. H. French. Boston: Butterworth.

Rought, B. G. 1984. "The southwestern salinity situation: The Rockies to the Mississippi River." In *Salinity in watercourses and reservoirs*, edited by R. H. French. Boston: Butterworth.

USACE (US Army Corps of Engineers). 1992. *Water resources development in Texas 1991*. Dallas: USACE, Southwestern Division.

USACE Fort Worth District. 1976. *Final environmental impact statement, Natural salt pollution control study, Brazos River basin, Texas*. Fort Worth, TX: USACE.

USACE Tulsa District. 1976. *Final environmental impact statement, Red River chloride control project*. Tulsa, OK: USACE.

USACE Tulsa District. 1995. *Final supplement to the final environmental impact statement, Red River chloride control project*. Tulsa, OK: USACE.

Wurbs, R. A. 2002. "Natural salt pollution control in the Southwest." *J. Am. Water Works. Assoc.* 94 (12): 58–57.

Wurbs, R. A. 2019a. *Daily water availability model for the Brazos River basin and San Jacinto-Brazos Coastal basin*. TR-513. College Station, TX: TWRI.

Wurbs, R. A. 2019b. *Daily water availability model for the Trinity River basin*. Contract No. 582-18-80410. Austin, TX: TCEQ.

Wurbs, R. A. 2020. *Daily water availability model for the Neches River basin*. Contract No. 582-18-80410. Austin, TX: TCEQ.

Wurbs, R. A. 2022. *Daily water availability model for the Colorado River basin*. Contract No. 582-18-80410. Austin, TX: TCEQ.

Wurbs, R. A. 2023a. *Daily water availability model for the Lavaca River basin*. Contract No. 582-18-80410. Austin, TX: TCEQ.

Wurbs, R. A. 2023b. *Daily water availability model for the Nueces River basin*. Contract No. 582-18-80410. Austin, TX: TCEQ.

Wurbs, R. A. 2024. *Incorporation of environmental flow standards and other water management complexities in water availability modeling*. TR-552. College Station, TX: TWRI.

Wurbs, R. A., and P. A. Carriere. 1988. *Evaluation of storage reallocation and related strategies for optimizing reservoir system operations*. TR-145. College Station, TX: TWRI.

Chapter 8

Synthesis, Elaboration, Summary, and Conclusions

Chapter 8 combines and builds on the preceding seven chapters. Fundamental concepts and lessons learned from the Texas experience in managing water in river and reservoir systems are discussed. This concluding chapter further elaborates on the book's exploration of water management in Texas along with summarizing and articulating conclusions.

Management of the water resources and constructed infrastructure of the river and reservoir systems of Texas is driven by extreme hydrologic variability, rapid population growth, depleting groundwater supplies, intensifying demands on surface water resources, and uncertainties about future hydrology and demands on water resources. A myriad of people and organizations work collaboratively to manage water within an institutional framework of laws, policies, programs, administrative procedures, professional disciplines, and water management practices. Management of water resources, agricultural and urban land use, electrical energy production, recreation facilities, ecosystems, and environmental resources are integrally interconnected.

Texas Paradigm for Comprehensive Water Management

Water management capabilities in Texas established over two centuries are still being expanded in response to intensifying demands and changing conditions. The Texas experience provides a comprehensive textbook case study of managing the water resources of river and reservoir systems that is relevant nationwide and worldwide. Endeavors and accomplishments in Texas are experiments in water management that can inform similar efforts throughout the United States and various other regions of the world. Diverse climate, hydrology, socioeconomic conditions, and water management practices are found in the various regions of Texas. Most regions of the nation and world are similar in relevant ways to some region of Texas. Water management infrastructure, policies, programs, and practices are crucial to the prosperity of the state and of necessity must be comprehensive and effective.

The general institutional framework for water management includes overlapping responsibilities and endeavors at the following levels: Local governmental, private sector, river basin, state, interstate, federal, and international. Water management roles have shifted significantly over the last several decades from local and federal governmental entities to an increased reliance on state agencies and state-level policies, programs, and funding.

Texas and this book may be viewed as the following two case studies in effective water management: (1) The Rio Grande, and (2) the rest of Texas. The rest of Texas experience outside of the international Rio Grande Basin can be viewed as a case study from the two perspectives of (1) innovative endeavors driven at the state level; and (2) nationwide activities at the federal, local, and private sector levels. The Texas case study provides both a general understanding of capabilities and practices established nationwide at the federal and local levels and newer innovative capabilities and practices created in Texas at the state level.

Hydrologic, institutional, economic, and water management differences between the Rio Grande Basin and the remainder of Texas are noted throughout the chapters of this book. The Rio Grande story is well over a hundred years of generally successful endeavors of two countries, several states, farming enterprises, and cities in managing water resources of a large dry river basin with water demands, largely agricultural irrigation, exceeding available reliable supplies. The rest of Texas case study spans a great diversity of climatic and hydrologic conditions, constructed infrastructure, and water needs, problems, opportunities, issues, and solutions.

The state of Texas is a notable leader in pioneering endeavors that can be replicated by water managers in many other places outside of Texas. Structured programs have been established in Texas over recent decades for regional and statewide water planning, water allocation, environmental flow protection, drought management, water availability modeling, and other state-level endeavors. Capabilities established by state agencies in Texas provide an experience base for implementing similar strategies and practices elsewhere.

Water management in the 15 major river basins and eight coastal basins of Texas is representative of water needs, problems, issues, constructed facilities, institutions, management strategies, and practices throughout the United States and other countries. The role of water users, cities and other local entities, private companies, and the federal government is similar in Texas and nationwide. Texas is a large state with a growing population and economy and resulting diverse water needs and issues. Water management needs and practices in rural areas, small communities, cities, large metropolitan areas, and private industry in Texas are illustrative of other similar places throughout the nation.

Texans have been major beneficiaries of federal water development and management programs and projects over the last century. Discussions of federal, nonfederal, and combined federal and nonfederal collaborative water resources development and management activities in this book are relevant nationwide.

River Basins of Texas

Texas encompasses all or portions of 15 major river basins and all of eight coastal basins located between the lower reaches of the major river basins. Each river basin has its own unique story, the basics of which are told briefly in Chapter 7. Aspects of the different river systems are highlighted throughout the book. The major river basins and their subbasins are diverse in size, climate, hydrology, population density, economic enterprises, and water management practices. Both the similarities and differences in water resources and water management between the different river and reservoir systems are notable.

The state of Texas encompasses an area of about 268,600 sq. mi. The 15 major river basins and eight coastal basins have an estimated total watershed drainage area of 263,200 sq. mi that contribute precipitation runoff to the river systems. Several of the larger river basins include upper-basin non-contributing land areas totaling about 5,400 sq. mi.

The Rio Grande Basin encompasses a land area of about 336,000 sq. mi, but only about 182,200 sq. mi of this area contributes runoff to the Rio Grande. Precipitation falling on the remainder of the arid land area within the basin is lost to playa lakes, infiltration, and evapotranspiration. About 27% of the 182,200 sq. mi of contributing area is in Texas with the remainder encompassing portions of New Mexico, Colorado, and five states in Mexico.

A third of the 93,450 sq. mi Red River Basin, a subbasin of the Mississippi River Basin, is in Texas and the remainder is in Oklahoma, New Mexico, Arkansas, and Louisiana. Although treated by Texas water agencies and this book as separate smaller Texas river basins, the upper Sulphur River and Cypress Creek Basins are part of the Red River Basin. Twenty-seven percent of the Canadian River Basin, also a subbasin of the Mississippi River Basin, is in Texas with the remainder in Colorado and Oklahoma. Seventy-eight percent of the Sabine Basin lies within Texas. Small arid portions of the upper Brazos and Colorado River Basins extending into New Mexico contribute almost no flow to the stream systems. The Brazos and Colorado are essentially totally Texas basins. The seven other major river basins and eight coastal basins are encompassed totally within Texas.

The 45,570 sq. mi Brazos River Basin is considered the largest basin entirely within Texas, though a 2,700 sq. mi. mostly non-contributing portion of this area is in New Mexico. The Brazos River Basin is larger in area than 16 of the 50 US states and over half the countries in the world. The Brazos River Authority was created by the Texas legislature in 1929 as the first government entity in the United States responsible for developing and managing the water resources of an entire river basin. BRA and USACE own and operate a system of 12 large multipurpose reservoirs and other facilities in collaboration with TWDB regional and statewide planning, TCEQ regulatory programs, entities managing recreational opportunities, and local municipal, industrial, and agricultural water managers and users.

The two smallest major rivers discharging into the Gulf of Mexico within Texas are the Lavaca and San Jacinto Rivers, with watershed areas of 2,310 and 3,940 sq. mi, respectively. The eight coastal basins, encompassing 7.3% of the area of the state, each discharge through multiple creeks and bayous into the Gulf of Mexico rather than through a single major river.

Half of the 30 million residents of Texas live in either the upper Trinity River Basin or the San Jacinto River Basin. The Dallas and Fort

Worth metropolitan area is encompassed by the upper Trinity River Basin. The Houston metropolitan area is in the much smaller San Jacinto River Basin. Lake Livingston, the largest reservoir in the Trinity River Basin, is a major water supply source for the city of Houston in the adjacent San Jacinto River Basin.

A total of about 20% of the people of Texas live in the third, fourth, and fifth most populous metropolitan areas, which are Greater San Antonio on the San Antonio River, the Austin metropolitan area on the Colorado River, and the El Paso metropolitan area on the Rio Grande. The remaining 30% of the population is scattered among many cities, communities, and rural areas throughout the state.

All the large cities and major metropolitan areas of the state have experienced annual population increases among the highest in the nation for comparably sized cities each year over many years. Many of the smaller cities are also experiencing rapid population growth.

River System Hydrology

Analyses of the characteristics of precipitation, reservoir surface evaporation rates, and river flow rates observed over many decades throughout Texas are presented in Chapter 5 with a focus on exploring variability and stationarity. Observed streamflow sequences adjusted computationally to represent natural conditions are also investigated in Chapter 5. Impacts of water development on river flows are further explored in Chapter 6. The effects of droughts on reservoir storage are investigated in Chapters 6 and 7.

Hydrologic Variability

The statewide average 1940 to 2022 mean annual precipitation and 1954 to 2022 mean annual reservoir evaporation are 28.03 and 59.43 in., respectively. Monthly precipitation during 1940 to 2022 spatially averaged statewide ranges from a January mean of 1.63 in. to a May mean of 3.32 in. Mean monthly reservoir evaporation during 1954 to 2022 averaged over the entire state ranges from 2.36 in. in January to 7.86 in. in July.

Hydrologic conditions in Texas vary dramatically from the dry western to wet eastern regions of the state. Mean annual precipitation varies from less than 9 in. per year at the city of El Paso on the Rio Grande in far West Texas to well over 50 in. per year in the Sulphur, Cypress, and Sabine River Basins in East Texas. Reservoirs in East Texas are full or almost full most of the time. Reservoirs in West Texas experience extended severe drawdowns. Major regions of West Texas and the Panhandle are not hospitable to reservoir development. Most of the large reservoirs in Texas are in Central and East Texas.

Rainfall and streamflow in Texas are extremely variable over time with continuous, storm event, seasonal, and multiple-year fluctuations reflecting extremes of droughts and floods as well as more frequent but less severe fluctuations. Reservoir evaporation rates vary greatly seasonally. Streamflow variability is driven by variability in precipitation and evaporation. Long-term mean river flow volumes are large, but most of the flow occurs during infrequent flood events or periods of high flows, separated by prolonged periods of fluctuating low-to-moderate flows that may include severe multiple-year droughts.

Hydrologic variability and associated water supply reliability and flood risk are fundamental to water management. Large volumes of reservoir storage are essential for developing supplies with acceptable levels of reliability and controlling flood flows to reduce economic damages and protect public safety.

Hydrologic Stationarity

Stationarity refers to constant long-term characteristics of time series variables with no permanent changes or trends. Stationarity or lack thereof of river flows and water availability for beneficial use may be affected by population growth, economic development, land use change, water resources development and use, climate change, other factors, and combinations thereof.

Monthly and annual precipitation and reservoir evaporation rates are climatic variables particularly relevant to river and reservoir systems. As discussed in Chapter 5, observed 1940 to 2022 sequences of 996 monthly and 83 annual precipitation depths in 92 grid cells covering Texas appear stationary. Nonstationarities, if they exist, are hidden in the great rainfall variability.

Annual time series of the maximum and minimum monthly precipitation in any month for each of the 83 years was also analyzed. Maximum monthly precipitation was much higher throughout the state in

2015, 2016, and 2018 than during the other years of 1940 to 2022. Major floods occurred in those years. Variations in instantaneous rainfall intensities during severe storms are obscured in aggregation of hourly or daily quantities to the monthly quantities explored in Chapter 5.

The literature on global warming indicates that evaporation rates are increasing worldwide. Analysis of stationarity or departures therefrom of 1954 to 2022 monthly and annual reservoir evaporation rates indicate a possible appearance of increasing evaporation rates. However, detection and measurement of changes in evaporation rates from the limited investigations presented in this book are difficult and inconclusive.

Permanent changes (nonstationarity) in river flow characteristics have resulted primarily from changes in water use accompanying population growth and construction of dams, reservoirs, conveyance facilities, and other infrastructure for storing, transporting, and using water. The impacts are significant, diverse, and vary with location. The impacts of water development and use on low flows are very different than on high flows. Regulation of rivers by dams reduces flood flows but may increase low flows at downstream locations. Changes in median flows are different than changes in average flows. The effects of a dam on flows just below the dam are much less evident further downstream.

Climate Change Uncertainties

Water management is driven by extreme hydrologic variability and uncertainties regarding the timing and characteristics of future floods and droughts, future economic and population growth, future changes in water needs and water management strategies, and various other future uncertainties. Climate change resulting from global warming likely increases hydrologic variability in various regions of the world and certainly adds to uncertainties regarding the future.

Global warming results in increases in reservoir evaporation, agricultural and urban irrigation demands, and other uses of water supplied by rivers and reservoirs. Global warming may increase rainfall intensities of extreme storms, worsening flooding.

Dams, reservoirs, conveyance facilities, other constructed infrastructure, and institutional water resources planning, allocation, and management capabilities are essential for dealing with hydrologic variability and future uncertainty even without long-term changes in climate resulting from global warming. These water management capabilities are likewise essential in dealing with the added complexities of global warming and climate change issues.

Surface Water and Groundwater Connections

Connections between surface and groundwater are basic components of the hydrologic cycle of importance in water management. Issues of coordination or lack thereof between surface and groundwater management are introduced in Chapter 1.

Many streams in Texas gain water from spring flow and seepage from groundwater and from wastewater return flows from water supplied from groundwater sources. Streams throughout Texas also lose water through seepage into the underlying ground. As discussed in Chapters 5, 6, and 7, although aquifer recharge is a significant loss of surface streamflow in several regions of the state, recharge of the Edwards Aquifer from flows of the Nueces River and its tributaries represents the extreme case. Most of the flow of the Nueces River and its tributaries is lost as the streams cross over the recharge zones of the Edwards and other aquifers.

Naturalized and Simulated Regulated Flows as a Percent of Precipitation

Most of the precipitation falling on the land area of river basins is lost through infiltration and evapotranspiration without ever reaching the river outlet under natural undeveloped conditions. River flows are further reduced by water development and use.

Full authorization WRAP/WAM simulations are based on the premise of all water users appropriating their full authorized streamflow amounts subject to water availability. Tables 5-5 and 6-7 of Chapters 5 and 6 relate naturalized and simulated regulated streamflow at basin outlets from the full authorization scenario WAMs to basin-wide precipitation in 11 major river basins encompasses 62.5% of the watershed area of Texas.

The mean annual naturalized flow volume at the basin outlet varies from 2.26% of the precipitation volume in the Nueces River Basin to 27.0% for the Sabine River Basin, with 13.4% of the precipitation falling on all 11 basins reaching their outlets. The extremely low long-term mean volume of flow of the Nueces River into Nueces Bay of 2.26% of the volume of precipitation falling on the Nueces River Basin reflects much of the river flow being lost to groundwater recharge as the river and its tributaries cross recharge zones.

The effects of water resources development and use on river flows vary greatly with location and between low flows versus more normal flows versus high flows. The long-term mean and median flows at the outlets of the 11 major rivers are compared in Table 6-7. Mean flow is significantly higher than median flow since the mean is affected more by flood flows than the median. The long-term mean of the simulated regulated flows expressed as a percentage of naturalized flows in Table 6-7 range from 24.3% for the Nueces River to 81.2% and 84.3%, respectively, for the Guadalupe and Lavaca Rivers. The median of the simulated regulated flows for these three rivers are 0.00%, 67.3%, and 64.7% of their naturalized flows.

With the premises and approximations reflected in the simulation model, the mean of the total simulated regulated flow at the outlets of the 11 major rivers is 46.1% of the flows under natural undeveloped conditions. The volume of the regulated flow at the outlets of the 11 rivers is estimated to be about 6.2% of the volume of the precipitation falling on their watersheds.

Reservoir Storage as a Drought Index

Storage contents and drawdowns of reservoirs are a practical measure of water availability that can also be viewed as a drought index. Reservoir storage depletions reflect cumulative past and recent weather and hydrology, water use, and water management. However, the basic weakness of using historical observed storage contents as an index for comparing past and current drought conditions is the nonstationarity resulting from the changes in water development and use that have accompanied population and economic growth over the last decades. Water needs and level of development during the 1950 to 1957 drought were much less than during the 2011 to 2014 drought. The WRAP/WAM modeling system deals with this issue by simulating the occurrence of a defined stationary condition of water resources development and use during a repetition of stationary historical natural hydrology.

Observed storage drawdowns from January 1998 through December 2023 are employed in Chapter 6 to investigate drought characteristics. Although population and water use have continued to grow, most of the currently existing reservoir storage capacity has been fully operational since before 1998. Using actual observed reservoir storage as a metric for measuring drought, the most severe drought since 1998 occurred throughout most of Texas during 2011 to 2014. The lowest statewide annual precipitation since 1940 of 13.6 in. occurred in 2011. Interestingly, 2015 had a statewide average precipitation of 39.9 in., exceeded since 1940 only by 40.6 in. in 1941.

Simulated reservoir storage contents is a useful measure of water availability that combines a long record of historical natural hydrology with a defined scenario of water resources development, allocation, management, and use. Simulated reservoir storage is also a meaningful drought index for investigating characteristics of droughts and their impacts on water supply capabilities. Storage contents for selected reservoirs or groups of reservoirs computed in WRAP/WAM full authorization scenario simulations are presented in Chapter 7. WAMs include sequences of naturalized streamflow extending from 1940 to near the present.

Figures 7-3, 7-5, and 7-16 are hydrologic period-of-analysis plots of the summations of simulated storage of all authorized reservoirs in the Brazos and Trinity River Basins and the only major reservoir in the Lavaca River Basin. These daily simulations begin on January 1, 1940, with all reservoirs full to conservation storage capacity. The most severe simulated reservoir drawdowns occur during 1950 to 1957 hydrologic conditions for these three river basins. Adopting full authorization scenario simulated storage as a measure of drought severity, the most hydrologically severe drought since 1940 occurred during 1950 to 1957 in these three river basins. The 1950 to 1957 drought began gradually during 1950 and ended with a major widespread flood in April–May 1957. All presently existing major reservoirs are included in the simulations. Many of these reservoirs did not yet actually exist during the 1950 to 1957 drought. Major population and economic growth and associated increases in water use have also occurred since 1957.

The simulations reflected in Figures 7-3, 7-5, and 7-16 support the conclusion that water managers and water users dependent on these reservoir/river systems have never experienced a drought as hydrologically severe as 1950 to 1957 with the population, economic development, water use, and supply capabilities reflected in the authorized use scenario WAMs. These reservoir and river systems supply the Dallas and Fort Worth metroplex (upper Trinity Basin reservoirs),

Houston (Lake Livingston), Corpus Christi (Lake Texana), and many other smaller cities and business enterprises in these regions.

The WAM full authorization scenario simulated daily storage contents of the six Highland Lakes on the Colorado River operated by LCRA are plotted in Figure 7-8. The maximum drawdown during the simulation reached a minimum storage content of 10.1% of conservation capacity on May 10, 2015. The second most severe drawdown depleted down to a minimum storage level of 24.7% of capacity on September 9, 1952. Droughts during 1950 to 1957 and 2011 to 2015 are more severe in terms of storage depletions in the Highlands Lakes than any other droughts occurring since 1940.

Total simulated storage contents of all reservoirs in the Neches River Basin WAM are plotted in Table 7-12. The lowest storage volume during the 1940 to 2020 simulation of 47.0% of the conservation storage capacity occurred on December 3, 2011. Drawdowns in the Neches River Basin in East Texas are less severe than drawdowns in the Trinity, Brazos, Colorado, and Lavaca River Basins to the west.

The storage levels in Lakes Choke Canyon and Corpus Christi in Figure 7-17 are generated in a full authorization Nueces River Basin WAM simulation. The dramatic storage depletions reflected in the simulation illustrate the extreme case of a greatly over-appropriated river basin. As discussed in Chapter 7, the city of Corpus Christi and other holders of water rights for use of water in the Nueces River Basin have developed alternative supplies including transport of water from the Lavaca and Colorado River Basins.

Storage in individual reservoirs fluctuate more than the summation of storage in multiple reservoirs due to the averaging-out of differences between reservoirs in the timing of drawdown and refilling cycles. WRAP/WAM simulations indicate that timing as well as magnitude of simulated reservoir drawdowns vary between different reservoirs and between different water use scenarios. However, for most large reservoirs in Texas, the minimum storage levels in simulations beginning in 1940 occur during either 1952 to 1957 or 2011 to 2015.

Using full authorization scenario simulated reservoir storage depletions as a drought index, the 1950 to 1957 and 2011 to 2014 droughts are the most hydrologically severe since before 1940 for most of Texas, with the 1950 to 1957 drought having a greater areal extent that includes the majority of the large reservoirs supplying water for the largest urban populations. The 2011 to 2014 drought was more economically costly due to population and economic growth since the 1950s.

Droughts more hydrologically and economically severe than those during past hydrologic periods of analysis are expected to occur in the future. The timing and other relevant characteristics of future severe droughts are unknown. However, the knowledge gained in investigations of historical past drought conditions is essential to the water management community in effectively preparing for the future.

Dams, Reservoirs, and Appurtenant Facilities

Reservoir projects include dams, outlet conduits through the dams, outlet intake structures, overflow spillways, hydraulic energy dissipation structures, water supply pumping stations, pipelines, channels, hydroelectric energy generation and transmission facilities, buildings, roads, parking lots, marinas, boat ramps, parks, camping areas, and other facilities. Reservoir projects in Texas are accompanied by major conveyance facilities comprised in several cases of over a hundred or even several hundred miles of pipeline and associated pumping stations to transport water to cities. River channels serve to transport reservoir releases to pumping plants at downstream diversion sites. Irrigation canals support farming operations in the Lower Rio Grande Valley and elsewhere in Texas. Most reservoir projects include parks, roads, and recreation facilities, and some have large additional areas of public land or wetlands. Hydroelectric power plants are located at 26 of the dams. Reservoir flood control operations may be integrated with downstream floodways comprised of modified river channels and/or levees.

Reservoir operations are central to comprehensive water management. Large conservation storage volumes are essential to provide reliable water supplies and provide opportunities for hydroelectric energy development, recreation, and fisheries. Flood control reservoir storage is an essential component of integrated structural and nonstructural flood risk mitigation.

Small to Very Large Reservoirs

Reservoirs vary greatly in size. Most of the storage capacity in thousands of storage facilities scattered throughout Texas is contained in a relatively small number of the largest reservoirs. The term *major* is adopted in this book to refer to reservoirs with storage capacities of at least 5,000 ac-ft. The 195 reservoirs listed with descriptive information in the two appendixes of this book have conservation storage capacities and/or flood control storage capacities of 5,000 ac-ft or larger. These major reservoirs do not include thousands of small farm ponds, ponds created for aesthetics, water supply storage facilities, urban stormwater detention basins, and rural flood-retarding structures which are smaller than 5,000 ac-ft.

The 195 major reservoirs have conservation, flood control, and total storage capacities of 58,872,709, 40,129,620, and 18,743,089 ac-ft. The 28 reservoirs in Texas with total flood control and conservation capacities exceeding 500,000 ac-ft are listed with descriptive information in Tables 1-3 and 1-4 and their locations shown in Figure 1-9. These 28 largest reservoirs contain 74.13%, 85.07%, and 77.61% of the conservation, flood control, and total storage capacities, respectively, of the 195 major reservoirs. The 69 reservoirs with at least 100,000 ac-ft of total storage capacity have 95.15% of the total storage capacity of the 195 major reservoirs.

Thirty-three of the 195 major reservoirs contain both flood control capacities of at least 5,000 ac-ft and conservation capacities of at least 5,000 ac-ft. Three other reservoirs have flood control capacities greater than 5,000 ac-ft but no conservation storage. Conversely, 192 of the 195 major reservoirs have only conservation storage with no flood control.

Lake Texoma on the Red River with flood control, conservation (hydropower, water supply, recreation), and total storage capacities of 2,877,000, 2,516,000, and 5,393,000 ac-ft, respectively, is the largest reservoir located partially or wholly in Texas and the 12th largest in the United States. Lake Texoma is owned and operated by the USACE Tulsa District. Toledo Bend Reservoir on the Sabine River is the largest conservation storage reservoir in Texas and the southern United States, with a conservation storage capacity of 4,477,000 ac-ft and corresponding water surface area of 181,600 ac, but no designated flood control pool. Toledo Bend Reservoir is owned and operated by the Sabine River Authorities of Texas and Louisiana.

The NID maintained by USACE as part of the NDSP contained 91,843 dams as of January 2024, of which 7,384 are located in Texas. The water rights system administered by TCEQ authorizes storage in more than 3,400 reservoirs with storage capacities of at least 200 ac-ft. Dams providing only flood protection are included in NID along with other types of dams but are not included in the more than 3,400 authorized reservoirs in the TCEQ-administered water rights system. Flood control storage is not included in water right authorizations. Thousands of small storage facilities are omitted from both databases.

Of the approximately 3,400 reservoirs with storage capacities authorized by water rights, which includes several authorized but not yet constructed projects, about 98% of the storage capacity is contained in the approximately 210 reservoirs that have storage capacities of at least 5,000 ac-ft. The remaining 2% of the storage volume is in reservoirs with capacities ranging from 200 to 5,000 ac-ft. The 81 reservoirs with storage capacities of at least 50,000 ac-ft account for about 92% of the total capacity of the approximately 3,400 storage facilities licensed by water rights.

All major reservoirs in Texas are impounded by constructed dams. The 129,000 ac-ft Caddo Lake on Cypress Bayou on the Texas and Louisiana border is the only natural lake in Texas with a storage capacity greater than 5,000 ac-ft. However, a dam was constructed by a private company in 1914 to raise the water level of the already existing natural lake and reconstructed by USACE in 1968 to 1971 motivated by dam safety concerns. Other than Caddo Lake, the oldest major reservoir in Texas is the 9,600 ac-ft Eagle Lake in the Colorado River Basin used for agricultural irrigation. Impoundment began in 1900. Lake Texoma with an initial impoundment date of 1943 is the oldest USACE project in Texas. Other USACE projects were completed shortly thereafter. Most major reservoir projects in Texas were constructed during the 1940s to 1980s.

Construction of the 180,000 ac-ft Lake Ralph Hall on the North Sulphur River and associated pipeline conveyance system began in June 2021 and is scheduled for completion in 2026. This municipal water supply reservoir, the latest of the few reservoir projects constructed over the last three decades, is being developed by the Upper Trinity Regional Water District.

Project Owners and Purposes

Of the 36 major reservoirs containing flood control storage, Olmos Reservoir owned and operated by the San Antonio River Authority is the smallest and the only totally nonfederal project. The flood control pools in the 35 other major flood control reservoirs were developed and are maintained at federal expense. Major reservoirs owned and operated by nonfederal entities do not include storage capacity designated for flood control primarily due to difficulty in obtaining nonfederal funding for large-scale flood control. With the exceptions of International Amistad and Falcon Reservoirs operated by IBWC, operations of officially designated reservoir flood control pools of 5,000 ac-ft or larger in Texas is the responsibility of the Corps of Engineers.

Except for Addicks and Barker Reservoirs in Houston operated by the USACE Galveston District, all reservoir projects in Texas constructed by USACE or USBR include water supply storage capacity subject to the nonfederal sponsor repayment provisions of the Water Supply Act of 1958. Nonfederal sponsors lease the federal storage capacity and generate revenue by selling water. USBR has constructed five reservoir projects in Texas, which were transferred to nonfederal sponsors to own and operate. Three of these reservoirs include flood control storage for which USACE retains oversight operational responsibility.

The majority of the major reservoirs in Texas are owned and operated for water supply by river authorities, water districts, cities, and electric power utility companies. The fewer but generally larger federal projects contain about half of the water supply storage capacity and most of the flood control storage capacity of the 195 major reservoirs. All costs of the federal reservoirs allocated to water supply are reimbursed to the federal government by nonfederal sponsors. The federal government covers all costs of federal reservoir projects allocated to flood control.

Hydroelectric power is produced at 26 dams, which include both federal and nonfederal projects. Hydroelectric energy generated at federal dams in Texas is marketed through the Western Area Power Administration and Southwestern Power Administration to nonfederal electric utilities. With some exceptions, the most notable being Lake Texoma, reservoir releases through hydroelectric power plants are limited to water that will be diverted downstream for municipal, industrial, or agricultural water supply.

Recreation is extremely popular at most of the major reservoirs. Boating, fishing, swimming, camping, hiking, and other recreational activities are common at most reservoirs. Federal and nonfederal reservoir owners, the Texas Parks and Wildlife Department, local communities, private marina operators, and other entities maintain parks and recreation facilities at the reservoirs.

Much of the shoreline of nonfederal reservoirs may be privately owned. Private homes and businesses may be located adjacent to the normal operating pool level in nonfederal reservoirs. Government-owned flood control pool lands of IBWC and USACE reservoirs are dry except during infrequent flood events. The publicly accessible land of IBWC and most USACE reservoirs also extends significant distances outside of the flood control pool. Government-owned land both within and outside of the flood control pools is publicly accessible providing recreational opportunities.

Reservoir System Operations

Reservoir operating procedures combine a framework of quantitative rules with experience-based operator judgment and may change over time with experience and changing conditions. Reservoir operations allocate river flows, reservoir storage capacity, downstream releases, and lakeside withdrawals between multiple reservoirs, project purposes, water users, and time periods. Operations under normal hydrologic conditions combine provisions for present water-related needs with maintenance of capabilities for responding to infrequent floods and droughts expected to occur at unknown times in the future. Flood damage reduction or drought management operations are activated during flood events and low flow or drought conditions. Modifications of reservoir operation objectives and procedures represent an important strategy for dealing with intensifying demands on limited water resources resulting from population and economic growth.

Water supply operations are often based simply on making withdrawals or releases as necessary to meet water demands subject to water rights and contractual agreements. Inflows pass through spillways and outlet conduits of water supply reservoirs if the reservoir is full. Predeveloped water conservation plans are in place for responding to supply depletion during infrequent severe droughts. Conversely, complex regulation plans guide operations of many reservoirs including large multiple-purpose, multiple-reservoir systems.

Reservoir operations employ designated trigger storage elevations to activate or alter various aspects of operating rules. Operational modifications such as permanent, seasonal, or temporary storage reallocations or strategies for managing water supply reliabilities involve revising storage trigger levels and/or other aspects of operating rules.

A designated top of conservation pool, which is also the bottom of flood control pool, elevation differentiates USACE flood control operations from nonfederal water supply operations in federal multiple-purpose reservoirs. Corps of Engineers flood control operations are in effect whenever water encroaches into the flood control pool. Nonfederal water supply entities contract for and control use of conservation pool storage capacity in USACE owned and maintained reservoirs. Reallocation of storage capacity between flood control and conservation (water supply, hydropower, recreation) purposes consists of modifying the designated top of conservation pool elevation.

Flood control operations vary depending on whether the top of flood control pool elevation is exceeded during a flood event. As long as flood control pool capacity is not exceeded, flood control operations are based on emptying flood control pools as expeditiously as practical without releases contributing to flows exceeding specified nondamaging or minimally damaging flow rates at downstream gauge sites. For some reservoirs, the designated controlling flow rates at downstream gauge sites vary depending on the height to which flood waters have encroached into flood control pools.

The top of flood control pool elevation of 13 of the 36 flood control reservoirs had never been overtopped as of January 2024 (Tables 6-4 and 6-5). The flood control capacities of 10 of the reservoirs have each been exceeded once. Multiple floods have encroached above the top of flood control pool into surcharge storage at each of the other 13 multipurpose projects.

Designated reservoir trigger levels can be employed to coordinate high reliability (firm or safe) water supply commitments with lower reliability (interruptible) water supply commitments. Interruptible water supply commitments from a reservoir system are curtailed if storage levels fall below specified trigger levels, along with activation of demand management or emergency water conservation plans and/or backup groundwater or other supply sources. This strategy is employed by LCRA in operations of the six Highland Lakes on the Colorado River and by BRA in operations of an 11-reservoir system in the Brazos River Basin, as discussed in Chapter 7. A water availability modeling (WRAP/WAM) system supports greater statewide implementation of reliability and frequency analysis-based water management strategies combining multiple water supply sources, commitments at varying levels of reliability, and/or conservation practices.

In many cases, single reservoirs are operated individually. Multiple-reservoir system operations are also common. In some cases, a single reservoir owner may own and operate multiple reservoirs as a system. In other cases, two or more agencies collaborate to develop a coordinated plan for managing their reservoirs as a system. Computer modeling studies quantify improvements in water supply reliability, flood risk reduction, hydroelectric energy generation, recreation, and/or other services achieved by coordinated operations of multiple reservoirs. Operating plans outline criteria and rules followed in balancing storage and releases between reservoirs.

Dam Safety and Rehabilitation

Maintenance, repair, and rehabilitation of a massive inventory of aging roads, bridges, buildings, water and wastewater treatment facilities, water distribution and wastewater collection systems, and other constructed infrastructure throughout the United States is a major national concern. Aging dams and appurtenant structures and associated safety issues are important components of the infrastructure maintenance and rehabilitation challenge. Dam failures can be catastrophic. Repairs of dams and appurtenant structures can be expensive. Monitoring and inspections are important.

FEMA coordinates NDSP in collaboration with USACE and other federal and state agencies. USACE, USBR, and other federal agencies are responsible for the safety of their own dams. The states are responsible for assuring the safety of nonfederal dams within national guidelines of the NDSP. Various resources are available to the states through NDSP. TCEQ administers the state nonfederal dam safety program for Texas.

Salinity Constraints

Water supply capabilities depend on water quality as well as quantity. The last section of Chapter 5 explores natural salt pollution originating from the Permian Basin geologic region in the upper Red, Canadian, Brazos, Colorado, and Pecos River Basins in Texas, New Mexico, Oklahoma, and Kansas. Salinity severely constrains water supply capabilities of a number of major reservoirs including some of the largest in Texas and neighboring states. Effects of salinity on water use and measures for dealing with Permian salt pollution in the different river basins are discussed in Chapter 7.

Saltwater intrusion from the Gulf of Mexico into the lower reaches of rivers is also a major concern in the coastal regions of Texas. Many cities and industries are located along the Texas coast. The Brazos River and the USACE/BRA reservoir system is a notable example of the dual problem of salinity entering the upper reaches of the river system from geologic formations in the Permian Basin Region and saltwater from the ocean extending upstream in the lower reach where streamflow is diverted to supply major water users.

Federal Roles

The Corps of Engineers of the Department of Army and Bureau of Reclamation of the Department of the Interior are the primary federal agencies that have constructed large multiple-purpose dam and reservoir projects and other associated river control structures in Texas and other states. The International Boundary and Water Commission is responsible for developing and managing the waters of the Rio Grande. The Natural Resources Conservation Service of the Department of Agriculture works with farmers and ranchers to reduce flooding from small rural watersheds, reduce land erosion, and protect natural resources. The US Environmental Protection Agency was created in 1970 to administer, in collaboration with state environmental agencies, the many environmental protection programs initially established during the 1960s to 1970s.

Navigation, Flood Control, and Multipurpose Water Development

USACE is as old as the nation, having constructed military works for the army led by General George Washington during the 1775 to 1783 Revolutionary War. The Civil Works Program of the Corps of Engineers places the United States in a unique situation, compared to other countries, of maintaining a large nationwide water resources development and management program within a military organization. However, except for top-level executive leadership, the large professional and administrative staff of USACE is comprised almost entirely of career civilian employees.

Congress assigned the Corps of Engineers responsibilities for navigation improvements in 1824. The rationale was that USACE was the foremost organization in the nation with civil engineering capabilities required to successfully address the challenges of making rivers navigable. The USACE Galveston District constructed during the mid-1900s and continues to maintain the Texas portion of the Gulf Intercoastal Waterway and port improvements along the Texas coast.

The Flood Control Act of 1936 established USACE responsibilities for flood control. Flood control is the federal project purpose justifying USACE construction and operation of multipurpose reservoir projects in Texas. All costs allocated to water supply in federal projects are repaid by local sponsors. More USACE multiple-purpose reservoirs with water supply storage are located in Texas than any other state. Other states have more USACE navigation facilities and single-purpose flood control reservoirs. The Corps of Engineers constructed the Addicks and Barker flood control dams in Houston, major levee and channel improvements in the Dallas and Fort Worth metroplex, major tunnels and channel improvements in downtown San Antonio, and local flood protection projects in other smaller cities.

USACE constructs and operates projects throughout the nation. USBR activities are limited to the 17 western states, which includes Texas. USACE is responsible for flood control operations at reservoirs constructed by USBR. Texas has been a major beneficiary of USACE programs. USBR has a greater role in California and other western reclamation states than in Texas.

USBR was created by the Reclamation Act of 1902 to plan and implement water projects to support population and economic growth in the arid western states. The original focus was irrigated agriculture which grew into comprehensive multipurpose water development. USBR has played a major role in other states in constructing and operating the largest reservoirs and water conveyance projects in the nation. Although USBR owns and operates many water projects in other states, the five reservoir projects and several irrigation canals constructed by USBR in the western, drier half of Texas were transferred to nonfederal entities to own and operate.

The US and Mexico sections of IBWC are federal agencies of the two nations. IBWC owns and operates International Amistad and Falcon Reservoirs on the Rio Grande. Initial impoundment in these two reservoirs constructed pursuant to a 1944 international treaty occurred in 1953 and 1968. The waters of the Rio Grande are allocated between the countries in accordance with the 1944 treaty. Water management in the Rio Grande Basin is notably different than the other river basins of Texas hydrologically, economically, and institutionally, as discussed throughout this book.

NRCS was created in 1935 as the Soil Conservation Service to work with farmers to reduce impacts of farming operations on land erosion. The name SCS was changed to NRCS in 1994 to reflect the agency's broadening responsibilities. The agency has constructed about 11,850 flood-retarding dams nationwide, including 2,040 in Texas, under provisions of the Watershed Protection and Flood Protection Act of 1954. The federal government funded construction costs for the small dams on private farms and ranches maintained by nonfederal entities at nonfederal expense. NRCS flood-retarding structures are not included in the 195 reservoirs in the appendices of this book due to their storage capacities being smaller than 5,000 ac-ft.

Most federal multipurpose reservoir, flood control levee and channel improvement, and coastal navigation projects in operation in Texas were constructed during the nationwide 1940s to 1980s construction era of large-scale water resources development. Most nonfederal water supply, steam-electric power plants, and hydropower reservoirs were also constructed during that period.

USACE played a dominant role in developing the larger reservoir, conveyance, and floodway projects. Congressionally authorized and funded comprehensive basin-wide planning studies performed by USACE and USBR during the 1930s to 1970s resulted in many federal multipurpose reservoirs and urban flood control channel improvement and levee projects being constructed in Texas during the 1940s to 1980s. Construction of large-scale water projects by the federal agencies transitioned during the 1970s to 1980s to operation and maintenance of a large inventory of constructed facilities concurrently with dramatic growth in regulatory programs to protect the environment and greater reliance on using water more efficiently and nonstructural strategies for reducing flood damages.

Environmental Protection

Creation of the EPA in 1970 and enactment of the NEPA also in 1970 followed the growing environmental consciousness of the 1960s. The Water Pollution Control Act of 1948, as amended by the Clean Water Acts of 1972 and 1977 and other legislation, established the NPDES, USACE Section 404 dredge and fill permit program, and other environmental regulatory programs. The Safe Drinking Water Act passed into law in 1974 and amended in 1986 and 1996 established programs administered by EPA and state environmental agencies to assure safe public drinking water.

Responsibilities for certain nationwide regulatory activities including NPDES and safe drinking water program were transferred to the states as state institutional and technical capabilities were established. EPA retained federal oversight authority. TCEQ administers NPDES, safe drinking water, and other programs in Texas in accordance with EPA rules and oversight.

Section 404 of the 1972 Clean Water Act established a program administered by USACE to regulate discharges of dredged or fill materials into the waters of the United States. Activities regulated under this program include but are not limited to urban and commercial development, dams and levees, roads and bridges, and mining operations. Proposed activities are regulated by USACE through a permit review process. No discharge of dredged or fill material is permitted if a practicable alternative exists that is less damaging to the aquatic environment or if the nation's waters would be significantly degraded. Issuance by USACE of a Section 404 permit to any federal or nonfederal entity including private developers is a federal action requiring adherence to all federal environmental laws, rules, and regulations.

State Agency Planning and Water Allocation

TWDB with an administrative and technical staff of about 400 employees is responsible for regional and statewide planning and administering grant and loan programs to assist local and regional entities in implementing water management strategies and projects. TCEQ with about 2,800 employees is the largest state-level environmental regulatory agency in the United States. TCEQ responsibilities include administration of the state's water rights system, dam safety program, and an array of other regulatory programs to protect water quality, public health, and environmental resources. The 3,500 employees of TPWD are responsible for managing 95 state parks, 51 wildlife management areas, eight fish hatcheries, and other related endeavors. TPWD owns land and recreation and fish and wildlife management facilities at many reservoirs.

TPWD, TWDB, and TCEQ collaborate in various ways in accomplishing their interrelated missions.

Statewide and Regional Planning

Creation of TWDB by the Texas legislature in 1957 was motivated by the devasting 1950 to 1957 drought which ended with a catastrophic flood in April to May 1957. A statewide water plan developed by TWDB staff in coordination with the public was completed in 1968 and formally adopted by the Board in 1969. Updates to the statewide water plan were adopted in 1984, 1990, and 1992. USACE and USBR coordinated federal planning studies for the river basins of Texas with TWDB during the 1950s to 1960s. TWDB and its regional planning group partners have been the dominant water planners for the state in recent decades. Cities, river authorities, and various other water management and use sectors are represented in the regional planning groups. Local and regional government agencies also perform their own water planning for their individual jurisdictions and water needs.

The central focus of the 1968 state water plan was construction of many large reservoir and conveyance projects by state, regional, and local entities to support the growing population and economy of the state. The plan recognized that USACE was dealing with flood control on a separate but parallel track at federal expense. Many of the projects recommended in the 1968 water plan were constructed by various entities between the 1968 water plan and the 1984 plan update. The 1984 and later versions of the statewide water plan included additional new reservoir and conveyance projects but shifted to a much greater emphasis on reducing water use by encouraging more efficient water use strategies and practices.

Senate Bill 1 enacted into law by the Texas legislature in 1997 established the current TWDB-led statewide water planning process that emphasizes integration of local stakeholder-guided consensus-based planning with statewide planning. Sixteen regional plans and a consolidated statewide plan are updated in five-year planning cycles, with a 50-year future planning horizon. Updated plans were completed in 2002, 2007, 2012, 2015, and 2022. The current cycle is scheduled for completion in 2027.

Senate Bill 8 enacted by the legislature in 2019 authorized creation of a state-level flood control planning process under the leadership of

TWDB. The new flood control planning process is patterned after the regional and statewide water supply planning process established pursuant to the 1997 SB1. The first statewide flood control plan is scheduled for completion in 2024.

TWDB also administers an array of grant and revolving fund loan programs to assist local and regional entities in financing water projects. The State Water Implementation Fund for Texas administered by TWDB provides low-interest loans with favorable terms for implementation of projects included in the most recently adopted state water plan.

Water Allocation

Water allocation mechanisms grow in importance as demands on limited water resources increase. IBWC administers the allocation of the waters of the Rio Grande between the two countries following provisions of a 1906 convention and 1944 treaty. TCEQ administers the Texas share of the water supplied by the Rio Grande in coordination with IBWC operation of the two international reservoirs. Five interstate river commissions administer compacts allocating water between Texas and neighboring states. TCEQ provides administrative and technical support for the Texas compact commissioners. The rules for allocating river flows between neighboring states are structured differently between the five interstate river compacts.

The severe drought during 1950 to 1957 that led to creation of TWDB and acceleration of reservoir construction efforts also motivated a massive lawsuit that resulted in judicial adjudication of water rights in the Lower Rio Grande Valley. The Water Rights Adjudication Act was enacted by the Texas legislature in 1967 to merge all historically claimed water rights into a structured prior appropriation permit system covering all of Texas except the Rio Grande downstream of the old Fort Quitman historic site 90 miles downstream of El Paso. TCEQ administers both the Rio Grande and the rest-of-Texas versions of the water rights system.

Water rights authorize diversion of specified amounts of water annually at specific locations, for specific purposes, and storage of streamflow in reservoirs of specified storage capacity. TCEQ approves applications for new permits or amendments to existing permits if unappropriated streamflow is available to supply proposed needs at acceptable levels of reliability; reliabilities of existing water rights are not impaired; efficient water conservation will be practiced; and proposed actions are consistent with the statewide and relevant regional water plans.

According to water availability modeling studies performed by TCEQ and TWDB staff, the author, and others, the several thousand existing water right holders have appropriated most of the flow of the streams of Texas occurring during normal and dry hydrologic conditions. Most remaining unappropriated flow identified in water availability modeling consists of high flows during floods. Large volumes of reservoir storage are required to store flood flows for later use as water supply. Many river reaches have practically no unappropriated flow still available. Western regions of the state have little or no unappropriated flow. Most of the remaining accessible unappropriated flow in the state is in the eastern river systems. Beneficial use of remaining unappropriated flow requires access to reservoir storage and for some cities, major pipeline conveyance facilities. Water is being transported between river basins, generally from east to west.

The quantities of water available to river authorities, water districts, cities, and other water management entities for storage in reservoirs and diversion for beneficial use are controlled by water rights. Monitoring of water storage and use authorized by water rights and associated water accounting systems are growing in importance as demands on limited water resources increase. Water master operations managing allocations of water in the Lower Rio Grande Valley supplied from International Amistad and Falcon Reservoirs reflect the most detailed water monitoring and accounting in the state. Water master operations are gradually being established or refined for the other river basins of the state.

Texas like many other states in the western United States has established over several decades a water rights permit system based on the prior appropriation doctrine. A fundamental concept of prior appropriation doctrine water rights is protection of existing water users from losing their water to new water users. Details of curtailment actions during droughts can be complicated. For example, a water right is typically assigned the same seniority (priority) for both refilling reservoir storage and water supply diversions. However, risks and potential consequences of curtailing refilling of depleted storage capacity of a half full reservoir will be very different than curtailing water supply diversions for present municipal, industrial, or agricultural use. The issue of curtailing refilling of reservoir storage is illustrative of various complexities that may lead to future refinements in rules for administering the water rights system.

Environmental Flow Standards

Establishment of EFS through the process created by Senate Bill 3 enacted into law by the Texas legislature in 2007 is a major commitment of the remaining unappropriated streamflow. The SB3 regulatory process results in EFS that are being incorporated by TCEQ into WAMs employed to evaluate water right permit applications. TCEQ will not approve a new water right permit or amendment to an existing water right that increases the amount of water to be stored or diverted to the detriment of any EFS. However, EFS do not constrain existing more senior water rights.

In the past, TCEQ has used special permit provisions to require consideration of instream uses, freshwater inflows to bays and estuaries, water quality, and fish and wildlife habitat. Instream flow requirements were defined in the past primarily in terms of minimum flow limits imposed on particular permits. More sophisticated EFS combining subsistence flow, base flow, and high flow pulse components that vary seasonally and with hydrologic condition are being established pursuant to the 2001 SB2 and 2007 SB3. SB2 enacted by the legislature in 2001 established the Texas Instream Flow Program jointly administered by TCEQ, TWDB, and TPWD to collect data and conduct studies to develop methods for determining flow conditions necessary to support a sound ecological environment. Recognizing the need to expeditiously establish requirements for protecting the environment, the legislature in its 2007 SB3 created an accelerated process for establishing EFS based on currently available information and best available science.

EFS were established during 2011 to 2014 by TCEQ based on recommendations of science teams and stakeholder committees appointed for selected priority river systems. The process established by the 2007 SB3 includes periodic reevaluations and improvements of EFS. The SB3 process is based on statewide agency oversight, scientific support by the science and engineering community, and public participation. Appointed teams of stakeholders and scientists develop recommendations regarding appropriate EFS for assigned river systems. TCEQ, TWDB, and TPWD provide administrative leadership and technical and funding support.

Sufficient unappropriated streamflow is not available for meeting all EFS requirements continuously. Unappropriated flow is limited by the prior appropriation practice of not applying the EFS requirements to existing senior water rights. Return flows from wastewater treatment plants, hydropower releases, and reservoir releases for water supply diversions many miles downstream contribute to instream flows and thus meeting EFS. Reservoir storage reduces flood flows and in many cases increases low flows. Developing and implementing management strategies involving existing water rights and stakeholders that further increase the protection of environmental flows while respecting senior water rights is a continuing challenge.

Water Availability Modeling

Assessments of hydrologic and institutional water availability and supply reliability are essential for effective water management. Water management is driven by the necessity to deal with extreme hydrologic variability and future uncertainty. Precipitation and streamflow are extremely variable. Water managers know that floods and droughts will occur in the future that are even more severe than experienced in the past. However, the timing and actual severity of future extreme hydrologic events are unknown. Uncertainties regarding future population growth, economic enterprises, needs of numerous water users, and various other factors also add to the complexity of water management. Water availability modeling and analysis tools support complex resource management decisions.

The WRAP modeling system and WRAP simulation input datasets called WAMs are employed by TCEQ staff in evaluating water right permit applications and administering water allocation processes and employed by permit applicants and consulting firms in developing applications for water use. The modeling system is also routinely applied by TWDB staff, regional planning groups, river authorities, consulting firms, and other entities in regional and statewide planning, operational planning, drought management, research studies, and other endeavors. The modeling system has been expanded and improved continuously since the late 1990s. Some of the more recent improvements have focused on modeling environmental instream flow requirements.

Development and implementation of the WRAP/WAM modeling system in Texas highlights two important institutional dimensions of employing computer modeling to support water management decisions.

1. The modeling system simulates institutional aspects of water management along with simulating natural river system hydrology and operation of constructed facilities.
2. The modeling system is implemented by a water management community within the institutional framework of water planning, allocation, and management responsibilities and endeavors.

The WRAP modeling system is generalized for application to river and reservoir systems located anywhere in the world. Applications range from simple to very complex. WRAP software, six lengthy manuals explaining use of the modeling system, other related publications, training courses, and example datasets are available free of charge at the TAMU WRAP website included in Table 4-1.

The TCEQ WAM website, also included Table 4-1, is interlinked with the TAMU WRAP website and provides WRAP simulation input datasets, called WAMs, for all Texas river basins and an array of other information supporting the water right permitting process. TWDB maintains versions of the WAM datasets modified for planning applications. Model users revise WAMs to add features of interest such as proposed increases in water use, new water development projects, or modifications to operations of existing facilities. Model users outside of Texas develop their own simulation input datasets for their river and reservoir systems of interest.

Simulations and analyses of simulation results generate a diverse array of information. Most applications of the modeling system focus on assessing water supply capabilities from the perspective of reliabilities or likelihoods of supplying specified needs or the demands that can be supplied at specified levels of reliability. Frequency statistics for reservoir storage and streamflow under various scenarios also provide useful information. Many water right permitting, planning, and research applications are very basic. Other water availability and reliability analysis endeavors have been quite complex.

In WRAP terminology, period reliability is the percentage of the days, months, or years of a simulation during which a specified percentage of the water supply or hydropower target of a water right or defined group of water rights is supplied during a simulation. Volume reliability is the volume of water supplied expressed as a percentage of the target volume. Firm yield is the maximum annual water supply or hydropower target that can be supplied completely without shortage in a simulation with period and volume reliabilities of 100%.

The WRAP/WAM modeling system provides an enhanced understanding of an array of considerations and complexities needed for employing more sophisticated water management strategies. Water supply capabilities of river and reservoir systems have been found to be greatly improved by management strategies requiring detailed analyses of relationships between quantities of water committed to certain water users and their associated supply reliabilities and the effects on supply reliabilities of other water users.

Water availability modeling studies have indicated that the volume of water supplied by river and reservoir systems can be significantly increased by accepting a small decrease in reliability, meaning an increased risk of not supplying the expected water commitment during infrequent extreme drought conditions. Predetermined reservoir trigger levels are employed to combine firm (high reliability) and interruptible (lower reliability) water supply commitments. Interruptible water supply diversions from a reservoir system are curtailed if storage levels fall below specified trigger levels established based on computer simulation studies. Predeveloped demand management or emergency water conservation plans or backup groundwater or other supply sources are activated based on predetermined reservoir storage levels.

Water availability modeling studies provide quantitative reliability metrics to support adoption of multiple-reservoir system operating plans and water use permits that recognize supply reliability increases achieved by operating multiple reservoirs as a coordinated system rather than as separate individual projects. The modeling system supports evaluations of interactions between operations for flood control, water supply, hydropower, recreation, and environmental flow requirements. WRAP/WAM capabilities also support operations during droughts that integrate multireservoir, multipurpose system operations, water conservation plans based on demand management measures, and various levels of curtailment actions.

Integrated Water Resources Management

Basic concepts of comprehensive river basin planning and development were established during the 1930s to 1960s. Multipurpose reservoir projects were constructed that combined flood control, municipal and industrial water supply, agricultural irrigation, hydroelectric power, recreation, and other project purposes. The Water Supply Act of 1958 formalized provisions for including water supply in federal multipurpose reservoirs with all cost allocated to water supply reimbursed to the federal government by nonfederal sponsors. Recreation was established as a primary reservoir purpose during the 1960s to 1970s with responsibility for development and maintenance of recreation facilities shared by multiple federal, state, and local entities.

The early national and Texas policy of comprehensive river basin planning and development was refined and expanded to a conceptual strategy commonly called integrated water resources management (IWRM) during the latter half of the twentieth century. The published literature on IWRM is extensive. IWRM is commonly defined as a process of coordinated development and management of water, land, and related resources to maximize the economic and social welfare in an equitable manner without compromising the sustainability of vital ecosystems. The water management strategy is based on principles of social equity, economic efficiency, and environmental sustainability. Early federal and state river basin planning and development and more recently state-level regional and statewide planning and water allocation in Texas demonstrate IWRM principles.

Water, land, energy, and environmental resources are integrally connected. The term *water–energy nexus*, emphasizing interconnections between water and energy systems, became popular during the 1990s to 2010s. Water is used extensively in various aspects of energy production, but particularly in generation of electricity. Water for producing and condensing (cooling) steam in thermal-electric power plants accounts for greater volumes of water withdrawn from lakes than any other water use sector, though most of the cooling water is returned to the lake. Addition of hydroelectric energy facilities at reservoir projects can contribute to the performance of electric power grids. Electrical energy is required to pump and otherwise deliver water of appropriate quality for human use and to treat wastewater. Oil and gas production, coal mining, and other energy-related activities also generate significant demands on water resources.

The cost of electricity for desalination operations has been the primary constraint to the use of seawater for supplying water needs in Texas. Growing demands on limited water resources is motivating desalination of seawater in the future for municipal and industrial supply in the Coastal Bend Region and other coastal areas of Texas.

Water Management Community

Most of the 30 million residents of Texas rely on the cities, towns, and communities in which they live for water supply, wastewater collection, and stormwater management services. A few Texans living on farms and ranches in rural areas are self-sufficient regarding water for domestic use, livestock, and irrigation of crops. Cities, counties, water districts, and river authorities play important foundational roles in water management within their local or regional jurisdictions. Water management responsibilities of TWDB, TCEQ, and TPWD are statewide. Electric power utilities and other private companies own and operate reservoirs and other facilities to supply steam-electric power plant steam production and cooling water.

Water management capabilities in Texas are the result of individual and collaborative private-sector, city, regional, river basin, state, federal, interstate, and international endeavors over many decades dating back to well before 1900. Shifts of responsibility from federal and local programs to greater reliance on state agencies and programs have occurred over the last several decades. Federal endeavors in Texas are representative of nationwide programs. Several successful major state programs have features uniquely developed within Texas but applicable elsewhere. Federal, state, and local collaborations became foundational to water planning and development during the 1930s to 1940s and have continued to grow in importance.

Water Conservation

The terms *conservation storage* and reservoir *conservation pool* are commonly used in the water management community and this book to refer to conserving streamflow by storage in reservoirs to facilitate its use for water supply, hydropower, recreation, and other purposes. Reservoir construction and operation entities have been referring to conservation storage since the 1940s and before. In more recent decades, river authorities, water districts, and cities develop water conservation plans to outline permanent strategies for using water efficiently and emergency strategies to deal with shortages during drought or conditions of inadequate supply. Water conservation plans are required for applications for water right permits submitted to TCEQ or requests for financial assistance submitted to TWDB.

Growing demands on limited water resources resulted in an increased emphasis during the 1970s to 1990s on developing strategies and methods for using water more efficiently. Water conservation measures include reduction in distribution losses and improving efficiencies in crop irrigation, urban landscaping, domestic use, and industrial use. Water conservation includes both long-term demand management strategies to reduce water use through improved efficiencies and short-term emergency plans activated only during shortages in supply.

As construction of additional water supply reservoirs became more difficult during the 1970s to 1980s, improving water use efficiency became a major strategy for dealing with growing demands on limited resources. Now after water demands have been significantly lowered through demand management strategies to more efficient levels of use, a shift back toward increasing supply to deal with still intensifying needs may be occurring.

Flood Risk Mitigation

The many federally funded flood control dams, levees, and channel improvements constructed nationwide and in Texas during the 1940s to 1980s are effective at partially controlling flood water and reducing damages. However, flood losses continued to increase despite large federal investments in flood control structures due to intensified urban development of floodplains. The national emphasis shifted during the 1960s to 1970s from construction of additional flood control projects to local regulation of floodplain land use, provision of flood insurance for property owners, and other nonstructural strategies for mitigating flood risk.

The National Flood Insurance Program was established during the 1960s to 1970s. NFIP is a partnership of the federal government, local communities, and the private sector. FEMA administers NFIP nationally. The 23,000 communities that participate in NFIP are responsible for restricting construction of residential and commercial properties in floodplains following standards set by FEMA. Civil engineering firms perform hydrologic and hydraulic engineering studies to delineate and update delineations of floodplains. A network of insurance companies coordinated by FEMA sells flood insurance to property owners.

Various strategies for reducing flood losses differ from the perspective of the relative relevance of different hydrologic metrics employed in planning, design, and management as well as institutional capabilities and mechanisms. Rural NRCS flood-retarding structures and urban stormwater detention basins constructed by cities and developers are designed primarily to reduce peak flow discharges from small watersheds. The intensity of short duration rainfall storm events is key factor in determining peak streamflow rates. Watershed land use is another governing concern. Floodplain management is based on maximum water surface elevation profiles along river reaches associated with given annual exceedance frequencies such as the 1.0% annual exceedance frequency.

Operation of the flood control pools of the major reservoirs highlighted in this book deal with large volumes of flood waters originating from large watersheds upstream of dams. No reservoir releases are made that contribute to downstream flooding unless the flood control pool fills. Typically at least multiple days and sometimes many days of flood inflows during extreme storm events are required to fill a flood control pool. Flood storage content may ratchet upward in response to multiple rainfall sequences occurring over many days. Severe flooding may occur downstream of either large reservoirs or smaller flood storage facilities due to rainfall runoff entering the stream system downstream of the dams.

Land and stream bank erosion are important aspects of rural small watershed flood protection and urban stormwater management. Reservoir sedimentation and shoreline and downstream channel erosion are important considerations in major reservoir flood control operations. Diverse aspects of water quality are important concerns in managing stormwater runoff, streamflows, and reservoir storage as well as the water supplied to municipal, industrial, and agricultural water users.

IWRM includes coordinated integration of the following types of flood damage reduction and other water management strategies and practices as illustrated by water management endeavors in Texas:

- Local stormwater management and drainage measures are integrated with measures for dealing with flood waters from overflowing major streams and rivers.
- Construction of flood control dams, levees, and channel improvements; regulation of floodplain land use; flood proofing of individual buildings to reduce susceptibility to flood damage; flood insurance; and computer-based flood forecasting, flood warning,

and decision-support systems are employed together in an integrated manner.

- Reservoir flood control pool land and adjacent project land are used for parks, playgrounds, athletic fields, hiking trials, and other public recreation opportunities between infrequent flood events. Stream floodplains are used for these same types of recreation activities as appropriate.
- Water supply reservoirs are operated during floods to mitigate flooding as much as possible.
- Reservoir releases from flood control pools as well as conservation pools are used to generate hydroelectric energy.
- Flood waters stored in reservoirs located upstream of groundwater recharge outcrop zones can be released at flow rates designed to optimally enhance aquifer recharge.
- Storage capacity has been reallocated between flood control and conservation pools of multipurpose reservoirs, in some cases permanently and sometimes seasonally or temporarily.
- Reservoir releases from flood control pools may be designed in the future to satisfy high pulse flow components of EFS. Likewise, multipurpose reservoir operations may include preserving subsistence, base, and high pulse components of downstream EFS along with other project purposes.

Appendix A

Major Reservoirs Listed by River Basin

Appendixes A and B provide information describing each of 195 reservoirs located entirely or partially in Texas that have conservation storage capacities of at least 5,000 ac-ft and/or flood control capacities of at least 5,000 ac-ft. The 195 major reservoirs are grouped by river basin and listed by age (fourth column) in Appendix A with the following information tabulated in seven columns for each reservoir. Reservoirs are numbered in Appendix B. The number in the last column of the Appendix A tabulation refers to the identifying number of the reservoir in Appendix B.

1. Name of reservoir.
2. Reservoir owner using abbreviations listed below.
3. Purposes served by the reservoirs using the notation defined below.
4. The year in which project construction was completed and/or impoundment began.
5. Height of the dam in feet above the stream bed.
6. Water surface area in acres at the top of conservation pool or normal operating level.
7. Connecting numerical identifier of reservoir in Appendix B.

The following abbreviations are used for reservoir owners.

IBWC—International Boundary and Water Commission

MWA—Municipal Water Authority
MWD—Municipal Water District
RA—River Authority
WCID—Water Control and Improvement District
WID—Water Improvement District
WSD—Water Supply District
USACE—US Army Corps of Engineers

The following abbreviations denote types of water use.

M—municipal and industrial use
I—industrial only without municipal
O—oil and gas and mining use
A—agricultural irrigation use
P—thermal-electric power water
H—hydroelectric power
R—recreation

Reservoir	Reservoir Owner	Project Purposes	First Year	Ht (ft)	Area (acres)	App B
	Rio Grande Basin					
San Esteban	William B. Blakemore	R	1911	68	760	156
Imperial	Pecos County WCID 2	A	1915	29	1,530	93
Balmorhea	Reeves County WCID 1	A	1917	46	510	11
Red Bluff	Red Bluff Water Power District	A,H	1936	102	11,700	151
Casa Blanca	Webb County	R	1949	76	1,660	36
Falcon	IBWC	F,M,A,H,R	1953	150	87,210	96
Amistad	IBWC	F,M,A,H,R	1968	254	64,900	95
	Canadian River Basin					
Rita Blanca	City of Dalhart	R	1941	75	520	20
Meredith	Canadian River MWD	F,M,R	1965	200	16,500	124
	Red River Basin					
Wichita	City of Wichita Falls	M,P,R	1901	23	2,200	192
Randall	City of Denison	M	1909	70	310	149

(Continued)

(Continued)

Reservoir	Reservoir Owner	Project Purposes	First Year	Ht (ft)	Area (acres)	App B
Kemp	Wichita Falls and Wichita County WID	M,P,F	1922	115	16,540	101
Crook	City of Paris	M	1923	38	1,230	53
Diversion	Wichita Falls and Wichita County WID	M	1924	55	3,420	59
Bivins	City of Amarillo	M	1926	48	380	18
Santa Rosa	W.T. Waggoner Estate	O	1929	41	1,500	157
Coffee Mill	US Forest Service	R	1938	27	650	47
Buffalo	US Fish and Wildlife Service	R	1938	37	1,900	30
Texoma	USACE Tulsa	F,H,M,R	1943	172	89,000	172
Kickapoo	City of Wichita Falls	M,P,R	1946	62	6,200	102
Baylor Creek	City of Childress	M,R	1949	66	610	15
Electra City	City of Electra	M	1950	40	660	64
Valley	Texas Power and Light	P	1960	55	1,080	183
Farmers Creek	North Montague County WSD	M,O	1961	77	1,470	67
NF Buffalo Creek	Wichita County WCID	M	1964	47	1,500	137
Hubert H. Moss	City of Gainesville	M	1966	93	1,130	129
Arrowhead	City of Wichita Falls	M	1966	62	16,200	7
Greenbelt	Greenbelt M&IWA	M	1966	110	1,990	82
Pat Mayse	USACE Tulsa	F,M,R	1967	96	5,990	122
Bonham	Bonham MWA	M	1969	70	1,020	21
Mackenzie	Mackenzie MWA	M	1974	174	900	119
Truscott	USACE Tulsa	Brine	1984	107	2,980	178
Bois d'Arc	North Texas MWD	M	2021	90	16,640	19
Sulphur River Basin						
River Crest	Texas Power and Light	P	1953	23	550	152
Wright Patman	USACE Fort Worth	F,M,R	1956	106	20,300	142
Sulphur Springs	Sulphur Springs WD	M	1973	44	1,340	168
Jim Chapman	USACE Fort Worth	F,M,R	1991	73	19,305	41
Cypress Creek Basin						
Ellison Creek	Lone Star Steel	P,I	1943	49	1,520	65
Lake O'the Pines	USACE Fort Worth	F,M,R	1957	97	1,060	107
Johnson Creek	Southwestern Electric	P	1961	60	650	100

(Continued)

(Continued)

Reservoir	Reservoir Owner	Project Purposes	First Year	Ht (ft)	Area (acres)	App B
Cypress Springs	Franklin County WD	M	1970	74	3,400	54
Caddo	USACE Fort Worth	M,R	1971	36	26,800	31
Monticello	Texas Utilities Company	P	1972	54	2,000	128
Welsh	Southwestern Electric	P,R	1975	60	1,360	188
Bob Sandlin	Titus County WSD	M,R	1978	69	9,460	155
Peacock Site 1A	Lone Star Steel	I	1983	121		143
Gilmer	City of Gilmer	M	2001	43	1,010	75
		Sabine River Basin				
Cherokee	Cherokee Water Company	M,R	1948	45	3,990	42
Gladewater	City of Gladewater	M,R	1952	48	800	76
Murvaul	Panola County WSD 1	M,R	1957	46	3,820	131
Tawakoni	Sabine River Authority	M,A,R	1960	85	36,700	170
Quitman	Wood County	F,R	1962	42	810	148
Holbrook	Wood County	F,R	1962	49	650	87
Winnsboro	Wood County	F,R	1962	45	810	193
Hawkins	Wood County	F,R	1962	58	780	85
Toledo Bend	Sabine River Authority	M,A,H,R	1966	112	181,600	174
Martin	Texas Utilities Service	P	1974	61	5,020	121
Lake Fork	Sabine River Authority	M	1980	82	27,690	70
Brandy Branch	Southwestern Electric	P	1983	80	960	23
		Neches River Basin				
Tyler	City of Tyler	M	1949	50	4,880	181
Steinhagen	USACE Fort Worth	F,M,R	1951	45	13,700	165
Striker Creek	Angelina–Nacogdoches WCID	M	1957	42	1,250	167
Jacksonville	City of Jacksonville	M,R	1957	72	1,320	98
Kurth	Southland Paper Mills	I	1961	37	770	105
Athens	Athens MWA	M	1962	67	1,799	8
Palestine	Upper Neches MWA	M,R	1962	75	25,560	140
Sam Rayburn	USACE Fort Worth	F,M,H,A	1965	120	114,500	150
Nacogdoches	City of Nacogdoches	M	1976	75	2,210	132
Pinkston	City of Center	M,R	1977	64	520	144
		Trinity River Basin				
White Rock	City of Dallas	R	1910	40	1,120	190
Anahuac	Chambers–Liberty Navigation District	A,O	1914	10	5,300	4

(Continued)

(Continued)

Reservoir	Reservoir Owner	Project Purposes	First Year	Ht (ft)	Area (acres)	App B
Worth	City of Fort Worth	M	1914	50	3,560	195
Halbert	City of Corsicana	M,R	1921	49	650	83
Trinidad	Texas Power and Light	P	1925	20	740	177
Eagle Mountain	Tarrant Regional Water District	M,A	1934	85	9,200	62
Mountain Creek	Dallas Power and Light	P	1937	47	2,710	130
Lost Creek	City of Jacksboro	M,A,R	1950	57	413	117
Grapevine	USACE Fort Worth	F,M,R	1952	137	7,380	81
Benbrook	USACE Fort Worth	F,M,R	1952	130	3,770	17
Lavon	USACE Fort Worth	F,M,R	1953	81	21,400	109
Lewisville	USACE Fort Worth	F,M,R	1954	125	23,280	112
New Terrell City	City of Terrell	M,R	1955	45	830	135
Amon G. Carter	City of Bowie	M	1956	71	20,050	35
Waxahachie	Ellis County WC&ID 1	M	1956	66	690	186
North	Dallas Power and Light	P	1957	65	800	136
Weatherford	City of Weatherford	M	1957	75	1,210	187
Arlington	City of Arlington	M	1957	83	2,270	6
Navarro Mills	USACE Fort Worth	F,M.R	1963	82	5,070	134
Bardwell	USACE Fort Worth	F,M,R	1965	82	3,570	12
Bridgeport	Tarrant Regional Water District	M	1965	91	33,750	26
Cedar Creek	Tarrant Regional Water District	M	1965	91	33,750	39
Houston County	Houston County WCID 1	M	1966	63	1,280	90
Ray Hubbard	City of Dallas	M	1968	68	22,740	92
Kiowa	Lake Kiowa Inc	R	1968	42	560	103
Livingston	Trinity River Authority and Houston	M,A,R	1968	100	82,600	114
Fairfield	Industrial Generating Service	P	1969	77	2,360	66
Forest Grove	Texas Utilities Service	P	1980	53	1,500	69
Joe Pool	USACE Fort Worth	F,M,R	1985	105	7,470	145
Ray Roberts	USACE Fort Worth	F,M,R	1987	141	29,350	154
Richland-Chambers	Tarrant Regional Water District	M	1987	120	38,850	153

(Continued)

(Continued)

Reservoir	Reservoir Owner	Project Purposes	First Year	Ht (ft)	Area (acres)	App B
	Trinity-San Jacinto Coastal Basin					
Cedar Bayou	Houston Power and Light	P	1972	10	2,600	37
	San Jacinto River Basin					
Sheldon	Texas Parks and Wildlife	R	1943	10	1,700	158
Barker	USACE Galveston	F	1945	365	0	13
Addicks	USACE Galveston	F	1948	49	0	2
Houston	City of Houston	M,A,R,O	1954	65	12,240	89
Lewis Creek	Gulf States Utilities	I	1969	54	1,010	111
Conroe	San Jacinto RA and Houston	M, O	1973	82	20,980	51
	San Jacinto-Brazos Coastal Basin					
Galveston County	Galveston County Water Authority	M,I	1949	14	810	72
	Brazos River Basin					
Mineral Wells	City of Mineral Wells	M	1920	74	650	127
Abilene	City of Abilene	M,R	1921	51	590	1
Cisco	City of Cisco	M	1923	96	440	44
Kirby	City of Abilene	M	1928	50	740	104
Sweetwater	City of Sweetwater	M	1930	50	630	169
Fort Phantom Hill	City of Abilene	M,K	1938	84	4,250	71
Possum Kingdom	Brazos River Authority	M,A,R,H,O	1941	189	14,440	146
William Harris	Dow Chemical Company	I	1947	12	1,660	84
Camp Creek	Camp Creek Water Company	M,R	1948	49	750	33
Daniel	City of Breckenridge	M	1948	60	520	55
Eagle Nest	T. L. Smith	A,R	1949	32	18,000	63
Whitney	USACE Fort Worth	F,M	1951	159	23,560	191
Lake Creek	Texas Power and Light	P	1952	50	560	106
Alcoa	Aluminum Company of America	I,R	1953	50	880	3
Stamford	City of Stamford	M	1953	78	4,690	164
Brazoria	Dow Chemical Company	I	1954	16	1,860	25
Leon	Eastland County WSD	M,I	1954	90	1,590	110
Belton	USACE Fort Worth	F,M,A,R	1954	192	12,300	16
Smithers	Houston Lighting and Power	P	1957	18	2,480	159

(Continued)

(Continued)

Reservoir	Reservoir Owner	Project Purposes	First Year	Ht (ft)	Area (acres)	App B
Graham	City of Graham	M	1958	82	2,550	78
Davis	League Ranch	A	1959	32	580	56
Mexia	Bristone MWSD	M	1961	50	1,200	125
Hubbard	West Central MWD	M,I,O	1962	112	16,250	91
Proctor	USACE Fort Worth	F,M,A,R	1963	86	4,610	147
White River	White River MWD	M,O	1963	84	1,810	189
Palo Pinto	Palo Pinto MWD 1	M	1964	96	2,660	141
Pat Cleburne	City of Cleburne	M	1964	78	1,560	45
Waco	USACE Fort Worth	F,M,R	1965	140	7,270	185
Somerville	USACE Fort Worth	F,M,A,R	1967	80	11,460	160
Stillhouse Hollow	USACE Fort Worth	F,M,A,R	1968	200	6,430	166
Tradinghouse	Texas Power and Light	P	1968	60	2,010	175
Granbury	Brazos River Authority	M,A,P	1969	84	1,300	79
Millers Creek	North Central Texas MWA	M	1974	75	1,900	126
Bryan Utilities	City of Bryan	P,R	1975	62	830	28
Squaw Creek	Texas Utilities Services	P	1977	159	3,230	163
Limestone	Brazos River Authority	M,A	1978	65	14,200	113
Granger	USACE Fort Worth	F,M,R	1980	115	4,400	80
Georgetown	USACE Fort Worth	F,M,R	1980	162	1,310	73
Gibbons Creek	Texas Municipal Power Agency	P	1981	50	2,490	74
Twin Oaks	Texas Power and Light	P	1982	56	1,460	180
Aquilla	USACE Fort Worth	F,M,R	1983	104	3,280	5
Alan Henry	City of Lubbock	M,A,R	1993	138	2,800	86
Colorado River Basin						
Eagle	Lakeside Irrigation Company	A	1900	6	1,200	61
Nasworthy	City of San Angelo	M,A,	1930	50	1,600	133
Brownwood	Brown County WID 1	M,A	1933	120	7,300	27
Buchanan	Lower Colorado RA	M,H,O	1937	146	23,060	29
Inks	Lower Colorado RA	M,A,H,O	1938	97	800	94
Austin	City of Austin	M,H	1939	85	1,830	9
Travis	Lower Colorado RA	F,M,A,H,O	1940	266	18,930	176
Ballinger	City of Ballinger	M	1947	76	500	10
Hords Creek	USACE Fort Worth	F,M,R	1948	91	500	88

(Continued)

(Continued)

Reservoir	Reservoir Owner	Project Purposes	First Year	Ht (ft)	Area (acres)	App B
Colorado City	Texas Electric Service	M,P	1949	85	1,610	50
Marble Falls	Lower Colorado RA	M	1951	99	780	120
J. B. Thomas	Colorado River WMD	M,K	1952	105	7,820	173
O. C. Fisher	USACE Fort Worth	F,M,A,R	1952	128	5,440	68
Oak Creek	City of Sweetwater	M	1953	95	2,380	138
Champion	Texas Electric Service	M	1959	114	1,560	40
Lady Bird	City of Austin	M,R,I	1960	64	465	108
L. B. Johnson	Lower Colorado RA	H	1961	118	6,380	99
Twin Buttes	City of San Angelo	F,M,A,R	1962	134	9,080	179
Winters	City of Winters	M	1963	65	640	194
Brady Creek	City of Brady	M	1963	104	2,020	22
Bastrop	Lower Colorado RA	I	1964	85	910	14
Coleman	City of Coleman	M	1966	90	2,000	48
Walter E. Long	City of Austin	M,R	1967	83	1,270	116
E. V. Spence	Colorado River MWD	M,O	1968	140	14,590	162
Clyde	City of Clyde	M	1969	63	450	46
Cedar Creek	Lower Colorado RA	P	1977	106	2,420	38
South Texas Project	Houston Lighting and Power	P	1981	51	7,000	161
O. H. Ivie	Colorado MWD	M	1990	148	19,150	97
Lavaca River Basin						
Texana	Lavaca–Navidad River Authority	M,A,R	1981	63	11,000	171
Guadalupe River Basin						
McQueeney	Gaudalupe–Blanco RA	H	1928	40	400	118
Dunlap	Guadalupe–Blanco RA	P	1928	41	410	60
Gonzales	Guadalupe–Blanco RA	H	1931	42	700	77
Canyon	USACE Fort Worth	F,M,R	1964	224	8,240	34
Coleto Creek	America Electric	P	1980	55	3,100	49
San Antonio River Basin						
Medina	Bexar–Medina–Atascosa Counties WID	A	1913	164	5,575	123
Olmos	City of San Antonio	F	1926	60	0-890	139
Victor Braunig	City of San Antonio	P	1962	80	1,350	24
Calaveras	City of San Antonio	P	1969	70	3,450	32

(Continued)

(Continued)

Reservoir	Reservoir Owner	Project Purposes	First Year	Ht (ft)	Area (acres)	App B
Nueces River Basin						
Upper Nueces	Zavala and Dimmit Counties WID 1	A	1948	60	320	182
Corpus Christi	Lower Nueces River WSD	M,A,R,O	1958	75	18,700	52
Choke Canyon	City of Corpus Christi	M,R	1982	128	26,000	43
Nueces-Rio Grande Coastal Basin						
Delta	Hidalgo–Willacy Counties WCID 1	A	1939	17	2,370	58
Valley Acres	Valley Acres WD	M,A	1947	15	910	184
Loma Alta	Brownsville Navigation District	M	1963	18	2,490	115
Barney M. Davis	Central Power and Light	P	1973	20	1,100	57

Appendix B

Information for Major Reservoirs

The 195 reservoirs located entirely or partially in Texas with conservation and/or flood control pools of at least 5,000 ac-ft are included in the two appendixes. This includes 192 reservoirs that have conservation storage capacities of 5000 ac-ft or larger, of which 33 also have flood control pools of at least 5,000 ac-ft, and Addicks, Barker, and Olmos Reservoirs that have flood control capacities greater than 5,000 ac-ft but no conservation storage capacity. The reservoirs are numbered in Appendix B. The last column in the table of Appendix A references the numbers. The reservoirs are listed in alphabetical order with the following information:

- Name of reservoir followed by name of dam which is often but not always the same.
- Former or other names for the reservoir and dam.
- River basin and stream location of the dam.
- County or counties in which the reservoir is located.
- Reservoir owner or owners and for USACE reservoirs nonfederal water supply sponsors.
- Primary purposes served by the reservoir.
- Date that construction was completed and/or impoundment of water began. Date of enlargement is also added if the reservoir has been enlarged after completion of the original project.
- Conservation and flood control storage capacity in acre-feet. Unless specified as flood control capacity, the storage volume in acre-feet is the conservation storage capacity for water supply,

hydropower, recreation, and other purposes. Storage capacities are approximate due mainly to sedimentation. The storage volumes obtained from various agency sources of information may or may not reflect measured or computed adjustments for sedimentation or the latest updated sediment survey data.

1. **Lake Abilene and Abilene Dam:** Brazos River Basin, Elm Creek; Taylor County; City of Abilene; municipal, industrial, recreation; impoundment began August 1921; 7,900 ac-ft.
2. **Addicks Reservoir and Addicks Dam:** San Jacinto River Basin, South Mayde Creek; Harris County; USACE Galveston District; flood control; completed 1948; 199,614 ac-ft flood control.
3. **Alcoa Lake and Alcoa Dam:** Also called Sandow Lake and Sandow Dam; Brazos River Basin, Sandy Creek; Milam County; Aluminum Company of America; industrial and recreation; impoundment began January 1953; 14,750 ac-ft.
4. **Anahuac Lake and Anahuac Dam:** Also called Turtle Bayou Reservoir; Trinity River Basin, Trinity River; Chambers County; Chambers-Liberty Counties Navigation District; industrial, irrigation and mining including oil production; impoundment began 1914, enlargement 1954; 35,300 ac-ft.
5. **Aquilla Lake and Aquilla Dam:** Brazos River Basin, Aquilla Creek; Hill County; operated by USACE Fort Worth District, Brazos River Authority has contracted for conservation storage; municipal, industrial, irrigation, recreation; impoundment began April 1983; 43,279 ac-ft conservation capacity; 102,721 flood control capacity.
6. **Lake Arlington and Arlington Dam:** Trinity River Basin, Village Creek; Tarrant County; City of Arlington; municipal, industrial; impoundment March 1957; 40,188 ac-ft.
7. **Lake Arrowhead and Lake Arrowhead Dam:** Red River Basin, Little Wichita River; Archer and Clay Counties; City of Wichita Falls; municipal; impoundment October 1966; 230,359 ac-ft.
8. **Lake Athens and Lake Athens Dam:** Formerly, Flat Creek Reservoir and Flat Creek Dam; Neches River Basin, Flat Creek; Henderson County; Athens Municipal Water Authority; municipal, recreation; impoundment began November 1962; 29,475 ac-ft.

9. **Lake Austin and Tom Miller Dam:** Colorado River Basin, Colorado River; Travis County; owned by City of Austin, operated by the Lower Colorado River Authority; municipal, industrial, hydropower; impoundment began 1939; 24,644 ac-ft.
10. **Ballinger City Lake and Dam:** Colorado River Basin, Valley Creek; Runnels County; City of Ballinger; water supply and recreation; completed 1947 and enlarged 1984; 8,215 ac-ft.
11. **Lake Balmorhea and Balmorhea Dam:** Rio Grande Basin, Sandia Creek; Reeves County; Reeves County Water Improvement District Number 1; irrigation; impoundment began September 1917; 6,350 ac-ft.
12. **Bardwell Lake and Bardwell Dam:** Trinity River Basin, Waxahachie Creek; Ellis County; owned and operated by USACE Fort Worth District, Trinity River Authority has contracted for conservation storage; flood control, municipal, industrial, recreation; impoundment began November 1965; 54,900 ac-ft conservation capacity; 85,100 ac-ft flood control capacity.
13. **Barker Reservoir and Barker Dam:** San Jacinto River Basin, Buffalo Bayou; Fort Bend and Harris Counties; USACE Galveston District; flood control; completed 1945; 209,600 ac-ft flood control.
14. **Lake Bastrop and Bastrop Dam:** Colorado River Basin, Spicer Creek; Bastrop County; Lower Colorado River Authority; industrial; impoundment began April 1964; 16,500 ac-ft.
15. **Baylor Creek Reservoir and Baylor Creek Dam:** Red River Basin, Baylor Creek; Childress County; City of Childress; municipal and recreation; impoundment began December 1949; 9,200 ac-ft.
16. **Belton Lake and Belton Dam:** Brazos River Basin, Leon River; Bell and Coryell Counties; USACE Fort Worth District, Brazos River Authority and Fort Hood have contracted for water supply storage; flood control, municipal, industrial, irrigation, recreation; impoundment began March 1954; 435,225 ac-ft conservation capacity; 662,375 ac-ft flood control capacity.
17. **Benbrook Lake and Benbrook Dam:** Trinity River Basin, Clear Fork Trinity River; Tarrant County; owned and operated by USACE Fort Worth District, City of Fort Worth and Benbrook Water and Sewage Authority have contracted for conservation storage; flood control, municipal, industrial,

recreation; impoundment September 1952; 85,648 ac-ft conservation capacity; 172,982 ac-ft flood control capacity.

18. **Bivins Lake and Bivins Dam:** Also known as Amarillo City Lake, Red River Basin, Palo Duro Creek; Randall County; City of Amarillo; municipal; impoundment began 1926; 5,120 ac-ft.
19. **Bois d´Arc Lake and Bois d´Arc Dam:** Red River Basin, Bois d´Arc Creek; Fannin County; North Texas Municipal Water District; municipal; impoundment April 2021; 367,609 ac-ft.
20. **Rita Blanca Lake and Rita Blanca Dam:** Canadian River Basin, Rita Blanca Creek, Hartley County; City of Dalhart, recreation; impoundment began September 1941; 12,100 ac-ft.
21. **Lake Bonham and Timber Creek Dam:** Red River Basin, Timber Creek; Fannin County; Bonham Municipal Water Authority; municipal; impoundment November 1969; 12,000 ac-ft.
22. **Brady Creek Reservoir and Dam:** Colorado River Basin, Brady Creek; McCulloch County; City of Brady; municipal; impoundment began May 1963; 30,430 ac-ft.
23. **Brandy Branch Cooling Pond and Dam:** Sabine River Basin, Brandy Branch; Harrison County; Southwestern Electric Power Company; steam-electric power; impoundment 1983; 29,500 ac-ft.
24. **Victor Braunig Lake and Victor Braunig Plant Dam:** Also called East Lake; San Antonio River Basin, Arroyo Seco; Bexar County; City Public Service Board of San Antonio; steam-electric power; impoundment began December 1962; 26,500 ac-ft.
25. **Brazoria Reservoir and Brazoria Dam:** Brazos River Basin, off-channel near Brazos River; Brazoria County; Dow Chemical Company; industrial; impoundment April 1954; 21,970 ac-ft.
26. **Bridgeport Reservoir and Bridgeport Dam:** Trinity River Basin, West Fork Trinity River; Wise County; Tarrant Regional Water District; municipal, industrial, recreation; impoundment began April 1932; 361,875 ac-ft.
27. **Brownwood Reservoir and Brownwood Dam:** Colorado River Basin, Pecan Bayou; Brown County; Brown County Water Improvement District No. 1; municipal, industrial, irrigation; impoundment began July 1933; 143,400 ac-ft.
28. **Bryan Utilities Lake and Bryan Utilities Lake Dam:** Brazos River Basin, unnamed creek; Brazos County; City of Bryan; steam-electric power and recreation; impoundment began 1975; 15,230 ac-ft.

29. **Lake Buchanan and Buchanan Dam:** Colorado River Basin, Colorado River; Burnet County; Lower Colorado River Authority; municipal, industrial, mining, hydropower; impoundment May 1937; 886,626 ac-ft.
30. **Buffalo Lake and Umbarger Dam:** Red River Basin, Tierra Blanca Creek; Randall County; US Fish and Wildlife Service; recreation; impoundment began June 1938; 18,150 ac-ft.
31. **Caddo Lake and Caddo Dam:** Cypress Creek Basin, Cypress Bayou; Harrison and Marion Counties; maintained by USACE Fort Worth District; municipal, industrial and recreation; natural lake, dam constructed in 1914 and reconstructed in 1971; 129,000 ac-ft.
32. **Calaveras Lake and Calaveras Creek Dam:** San Antonio River Basin, Calaveras Creek; Bexar County; City of San Antonio; steam-electric power; impoundment began January 1969; 61,800 ac-ft.
33. **Camp Creek Lake and Camp Creek Dam:** Brazos River Basin, Camp Creek; Robertson County; Camp Creek Water Company; municipal and recreation; impoundment November 1948; 8,500 ac-ft.
34. **Canyon Reservoir and Canyon Dam:** Guadalupe River Basin, Guadalupe River; Comal County; constructed and operated by USACE Fort Worth District, Guadalupe-Blanco River Authority has contracted for conservation storage; flood control, municipal, industrial, recreation; impoundment June 1965; 378,852 ac-ft conservation; 362,048 ac-ft flood control.
35. **Lake Amon G. Carter and Amon G. Carter Dam:** Also called Bowie Lake and Bowie Lake Dam; Trinity River Basin, Big Sandy Creek; Montague County; City of Bowie; municipal, industrial; impoundment began May 1956; 29,000 ac-ft.
36. **Casa Blanca Lake and Country Club Dam:** Rio Grande Basin, Chacon Creek; Webb County; recreation; impoundment began 1949; 20,000 ac-ft.
37. **Cedar Bayou Cooling Reservoir:** Trinity-San Jacinto Coastal Basin, Cedar Bayou; Chambers County; Houston Lighting and Power Company; steam-electric power; 20,000 ac-ft.
38. **Cedar Creek Reservoir and Cedar Creek Dam:** Colorado River Basin, Cedar Creek; Fayette County; Lower Colorado River Authority; steam-electric power; impoundment 1977; 71,400 ac-ft.

39. **Cedar Creek Reservoir and Joe B. Hoggsett Dam:** Also called Joe B. Hoggsett Lake; Trinity River Basin, Cedar Creek; Henderson and Kaufman Counties; Tarrant Regional Water District; industrial; impoundment began July 1965; 644,785 ac-ft.
40. **Champion Creek Reservoir and Champion Creek Dam:** Colorado River Basin, Champion Creek; Mitchell County; Texas Electric Service Company; municipal, steam-electric power; impoundment began February 1959; 41,600 ac-ft.
41. **Jim Chapman Lake and Dam:** Also known as Cooper Lake and Dam; Sulphur River Basin, South Sulphur River; Delta and Hopkins Counties; USACE Fort Worth District, water supply for Sulphur Springs, Commerce, North Texas Municipal Water District, and Sulphur River Municipal Water District; completed 1991; 298,930 ac-ft conservation and 130,405 ac-ft flood control capacity.
42. **Lake Cherokee and Cherokee Dam:** Sabine River Basin, Cherokee Bayou; Gregg and Rusk Counties; Cherokee Water Company; municipal, industrial, and recreation; impoundment began October 1948; 46,700 ac-ft.
43. **Choke Canyon Reservoir and Choke Canyon Dam:** Nueces River Basin, Frio River; Live Oak and McMullen Counties; Constructed by Bureau of Reclamation; operated and maintained by City of Corpus Christi; municipal, industrial, recreation; impoundment 1982; 662,821 ac-ft.
44. **Lake Cisco and Williamson Dam:** Brazos River Basin, Sandy Creek; Eastland County; City of Cisco; municipal; impoundment began 1923; 8,800 ac-ft capacity.
45. **Lake Pat Cleburne and Cleburne Dam:** Brazos River Basin, Nolan River; Johnson County; City of Cleburne; municipal; impoundment began August 1964; 25,300 ac-ft.
46. **Lake Clyde and Upper Pecan Bayou Dam Site 7:** Colorado River Basin, North Prong Pecan Bayou; Callahan County; City of Clyde; municipal; impoundment November 1969; 5,748 ac-ft.
47. **Coffee Mill Lake and Coffee Mill Dam:** Red River Basin, Coffee Mill Creek; Fannin County; US Forest Service; recreation; impoundment began 1938; 8,000 ac-ft.
48. **Coleman Lake and Dam:** Colorado River Basin, Jim Ned Creek; Coleman County; City of Coleman; municipal; impoundment April 1966; 38,100 ac-ft.

49. **Coleto Creek Reservoir and Dam:** Guadalupe River Basin, Coleto Creek; Victoria County; Guadalupe-Blanco River Authority; steam-electric power; construction completed 1980; 35,080 ac-ft.
50. **Colorado City Lake and Colorado City Dam:** Colorado River Basin, Morgan Creek; Mitchell County; Texas Electric Service Company; municipal, industrial and hydropower; impoundment began April 1949; 30,800 ac-ft.
51. **Lake Conroe and Conroe Dam:** Also called Honea Reservoir and Honea Dam; San Jacinto River Basin, West Fort of San Jacinto River; San Jacinto River Authority and City of Houston; municipal, industrial and mining including oil production; impoundment January 1973; 429,890 ac-ft.
52. **Lake Corpus Christi and Wesley E. Seale Dam:** Also known as Lake Mathis; Nueces River Basin, Nueces River; San Patricio, Live Oak, and Jim Wells Counties; Lower Nueces River Water Supply District; municipal and recreation; impoundment began April 1958; 254,732 ac-ft. Lake Corpus Christi inundates the 10,000 ac-ft Lake Lovenskiod and La Fruta Dam constructed during 1927 to 1929.
53. **Lake Crook and Crook Dam:** Red River Basin, Pine Creek, Lamar County; City of Paris; municipal; impoundment began 1923; 9,964 ac-ft.
54. **Lake Cypress Springs and Franklin County Dam:** Formerly, Franklin County Lake; Cypress Creek Basin, Big Cypress Creek; Franklin County; Franklin County Water District; municipal and industrial; impoundment July 1970; 72,800 ac-ft.
55. **Lake Daniel and Gonzales Creek Dam:** Brazos River Basin, Gonzales Creek; Stephens County; City of Breckenridge; municipal and industrial; impoundment 1948; 9,515 ac-ft.
56. **Davis Lake and Davis Dam:** Brazos River Basin, Double Dutchman Creek; Knox County; League Ranch; irrigation; impoundment began 1959; 5,395 ac-ft.
57. **Barney M. Davis Cooling Reservoir:** Nueces-Rio Grande Coastal Basin, off-channel of Laguna Madre; Nueces County; Central Power and Light; steam-electric power; impoundment began 1973; 6,600 ac-ft.
58. **Delta Lake and Delta Dam (Reservoir Unit 1 and Unit 2):** Formerly, Monte Alto Reservoir and Monte Alto Dam; Nueces-Rio Grande Coastal Basin, off-channel from the Rio Grande;

Hidalgo County; Hildalgo-Willacy Counties Water Control and Improvement District No. 1; irrigation; impoundment began 1939; 25,000 ac-ft.

59. **Lake Diversion and Lake Diversion Dam:** Red River Basin, Wichita River; Acher and Baylor Counties; City of Wichita Falls and Wichita County Water Improvement District No. 2; municipal and industrial; impoundment began 1924; 40,000 ac-ft.
60. **Lake Dunlap and TP-1 Dam:** Guadalupe River Basin, Guadalupe River; Guadalupe County; Guadalupe-Blanco River Authority; hydropower; impoundment began 1928; 5,900 ac-ft.
61. **Eagle Lake and Eagle Lake Dam:** Colorado River Basin, off-channel from Colorado River; Colorado County; Lakeside Irrigation Company; irrigation; impoundment 1900; 9,600 ac-ft.
62. **Eagle Mountain Reservoir and Eagle Mountain Dam:** Trinity River Basin, West Fort Trinity River; Tarrant County; Tarrant Regional Water District; municipal, industrial, irrigation; impoundment began February 1934; 179,880 ac-ft.
63. **Eagle Nest Lake:** Brazos River Basin, off-channel reservoir near Brazos River; Brazoria County; T. L. Smith; irrigation; completed 1949; 18,000 ac-ft.
64. **Electra City Lake and Electra City Dam:** Red River Basin; Beaver and Camp Creeks; Wilbarger County; City of Electra; municipal and industrial; impoundment 1950; 8,055 ac-ft.
65. **Ellison Creek Reservoir and Ellison Creek Dam:** Also called Lone Star Dam; Cypress Creek Basin, Ellison Creek; Ellison County; Lone Star Steel Company; steam-electric power and industrial; impoundment began January 1943; 24,700 ac-ft.
66. **Fairfield Lake and Fairfield Dam:** Formerly, Big Brown Creek Reservoir and Big Brown Creek Dam; Trinity River Basin, Big Brown Creek; Freestone County; Industrial Generating Company; steam-electric power; impoundment began December 1969; 49,750 ac-ft.
67. **Farmers Creek Reservoir and Farmers Creek Dam:** Also known as Lake Nocona; Red River Basin, Farmers Creek; Montague County; North Montague County Water Supply District; municipal, industrial, mining; impoundment began Spring 1961; 25,400 ac-ft.
68. **O. C. Fisher Lake and Dam:** Formerly, San Angelo Reservoir and Dam; Colorado River Basin, Concho River; Tom Green County; constructed and operated by USACE Fort Worth

District, Upper Colorado River Authority has contracted for conservation storage; flood control, municipal, industrial, irrigation and recreation; impoundment began February 1952; 115,743 ac-ft conservation; 276,944 ac-ft flood control.

69. **Forest Grove Reservoir and Forest Grove Dam:** Trinity River Basin, Caney Creek; Henderson County; Texas Utilities Service Company; steam-electric power; impoundment 1980; 20,040 ac-ft.
70. **Lake Fork Reservoir and Lake Fork Dam:** Sabine River Basin, Lake Fork Creek; Rains and Wood Counties; Sabine River Authority; municipal and industrial; impoundment 1980; 636,504 ac-ft.
71. **Fort Phantom Hill Reservoir and Fort Phantom Hill Dam:** Brazos River Basin, Elm Creek; Jones County; City of Abilene; municipal and recreation; impoundment October 1938; 74,310 ac-ft.
72. **Galveston County Industrial Water Reservoir:** San Jacinto-Brazos Basin, off-channel of Dickinson Bayou; Galveston County; Galveston County Water Authority; impoundment began 1949; municipal and industrial; 7,308 ac-ft.
73. **Georgetown Lake and Georgetown Dam:** Formerly, North Fork Lake and North Fork Dam; Brazos River Basin, San Gabriel River; Williamson County; owned and operated by USACE Fort Worth District, Brazos River Authority has contracted for conservation storage; flood control, municipal, industrial, and recreation; impoundment began January 1980; 29,200 ac-ft conservation capacity; 87,600 ac-ft flood control.
74. **Gibbons Creek Reservoir and Gibbons Creek Dam:** Brazos River Basin, Gibbons Creek; Grimes County; Texas Municipal Power Agency; steam-electric power; impoundment 1981; 26,824 ac-ft.
75. **Lake Gilmer and Gilmer Dam:** Red River Basin, Kelsey Creek tributary of Big Cypress River; Upshur County; City of Gilmer; municipal; impoundment began 2001; 12,700 ac-ft.
76. **Lake Gladewater and Gladewater Dam:** Sabine River Basin, Glade Creek; Upshur County; City of Gladewater, municipal and recreation, impoundment began August 1952; 6,950 ac-ft.
77. **Lake Gonzales and H-4 Dam:** Also known as H-4 Reservoir; Guadalupe River Basin, Guadalupe River; Gonzales County; Guadalupe-Blanco River Authority; hydropower; impoundment began 1931; 5,200 ac-ft.

78. **Lake Graham and Eddleman and Graham Dam**: Also called Eddleman and Salt Creek Lakes; Brazos River Basin, Flint and Salt Creek; Young County; City of Graham; municipal and industrial; impoundment began 1929 (Eddleman Dam) and 1958 (Graham Dam); 45,000 ac-ft.
79. **Lake Granbury and De Cordova Bend Dam:** Also called De Cordova Bend Reservoir; Brazos River Basin, Brazos River; Hood and Parker Counties; Brazos River Authority; municipal, industrial, irrigation, power; impoundment began September 1970; 153,000 ac-ft.
80. **Granger Lake and Granger Dam:** Formerly, Laneport Lake and Laneport Dam; Brazos River Basin; San Gabriel River; Williamson County; owned and operated by USACE Fort Worth District; Brazos River Authority has contracted for conservation storage; flood control, municipal, and industrial; impoundment January 1980; 37,900 ac-ft conservation capacity; 162,200 ac-ft flood storage capacity.
81. **Grapevine Lake and Grapevine Dam:** Trinity River Basin, Denton Creek; Tarrant County; owned and operated by USACE Fort Worth District, Cities of Grapevine and Dallas have contracted for conservation storage; municipal, industrial; impoundment July 1952; 163,064 ac-ft conservation capacity; 245,136 ac-ft flood control capacity.
82. **Greenbelt and Greenbelt Dam:** Red River Basin, Salt Fork Red River, Donley County; Greenbelt Municipal and Industrial Water Authority; municipal and industrial; impoundment began December 1966; 58,200 ac-ft.
83. **Lake Halbert and Halbert Dam:** Trinity River Basin, Elm Creek; Navarro County; City of Corsicana; municipal, industrial, and recreation; impoundment began 1921; 7,420 ac-ft capacity.
84. **William Harris Reservoir and William Harris Dam:** Brazos River Basin, off-channel between Brazos River and Oyster Creek; Brazoria County; Dow Chemical Company; industrial; impoundment began 1947; 12,000 ac-ft.
85. **Lake Hawkins and Wood County Dam No. 3:** Sabine River Basin, Little Sandy Creek; Wood County; owned by Wood County; recreation and flood control; impoundment August 1962; 11,570 ac-ft.
86. **Lake Alan Henry and John T. Montford Dam:** Brazos River Basin, South Fork of Double Mountain Fork of Brazos River;

Garza and Kent Counties; City of Lubbock; municipal, irrigation, recreation; impoundment began 1993; 96,210 ac-ft.

87. **Lake Holbrook and Wood County Dam No. 2:** Sabine River Basin, Keys Creek; Wood County; recreation and flood control; impoundment began September 1962; 7,700 ac-ft.
88. **Hords Creek Reservoir and Hords Creek Dam:** Colorado River Basin, Hords Creek; Coleman County; operated by USACE Fort Worth District, Central Colorado River Authority has contracted for conservation storage; flood control, municipal, recreation; construction completed June 1948; 7,959 ac-ft conservation capacity; 16,678 ac-ft flood control capacity.
89. **Lake Houston and Lake Houston Dam:** San Jacinto River Basin, San Jacinto River; Harris County, City of Houston; municipal, industrial, irrigation, recreation, and mining, including oil production; impoundment began April 1954; 140,520 ac-ft.
90. **Houston County Lake and Houston County Dam:** Trinity River Basin, Little Elkhart Creek; Houston County; Houston County Water Control and Improvement District No. 1; municipal and industrial; impoundment began November 1966; 19,500 ac-ft.
91. **Hubbard Creek Reservoir and Hubbard Creek Dam:** Brazos River Basin, Hubbard Creek; Shackleford and Stephens Counties; West Central Municipal Water District; municipal, industrial and mining including oil production; impoundment December 1962; 314,280 ac-ft.
92. **Lake Ray Hubbard and Rockwall Forney Dam:** Formerly, Forney Reservoir; Trinity River Basin, East Fork Trinity River; Dallas and Kaufman Counties; City of Dallas; municipal; impoundment began December 1968; 439,559 ac-ft.
93. **Imperial Reservoir and Imperial Dam:** Rio Grande Basin, Pecos River; Pecos and Reeves Counties; Pecos County Water Control and Improvement District No. 2; irrigation; impoundment began 1915; 6,000 ac-ft.
94. **Inks Lake and Roy Inks Dam:** Colorado River Basin, Colorado River; Burnet County; Lower Colorado River Authority; municipal, irrigation, mining and hydropower; impoundment began 1938; 17,540 ac-ft.
95. **International Amistad Reservoir and Dam:** Also called Diablo Reservoir and Diablo Dam; Rio Grande Basin, Rio Grande; Val Verde County; International Boundary and Water Commission; conservation, recreation, irrigation, hydropower and flood

control; impoundment May 1968; 3,275,530 ac-ft conservation; 1,734,400 ac-ft flood control.

96. **International Falcon Reservoir and Dam:** Rio Grande Basin, Rio Grande; Starr County, Texas and Estado de Tamaulipas, Mexico, International Boundary and Water Commission; municipal, industrial, irrigation, flood control, hydropower; impoundment August 1953; 2,264,810 ac-ft conservation; 501,330 flood control.
97. **O. H. Ivie Lake and S. W. Freese Dam:** Formerly called Stacey Lake and Dam; Colorado River Basin, Colorado River; Colorado River Municipal Water District; impoundment 1990; 554,340 ac-ft.
98. **Lake Jacksonville and Buckner Dam:** Neches River Basin, Gum Creek; Cherokee County; City of Jacksonville; municipal and recreation; impoundment began June 1957; 30,500 ac-ft.
99. **Lake Lyndon B. Johnson and Alvin Wirtz Dam:** Formerly, Granite Shoals Lake; Colorado River Basin, Colorado River; Burnet and Llano Counties; Lower Colorado River Authority; hydroelectric power; impoundment began May 1951; 138,500 ac-ft.
100. **Johnson Creek Reservoir and Johnson Creek Dam:** Cypress River Basin, Johnson Creek; Marion County; Southwestern Power Company; cooling water for Wilkes Power Plant; 10,100 ac-ft.
101. **Lake Kemp and Lake Kemp Dam:** Red River Basin, Wichita River; Baylor County; City of Wichita Falls and Wichita County Water Improvement District No. 2; municipal, steam-electric power, irrigation; impoundment October 1922, reconstruction by USACE Tulsa District; impoundment 1974; 245,430 ac-ft conservation; 322,470 ac-ft flood control capacity.
102. **Lake Kickapoo and Lake Kickapoo Dam:** Red River Basin, North Fork Little Wichita River; Archer County; City of Wichita Falls; municipal; impoundment February 1946; 106,000 ac-ft.
103. **Lake Kiowa and Kiowa Dam:** Trinity River Basin, Indian Creek; Cooke County; Lake Kiowa, Inc.; recreation; impoundment began March 1970; 7,000 ac-ft.
104. **Kirby Lake and Kirby Dam:** Brazos River Basin, Cedar Creek; Taylor County; City of Abilene; municipal; impoundment began 1928; 7,620 ac-ft.
105. **Lake Kurth and Kurth Dam:** Also called Southland Paper Mills Reservoir; Neches River Basin, off-channel near Angelina River; Angelina County; Southland Paper Mills; industrial; impoundment began September 1961; 16,200 ac-ft.

106. **Lake Creek Lake and Lake Creek Dam:** Brazos River Basin, Manos Creek; McLennan County; Texas Power and Light Company; steam-electric power; impoundment began June 1952; 8,400 ac-ft.
107. **Lake O' the Pines and Ferrells Bridge Dam:** Also called Ferrells Bridge Dam and Reservoir, Cypress Creek Basin, Cypress Creek; Camp, Harrison, Marion, Morris, and Upshur Counties, operated by USACE Fort Worth District, Northeast Texas MWD has contracted for conservation storage; municipal, industrial, recreation, and flood control; impoundment August 1957; 241,363 ac-ft conservation capacity; 600,737 ac-ft flood control capacity.
108. **Lady Bird Lake and Longhorn Dam**; Formerly, Town Lake; Colorado River Basin, Colorado River; Travis County; City of Austin and Lower Colorado River Authority; water supply, flood control, wildlife protection, recreation; impoundment 1960; 7,150 ac-ft.
109. **Lavon Lake and Lavon Dam:** Trinity River Basin, East Fork Trinity River; Collin County; owned and operated by USACE Fort Worth District, North Texas MWD has contracted for conservation storage; flood control, municipal, industrial; impoundment began September 1953, project enlarged 1979; 409,300 ac-ft conservation; 338,900 ac-ft flood control.
110. **Leon Lake and Dam:** Brazos River Basin, Leon River; Eastland County; Eastland County Water Supply District; municipal and industrial; impoundment April 1954; 28,040 ac-ft.
111. **Lewis Creek Reservoir and Lewis Creek Dam:** San Jacinto River Basin, Lewis Creek; Montgomery County; Gulf States Utilities Company; steam-electric power; impoundment began February 1969; 16,400 ac-ft.
112. **Lewisville Lake and Lewisville Dam:** Also called Lake Dallas and Lake Dallas Dam and Garza-Little Elm Lake and Garza-Little Elm Dam; Trinity River Basin, Elm Fork Trinity River; Denton County; owned and operated by USACE Fort Worth District, Cities of Dallas and Denton contracted for conservation storage; flood control, municipal, industrial, recreation; impoundment November 1954; 598,902 ac-ft conservation; 382,861 ac-ft flood control.
113. **Limestone Lake and Limestone Dam:** Brazos River Basin, Navasota River; Leon, Limestone, and Robertson Counties;

Brazos River Authority; municipal, industrial, irrigation; impoundment began 1978; 225,400 ac-ft.

114. **Lake Livingston and Livingston Dam:** Trinity River Basin, Trinity River; Polk, San Jacinto, Trinity, and Walker Counties; City of Houston and Trinity River Authority; municipal, industrial and irrigation; impoundment began October 1968; 1,742,000 ac-ft.
115. **Loma Alta Reservoir and Loma Alta Dam:** Nueces-Rio Grande Coastal Basin, off-channel from the Rio Grande; Cameron County; Brownsville Navigation District; municipal, industrial; impoundment began 1963; 26,500 ac-ft.
116. **Walter E. Long Lake and Dam**: Formerly, Decker Lake; Colorado River Basin, Decker Creek; Travis County; City of Austin; municipal, industrial, recreation; impoundment began January 1967; 33,940 ac-ft.
117. **Lost Creek Reservoir and Lost Creek Dam**: Also known as Lake Jacksboro and Jacksboro Dam; Trinity River Basin, Little Cleveland Creek; Jack County; City of Jacksboro; municipal, irrigation, recreation; completed in 1950 and enlarged in 1990; 11,950 ac-ft.
118. **Lake McQueeney and Abbott Dam:** Guadalupe River Basin, Guadalupe River; Guadalupe County; Guadalupe-Blanco River Authority; hydropower; impoundment 1928; 5,000 ac-ft.
119. **Mackenzie Reservoir and Mackenzie Dam:** Red River Basin, Tele Creek; Swisher and Brisco Counties; Mackenzie Municipal Water Authority; municipal, impoundment April 1974; 46,250 ac-ft.
120. **Marble Falls Lake and Max Starcke Dam:** Also called Max Starcke Lake; Colorado River Basin, Colorado River; Burnet County; Lower Colorado River Authority; hydropower, impoundment began July 1951; 8,760 ac-ft.
121. **Martin Lake and Martin Dam:** Sabine River Basin, Martin Creek; Panola and Rusk Counties; Texas Utilities Services; steam-electric power; impoundment began 1974; 77,619 ac-ft.
122. **Pat Mayse Lake and Pat Mayse Dam:** Red River Basin, Sanders Creek; Lamar County; operated by USACE Tulsa District, City of Paris has contracted for conservation storage; municipal, industrial and flood control; impoundment September 1967; 189,100 ac-ft conservation capacity; 71,256 ac-ft flood control capacity.

123. **Medina Lake and Medina Dam:** San Antonio River Basin, Medina River; Medina and Bandera Counties; Bexar-Medina-Atascosa Counties Water Improvement District No. 1; irrigation; impoundment began May 1913; 254,000 ac-ft capacity.
124. **Lake Meredith and Sanford Dam:** Also called Lake Sanford; Canadian River Basin, Canadian River; Hutchison, Moore, and Potter Counties; constructed by Bureau of Reclamation, owned and operated by Canadian River Municipal Water Authority; municipal, flood control, recreation; impoundment 1965; 817,970 ac-ft conservation; 589,630 ac-ft flood control.
125. **Lake Mexia and Bristone Dam:** Brazos River Basin, Navasota River; Limestone County; Bristone Municipal Water Supply District; municipal and industrial; impoundment began June 1961; 10,000 ac-ft.
126. **Millers Creek Reservoir and Millers Creek Dam:** Brazos River Basin, Millers Creek; Baylor and Throckmorton Counties; North Central Texas Municipal Water Authority; municipal; impoundment began 1974; 25,520 ac-ft.
127. **Lake Mineral Wells and Mineral Wells Dam:** Brazos River Basin, Rock Creek; Parker County; City of Mineral Wells; municipal; impoundment began 1920; 16,760 ac-ft.
128. **Monticello Reservoir and Monticello Dam:** Cypress Creek Basin, Blundell Creek; Titus County; Texas Utilities Generating Company; steam-electric power; impoundment August 1972; 40,100 ac-ft.
129. **Hubert H. Moss Lake and Fish Creek Dam:** Red River Basin, Fish Creek; Cooke County; City of Gainsville; municipal and industrial; impoundment April 1966; 23,210 ac-ft.
130. **Mountain Creek Lake and Mountain Creek Dam:** Trinity River Basin, Mountain Creek; Dallas County; Dallas Power and Light Company; industrial; impoundment January 1937; 22,840 ac-ft.
131. **Murvaul Lake and Murvaul Dam:** Also called Panola Lake and Panola Dam; Sabine River Basin, Murvaul Bayou; Panola County; Panola County Fresh Water Supply District No.1; municipal, industrial, recreation; impoundment began 1957; 45,810 ac-ft.
132. **Nacogdoches Lake and Nacogdoches Lake Dam:** Neches River Basin, Bayo Loco Creek; Nacogdoches County; City of Nacogdoches; municipal; impoundment May 1976; 45,810 ac-ft.

133. **Lake Nasworthy and Nasworthy Dam:** Colorado River Basin, South Concho River; Tom Green County; City of San Angelo; municipal, industrial, irrigation; impoundment began March 1930; 12,390 ac-ft.
134. **Navarro Mills Lake and Navarro Mills Dam:** Trinity River Basin, Richland Creek; Navarro and Hill Counties; owned and operated by USACE Fort Worth District, Trinity River Authority has contracted for conservation storage; municipal and flood control; impoundment March 1963; 49,827 ac-ft conservation capacity; 149,403 ac-ft flood storage capacity.
135. **New Terrell City Lake and Terrell Dam:** Trinity River Basin, Muddy Cedar Creek; Kaufman County; City of Terrell; municipal; construction completed November 1955; 8,590 ac-ft.
136. **North Lake and North Lake Dam:** Trinity River Basin, South Fork Grapevine Creek; Dallas County; Dallas Power and Light Company; steam-electric power; impoundment began March 1957; 17,000 ac-ft.
137. **North Fork Buffalo Creek Reservoir and North Fork Buffalo Creek Dam:** Red River Basin, North Fork Buffalo Creek; Wichita County; Wichita County Water Control and Improvement District No. 3; municipal; impoundment began November 1964; 15,400 ac-ft.
138. **Oak Creek Reservoir and Oak Creek Dam:** Colorado River Basin, Oak Creek; Coke and Nolan Counties; City of Sweetwater; municipal, industrial; impoundment May 1953; 39,360 ac-ft.
139. **Olmos Reservoir and Olmos Dam:** San Antonio River Basin, Olmos Creek; Bexar County; San Antonio River Authority; flood control; completed 1926; 12,600 ac-ft flood control capacity.
140. **Lake Palestine and Blackburn Crossing Dam:** Also called Blackburn Crossing Lake; Neches River Basin, Neches River; Anderson, Cherokee, Henderson, and Smith counties; Upper Neches River Municipal Water Authority; municipal, industrial, recreation; impoundment May 1962; 367,312 ac-ft.
141. **Lake Palo Pinto and Palo Pinto Creek Dam:** Brazos River Basin, Palo Pinto Creek; Palo Pinto County; Palo Pinto County Municipal Water District No. 1 and City of Mineral Wells; municipal and industrial; impoundment began April 1964; 42,200 ac-ft.
142. **Wright Patman Lake and Wright Patman Dam:** Formerly, Texarkana Lake and Dam; Sulphur River Basin, Sulphur

River; Bowie and Cass Counties; USACE Fort Worth District; City of Texarkana has contracted for conservation storage; flood control, municipal, industrial and recreation; impoundment June 1956; 97,927 ac-ft conservation; 2,556,048 ac-ft flood control.

143. **Peacock Site 1A Tailings Reservoir:** Cypress River Basin, Peacock Creek; Lone Star Steel Company; industrial; completed 1983; 11,250 ac-ft.
144. **Pinkston Reservoir and Pinkston Dam:** Formerly, Sandy Creek Reservoir; Neches River Basin, Sandy Creek; Shelby County; City of Center; municipal; impoundment began 1976; 7,380 ac-ft.
145. **Joe Pool Reservoir and Joe Pool Dam:** Formerly, Lakeview Lake and Lakeview Dam; Trinity River Basin, Mountain Creek; Dallas, Ellis and Tarrant Counties; constructed and owned by USACE Fort Worth District, Trinity River Authority has contracted for conservation storage; flood control, municipal, and recreation; impoundment December 1985, 176,900 ac-ft conservation capacity; 127,100 ac-ft flood storage capacity.
146. **Possum Kingdom Lake and Morris Sheppard Dam:** Brazos River Basin, Brazos River; Palo Pinto, Stephens, Young Counties; Brazos River Authority; municipal, industrial, irrigation, recreation, power, mining including oil production; impoundment March 1941; 538,139 ac-ft.
147. **Proctor Lake and Proctor Dam:** Brazos River Basin, Leon River; Comanche County; owned and operated by USACE Fort Worth District; Brazos River Authority has contracted for conservation storage; flood control, municipal, industrial, irrigation, and recreation; impoundment September 1963; 54,762 ac-ft conservation; 319,438 ac-ft flood control.
148. **Lake Quitman and Wood County Dam No. 1:** Sabine River Basin, Dry Creek; Wood County; recreation and flood control, impoundment began May 1962; 7,440 ac-ft.
149. **Lake Randall and Randall Dam:** Red River Basin, Shawnee Creek; Grayson County; City of Denison; municipal; impoundment began 1909, 6,290 ac-ft.
150. **Sam Rayburn Reservoir and Sam Rayburn Dam:** Formerly, McGee Bend Reservoir and Dam; Neches River Basin, Angelina River; Angelina, Jasper, Nacogdoches, Sabine, San Augustine Counties; operated by USACE Fort Worth District, Lower Neches Valley Authority has contracted for conservation

storage; municipal, industrial, irrigation, hydroelectric power, recreation, and flood control; impoundment March 1965; 2,876,033 ac-ft conservation; 1,121,567 ac-ft flood control.

151. **Red Bluff Reservoir and Red Bluff Dam:** Rio Grande Basin, Pecos River; Reeves and Loving Counties; Red Bluff Water Power Control District; irrigation and hydropower; impoundment began September 1936; 307,000 ac-ft.
152. **River Crest Lake and River Crest Levee:** Sulphur River Basin, off-channel near Sulphur River; Red River County; Texas Power and Light Company; steam-electric power; impoundment began November 1953; 7,000 ac-ft.
153. **Richland-Chambers Reservoir and Dam:** Trinity River Basin, Richland and Chambers Creeks; Freestone and Navarro Counties; Tarrant Regional Water District; municipal; impoundment July 1987; 1,112,760 ac-ft.
154. **Ray Roberts Lake and Ray Roberts Dam:** Formerly, Aubrey Lake and Dam; Trinity River Basin, Elm Fork Trinity River; Cooke, Denton, and Grayson Counties; constructed and owned by USACE Fort Worth District, Cities of Dallas and Denton have contracted for conservation storage; flood control, municipal, industrial, recreation; impoundment June 1987; 788,490 ac-ft conservation capacity; 276,110 ac-ft flood control capacity.
155. **Bob Sandlin Lake and Fort Sherman Dam:** Also called Cherokee Trails Lake; Cypress Creek Basin, Big Cypress Creek; Camp, Franklin, Titus, and Wood Counties; Titus County Fresh Water Supply District No. 1; impoundment began 1978; 203,148 ac-ft.
156. **San Esteban Lake and San Esteban Dam:** Rio Grande Basin, Alamito Creek; Presidio County; owned by William B. Blakemore; recreation; impoundment began 1911; 18,770 ac-ft.
157. **Santa Rosa Lake and Santa Rosa Dam:** Red River Basin, Beaver Creek, Wilbarger County; W.T. Waggoner Estate; mining; impoundment began 1929; 11,570 ac-ft.
158. **Sheldon Reservoir and Sheldon Dam:** San Jacinto River Basin, Carpenters Bayou; Harris County; Texas Parks and Wildlife Department; recreation and fish hatchery; impoundment began December 1943; 5,420 ac-ft.
159. **Smithers Lake and Smithers Lake Dam:** Also called Lake George; Brazos River Basin, Dry Creek; Fort Bend County;

Houston Lighting and Power; steam-electric power; impoundment began October 1957; 18,700 ac-ft.

160. **Somerville Lake and Somerville Dam:** Brazos River Basin, Yegua Creek; Burleson, Lee, and Washington Counties; owned and operated by USACE Fort Worth District; Brazos River Authority has contracted for conservation storage; impoundment January 1967; 150,293 ac-ft conservation capacity; 347,400 ac-ft flood control capacity.
161. **South Texas Project Reservoir:** Colorado River Basin, off-channel near Colorado River; Matagorda County; Houston Lighting and Power; steam-electric power; 187,000 ac-ft.
162. **E. V. Spence Reservoir and Robert Lee Dam:** Also called Robert Lee Lake; Colorado River Basin, Colorado River; Coke County; Colorado River Municipal Water District; municipal, industrial, and mining; impoundment began December 1968; 484,800 ac-ft.
163. **Squaw Creek Reservoir and Squaw Creek Dam:** Brazos River Basin, Squaw Creek; Hood and Somerville Counties; Texas Utilities Services Company; steam-electric power; impoundment began 1977; 151,047 ac-ft.
164. **Lake Stamford and Stamford Dam:** Brazos River Basin, Paint Creek; Haskell County; City of Stamford; municipal and industrial; impoundment began June 1953; 52,700 ac-ft.
165. **B.A. Steinhagen Lake and Town Bluff Dam:** Also known as Town Bluff Reservoir and Dam B; Neches River Basin, Neches River; Jasper and Tyler Counties; owned and operated by USACE Fort Worth District; Lower Neches Valley Authority has contracted for conservation storage; municipal, industrial, recreation; impoundment April 1951; 66,972 ac-ft conservation capacity; 57,728 ac-ft flood control capacity.
166. **Stillhouse Hollow Lake and Stillhouse Hollow Dam:** Also called Lampasas Reservoir and Lampasas Dam; Brazos River Basin, Lampasas River; Bell County; owned and operated by USACE Fort Worth District, Brazos River Authority has contracted for conservation storage; flood control, municipal, industrial, irrigation, recreation; impoundment February 1968; 227,825 ac-ft conservation capacity; 402,575 ac-ft flood control capacity.
167. **Striker Creek Reservoir and Striker Creek Dam:** Neches River Basin, Striker Creek; Cherokee and Rusk Counties; Angelina and

Nacogdoches Counties Water Control and Improvement District No. 1; municipal and industrial; impoundment began May 1957; 26,960 ac-ft.

168. **Sulphur Springs Lake and Sulphur Springs Lake Dam:** Formerly, White Oak Creek Reservoir and White Oak Creek Dam; Sulphur River Basin, White Oak Creek; Hopkins County; Sulphur Springs Water District; municipal; impoundment began July 1973; 13,520 ac-ft.
169. **Sweetwater Lake and Sweetwater Dam:** Brazos River Basin, Bitter and Cottonwood Creeks; Nolan County; City of Sweetwater; municipal and industrial; impoundment 1930; 11,900 ac-ft.
170. **Lake Tawakoni and Iron Bridge Dam:** Also called Iron Bridge Dam and Lake; Sabine River Basin, Sabine River; Hunt, Rains, and Van Zandt Counties; Sabine River Authority; municipal, industrial, irrigation, recreation; impoundment October 1960; 871,695 ac-ft.
171. **Texana Lake and Texana Dam:** Formerly, Palmetto Bend Reservoir and Dam; Lavaca River Basin, Navidad River and Sandy Creek; Jackson County; constructed by Bureau of Reclamation, owned and operated by Lavaca-Navidad River Authority; municipal and irrigation; impoundment began 1981; 157,900 ac-ft capacity.
172. **Lake Texoma and Denison Dam:** Red River Basin, Red River; Cooke and Grayson Counties; owned and operated by USACE Tulsa District; hydropower, flood control, conservation, and recreation; impoundment began October 1943; 2,516,000 ac-ft conservation capacity; 2,877,000 ac-ft flood control capacity.
173. **J. B. Thomas Lake and Colorado River Dam:** Colorado River Basin, Colorado River; Scurry and Borden Counties; Colorado River Municipal Water District; municipal, industrial, and recreation; impoundment began July 1952; 202,300 ac-ft.
174. **Toledo Bend Reservoir and Toledo Bend Dam:** Sabine River Basin, Sabine River; Newton, Sabine and Shelby Counties; Sabine River Authorities of Texas and Louisiana; municipal, industrial, irrigation, hydropower, recreation; impoundment October 1966; 4,477,000 ac-ft.
175. **Tradinghouse Creek Reservoir and Tradinghouse Creek Dam:** Brazos River Basin, Tradinghouse Creek; McLennan County;

Texas Power and Light Company; steam-electric power; impoundment began July 1968; 35,124 ac-ft.

176. **Lake Travis and Mansfield Dam:** Also called Marshall Ford Dam; Colorado River Basin, Colorado River; Travis and Burnet Counties; Lower Colorado River Authority; municipal industrial, irrigation, mining, hydropower, flood control, and recreation; impoundment began September 1940; 1,134,956 ac-ft conservation capacity; 786,775 ac-ft flood control capacity.

177. **Trinidad Lake and Trinidad Levee:** Trinity River Basin, off-channel of Trinity River; Henderson County; Texas Power and Light Company; hydropower; impoundment 1925; 7,450 ac-ft.

178. **Truscott Brine Lake and Truscott Brine Lake Dam:** Red River Basin, Bluff Creek; Knox County; constructed by USACE Tulsa District; permanent impoundment of naturally high-salinity flows to decrease salinity downstream; impoundment began 1984; 107,000 ac-ft.

179. **Twin Buttes Reservoir and Twin Buttes Dam:** Colorado River Basin, South Concho River, Spring Creek and Middle Concho River; Tom Green County; constructed by Bureau of Reclamation, conservation storage maintained and operated by the City of San Angelo, flood control storage operated by USACE Fort Worth District; municipal, industrial, flood control, irrigation, and recreation; impoundment December 1962; 186,000 ac-ft conservation; 446,000 ac-ft flood control.

180. **Twin Oaks Reservoir and Twin Oaks Dam:** Brazos River Basin, Duck Creek; Robertson County; Texas Power and Light Company; steam-electric power; impoundment 1982; 30,319 ac-ft.

181. **Lake Tyler and Mud Creek Dam and Whitehouse Dam:** Also called Mud Creek Dam Lake; Neches River Basin, Mud Creek and Prairie Creek; Smith County; City of Tyler; municipal and industrial; impoundment began January 1949 (Whitehouse) and November 1966 (Mud Creek); 73,700 ac-ft.

182. **Upper Nueces Reservoir and Upper Nueces Dam:** Nueces River Basin, Nueces River; Zavala County; Zavala and Dimmit Counties Water Improvement District Number 1; irrigation; impoundment began March 1948; 7,590 ac-ft.

183. **Valley Lake and Valley Dam:** Formerly, Brushy Creek Reservoir and Dam; Red River Basin, Brushy Creek; Fannin and Grayson Counties; Texas Power and Light Company; steam-electric

power; impoundment December 1960; main water supply is pumped from Red River; 16,400 ac-ft.

184. **Valley Acres Lake and Valley Acres Dam:** Nueces-Rio Grande Coastal Basin, off-channel from the Rio Grande; Hidalgo County; Valley Acres Water District; irrigation and municipal; impoundment began 1947; 7,840 ac-ft.
185. **Waco Lake and Waco Dam:** Brazos River Basin, Bosque River; McLennan County; owned and operated by USACE Fort Worth District; Brazos River Authority and City of Waco contracted for conservation storage; flood control, municipal, industrial, and recreation; impoundment February 1965; 189,773 ac-ft conservation capacity; 536,587 ac-ft flood control capacity. The 1965 Lake Waco completely submerged the old Lake Waco constructed by the City of Waco in 1928 to 1929. The top of conservation pool was raised 7.0 ft in in 2003.
186. **Lake Waxahachie and South Prong Dam:** Trinity River Basin, Clear Fork Trinity River; Parker County; City of Weatherford; municipal and industrial; impoundment November 1956; 19,470 ac-ft.
187. **Weatherford Lake and Weatherford Dam:** Trinity River Basin, White River; Crosby County; White River Municipal Water District; municipal, industrial, and mining including oil production; impoundment began March 1957; 37,950 ac-ft.
188. **Welsh Reservoir:** Cypress River Basin, Swauano Creek; Titus County; Southwestern Power Company; cooling water for Welch Power Plant; construction completed 1975; 20,240 ac-ft.
189. **White River Lake and White River Dam:** Brazos River Basin, White River; Crosby County; White River Municipal Water District; municipal, industrial, and mining including oil production; impoundment began October 1963; 37,950 ac-ft.
190. **White Rock Lake and White Rock Dam:** Trinity River Basin, White Rock Creek; Dallas County; City of Dallas; recreation; impoundment began 1910; 10,740 ac-ft.
191. **Whitney Lake and Whitney Dam:** Brazos River Basin, Brazos River; Bosque, Hill and Johnson Counties; owned and operated by USACE Fort Worth District; Brazos River Authority has contracted for 50,000 ac-ft water supply storage; flood control, municipal, hydroelectric power; impoundment December 1951; 554,203 ac-ft conservation; 1,445,297 ac-ft flood control.

192. **Lake Wichita and Lake Wichita Dam:** Red River Basin, Holiday Creek; Archer and Wichita Counties; City of Wichita Falls; municipal, steam-electric power, and recreation; impoundment began 1901; 14,000 ac-ft.
193. **Lake Winnsboro and Wood County Dam No. 4:** Sabine River Basin, Big Sandy Creek; Wood County; recreation and flood control; impoundment began July 1962; 8,100 ac-ft.
194. **Winters Lake and Winters Dam:** Colorado River Basin, Elm Creek; Runnels County; City of Winters; municipal; dam completed in 1945 replaced with new dam completed in 1983; 8,370 ac-ft.
195. **Lake Worth and Lake Worth Dam:** Trinity River Basin, West Fork Trinity River; Tarrant County; City of Fort Worth; municipal; impoundment began October 1914; 17,000 ac-ft.

Index

Page numbers followed by *f* and *t* indicate figures and tables.

About the Author

Ralph A. Wurbs has participated in managing water in rivers and reservoirs since May 1971 when, upon graduation from Texas A&M University (TAMU) with a degree in civil engineering, he joined the water resources development program of the US Army Corps of Engineers (USACE), Fort Worth District. During his nine-year full-time USACE employment as a civil engineer and program manager, he also completed, with USACE support, an M.S. degree in civil engineering (Water Resources Engineering) at the University of Texas at Arlington in 1974 and a Ph.D. in civil engineering (Water Resources Systems Planning and Management) at Colorado State University in 1978, along with his responsibilities in planning and design of multipurpose reservoir and local flood control projects. After leaving USACE in August 1980 to join the TAMU Civil Engineering Department faculty, his university research endeavors were sponsored for many years by the USACE Engineering Research and Development Center, Hydrologic Engineering Center, Institute for Water Resources, and Fort Worth District.

Professor Wurbs has completed many research and consulting projects over the past four decades sponsored by an array of federal and state agencies and other organizations. He has worked closely with the water management community of Texas, most notably under the auspices of the Texas Commission on Environmental Quality, which has sponsored his research at TAMU in modeling and analysis of management of water in river and reservoir systems continuously since 2005.

Professor Wurbs has taught many undergraduate and graduate courses at TAMU, short courses for practicing engineers and water managers in Texas and abroad, and a graduate water management course several semesters as a visiting professor at the Catholic University of Leuven in Belgium. Responsibilities at TAMU have included serving as head of the department's Environmental and Water Resources Engineering Division, associate director for engineering of the Texas Water Resources Institute, and member of the Executive Committee of the Interdepartmental Graduate Program in Water Management and Hydrologic Science.

His publications include three books from the Prentice Hall Division of Pearson: *Water Management Models* (1995), *Modeling and Analysis of Reservoir System Operations* (1996), and as coauthor of the textbook *Water Resources Engineering* (Wurbs and James 2002) used by thousands of students since 2002 in courses at TAMU and other universities. He has authored chapters in 14 other books and published 117 papers, as well as numerous published and unpublished technical reports sponsored by water agencies. His work has been recognized by an endowed professorship, several teaching awards, and various other awards including the Outstanding Research and Innovation Award of the American Academy of Water Resources Engineers, an ASCE affiliate. Professor Wurbs is a Founding and Honorary Diplomate of AAWRE and Fellow and Life Member of ASCE.

A group of TAMU former students endowed the Ralph A. Wurbs Endowed Graduate Fellowship in Water Resources in 2024, named in his honor. Also in 2024, he was honored by the *Texas Water Journal*, a Texas Water Resources Institute affiliate, as the inaugural recipient of the Texas Water Lifetime Achievement Award. This book is based largely on his career as recognized by these 2024 honors.